# The Geometer's Sketchpad
# Reference Center

# The Geometer's Sketchpad
## Reference Center

# Sketchpad Reference Center



**Project Design**

*Nicholas Jackiw*

**Implementation**

*Nicholas Jackiw*
*Scott Steketee*
*Matt Litwin*

**Engineering Support**

*Jon Brooks*
*Scott Johnson*
*Kirk Swenson*

**Special Thanks to**

*Eugene Klotz*
*Kendra Lockman*
*Tawnia Litwin*
*Vishakha Parvate*
*Steven Rasmussen*
*Doris Schattschneider*
*Daniel Scher*
*Nathalie Sinclair*
*and to all of our many field testers!*

*Portions of this work were funded by grants from the National Science Foundation to KCP
Technologies, Key Curriculum Press, and the Visual Geometry Project at Swarthmore College.*



ISBN: 978-1-60440-101-1

# Table of Contents

# Objects

# 1　Objects

From one perspective, mathematics is the art of creating knowledge by finding new and interesting relationships among existing mathematical objects. Sketchpad provides you with a rich set of such mathematical objects and with many ways to connect them. It's up to you to create those objects, set up the connections between them, and determine their attributes [77]. Then you can investigate their behavior, find new relationships, discover symmetry and patterns, and display and present your results.

Here's a list of the objects that Sketchpad makes available:

| **Geometric Objects:** | **Numeric and Algebraic Objects:** | **Presentation Objects:** |
|---|---|---|
| • Points [2] | • Measurements [36] | • Captions [56] |
| • Segments, Rays, and Lines [5] | • Parameters [36] | • Angle Markers [60] |
| • Circles [6] | • Calculations [39] | • Tick Marks [63] |
| • Arcs [7] | • Tables [39] | • Action Buttons [66] |
| • Polygons and Other Interiors [8] | • Coordinate Systems [41] | |
| • Loci [10] | • Functions [45] | |
| • Pictures and Drawings [18] | • Function Plots [52] | |
| • Iterations [24] (some iterations are numeric) | • Parametric Plots [54] | |

### Object Categories:

Certain categories of objects can be used for special purposes.

- Path Objects [89] are objects on which you can construct and animate points.

- Straight Objects [90] are path objects that you can use to construct parallels and perpendiculars. You can measure the slope of a straight object and the distance from a point to a straight object.

- Layered Objects [91] occupy two-dimensional area and appear in a layered order when they overlap.

- Translucent Objects [91] have adjustable opacity, allowing similar objects in layers below them to show through.

- Values [92] are objects that you can use in situations when you need a numeric value.

- Plots [95] are displayed by combining many small elements (called *samples*) that make up the full object. These objects include loci [10] and function plots [52], among others.

- Text Objects [99] include captions [56], values [92], functions [45], and action buttons [66].

## 1.1　Points

Points are the fundamental building blocks of classical geometry, and geometric figures such as lines and circles are defined in terms of points. All of Sketchpad's geometric constructions begin with points.

There are three kinds of points in Sketchpad.

| | |
|---|---|
| An *independent point* has no parents [80], and so does not depend on any other object. An independent point is free to move anywhere on the sketch plane. | |
| A *point on a path* is constructed on a path object [89] such as a line or circle. A point on a path is free to move along its path, but cannot leave that path. | |
| A *dependent point* — such as a point of intersection — is constructed so that its position is completely determined by its parents. A dependent point cannot move by itself; it can move only if at least one of its parent objects also moves. Thus, a point at the intersection of two segments cannot move unless one or both of the intersecting segments are moved, and the reflected [207] image of a point cannot move unless either the pre-image or the mirror moves. | |

You can change the object properties [79], label properties [82], and display attributes [86] of points.

Sketchpad provides a number of ways to create, manipulate, and use points.

## ▼ Construct a Point

There are many different ways to construct points in a sketch.

- Click the **Point** [112] tool to construct a point.

- Click the **Straightedge** [115] tool or the **Compass** [113] tool in empty space to construct a point that determines a straight object or circle. Many **Custom** [125] tools also construct points.

- Choose **Construct | Point on Object** [176] to construct a point on each selected path object [89].

- Choose **Construct | Midpoint** [176] to construct a midpoint on each selected segment.

- Choose **Construct | Intersection** [177] to construct the intersection point of two selected objects. Each object must be a straight object [5], a circle [6], an arc [7], a point locus [10], or a function plot [52].

- Choose **Graph | Plot Value on Axis** [237] to plot a point on an axis or other path [89], at a position defined by a value [92] or number.

- Choose **Graph | Plot Points** [239] to plot a point with coordinates defined by two values [92] or numbers.

You can also construct a point by using commands from the Transform menu [192] to create a transformed image of an existing point.

## ▼ Drag, Animate, or Move a Point

- Drag an independent point or a point on a path using the Translate Arrow [104] tool, the Rotate Arrow [104] tool, or the **Dilate Arrow** [104] tool.

- Animate an independent point or a point on a path using either the **Display | Animate** [171] command or using an Animation button [149].

- Move an independent point or a point on path toward a destination using a Movement button [150].

When you drag or animate any other geometric object, it moves by dragging or animating the parents [80] on which the object depends.

## ▼ Split or Merge a Point

There may be times when you want to attach an independent point to another object.

Similarly, there may be times when you want to separate a point from its parents to make it independent.

The **Split/Merge** [153] commands allow you to make such changes in the family tree [80] of your sketch.

- Choose **Edit | Merge** [153] to merge an independent point with any other point or with a path [89], provided that the object to which it is being merged does not depend on the independent point.

- Choose **Edit | Split** [153] to separate any point on a path, midpoint, or intersection from its parents, making it into an independent point.

- Choose **Edit | Split Plotted Point from Coordinate System** [153] to split a plotted point from its defined coordinate values on its coordinate system.

- Hold Shift and choose **Edit | Split from Definition** [153] to split a transformed image point from its pre-image.

*See also:*
*How to Use Split and Merge to Explore Constructions* [156]

## ▼ Transform a Point

Transform a point by selecting it and choosing one of the following commands:

- **Transform | Translate** [200]
- **Transform | Rotate** [203]
- **Transform | Dilate** [206]
- **Transform | Reflect** [207]

## ▼ Use Points to Define Transformations

Use points in several ways to help specify how other objects will be transformed.

- Choose **Transform | Mark Center** [194] to designate a center point. The center point can be used for rotation or dilation.

- Choose **Transform | Mark Angle** [195] to designate an angle formed by three points. The angle can be used for translation or rotation.
  You can also designate an angle using two coterminal rays or segments.

- Choose **Transform | Mark Ratio** [197] to designate a point's value on a path or to designate a ratio defined by three collinear points. The ratio can be used for dilation.

## ▼ Measure a Point

Several Measure [216] menu commands apply to selected points or to combinations of points.

- Choose **Measure | Distance** [218] to measure the distance between two points.

- Choose **Measure | Angle** [218] to measure the angle formed by three points.

- Choose **Measure | Ratio** [221] to measure the ratio defined by three collinear points.

- Choose **Measure | Value of Point** [221] to measure the relative position of a point on a path.

- Choose **Measure | Coordinates** [223] to measure the coordinates of a point.

- Choose **Measure | Abscissa (x)** [223] to measure the $x$-coordinate of a point.

- Choose **Measure | Ordinate (y)** [224] to measure the $y$-coordinate of a point.

## 1.2   Segments, Rays, and Lines

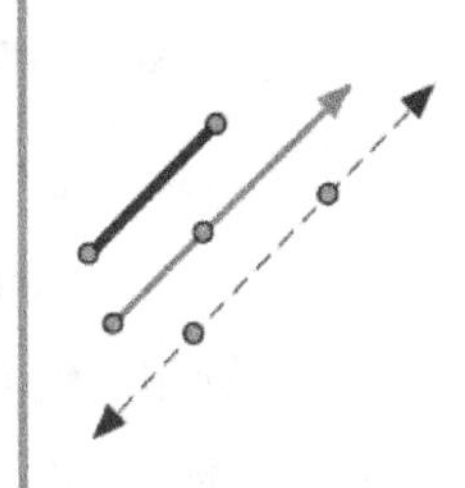

Segments, rays, and lines are fundamental objects in Euclidean geometry. In classical geometry constructions, you use a straightedge to construct these objects.

    An axis [41] acts as a line for most purposes.

A straight object is a path object [89].

You can change the object properties [79], label properties [82], and display attributes [86] of straight objects.

### ▼ Construct a Straight Object

Sketchpad provides several ways to construct straight objects.

- Use a **Straightedge** [115] tool to construct a straight object based on two points.Use existing points, create new independent points, or create points on paths or intersections.

  Constrain the straight object [115] to be horizontal, vertical, or at other common angles by pressing and holding the Shift key while you drag.

- Use the **Construct | Segment** [178], **Construct | Ray** [178], or **Construct | Line** [178] command to construct a straight object using two selected points.

- Use the **Construct | Perpendicular Line** [180] or **Construct | Parallel Line** [179] command to construct a line parallel or perpendicular to a selected straight object.

- Use the **Construct | Angle Bisector** [181] command to construct the ray bisecting the angle formed by three selected points.

*See also:*
*Construct a Segment of Fixed Length* [182]
*Construct an Angle of Fixed Measure* [204]
*Construct Congruent Segments* [184]
*Construct Congruent Angles or Triangles* [204]

### ▼ Use a Straight Object

- Construct a point on a straight object using the **Point** [112] tool or the **Construct | Point on Object** [176] command. Then animate [267] the point along the straight object or use the point as the driver for a locus [10].

- Attach an independent point to the straight object by using the **Edit | Merge** [153] command.

- Use a straight object as a mirror for reflections by choosing the **Transform | Mark Mirror** [195] command.

- Use **Construct | Midpoint** [176] to construct a segment's midpoint.

- Use **Construct | Intersection** [177] to construct a straight object's intersection with another straight object, a circle, an arc, a point locus, or a function plot.

- Use **Construct | First Intersection** [177] to construct a ray's first intersection with a polygon. (The first intersection is the intersection closest to the ray's endpoint.)

- Use **Graph | Define Unit Distance** [233] to define a coordinate system with a unit distance defined by one or two selected segments. (The unit of the coordinate system is defined by the length of the segment.)

   If you use two selected segments, the first defines the unit for *x*, and the second for *y*.

### ▼ Measure a Straight Object

Use the Measure [216] menu to find:

- the **Length** [217] of a segment

- the **Ratio** [221] of two segments' lengths

- the **Equation** [225] of a line

- the **Slope** [224] of any straight object

## 1.3    Circles

Circles are fundamental objects in Euclidean geometry. In classical geometric constructions, you use a compass to construct circles.

A circle is a path object [89].

You can change the object properties [79], label properties [82], and display attributes [86] of circles.

### ▼ Construct a Circle

Sketchpad provides several different ways to construct a circle.

Use the **Compass** [113] tool to construct a circle using a center point and another point which defines the radius.

Use the **Construct | Circle by Center+Point** [182] command to construct a circle using a center point and another point which defines the radius.

Use the **Construct | Circle by Center+Radius** 183 command to construct a circle using a center point and either a segment or a distance measurement to define the radius.

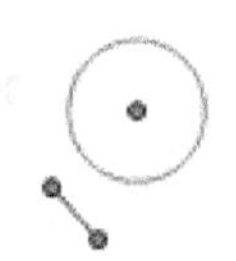

### ▼ Use a Circle

- Construct a point on a circle using the **Point** 112 tool or the **Construct | Point on Object** 176 command. Then animate 267 the point around the circle or use the point as the driver for a locus 10.

- Attach an independent point to the circle by using the **Edit | Merge** 153 command.

- Use **Construct | Interior** 186 to construct the interior of a circle.

- Use **Construct | Intersection** 177 to construct a circle's intersection with a straight object, another circle, an arc, a point locus, or a function plot.

- Use **Graph | Define Unit Circle** 233 to define a coordinate system that uses the selected circle as its unit circle.

### ▼ Measure a Circle

Use the Measure 216 menu to find:

- the **Circumference** 218 of a circle

- the **Radius** 221 of a circle

- the **Area** 219 of a circle

- the **Equation** 225 of a circle

## 1.4    Arcs

Arcs are fundamental objects in Euclidean geometry. In Sketchpad, use the Construct 174 menu to create arcs.

An arc is a path object 89.

You can change the object properties 79, label properties 82, and display attributes 86 of arcs.

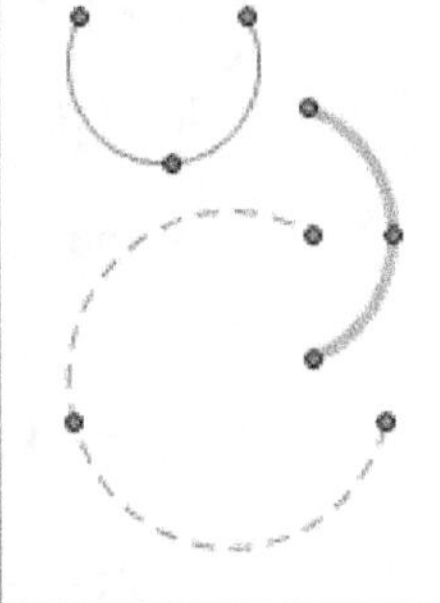

 Hint

If you construct an arc through three points 186 and then drag the three points so they are collinear, the second point determines the appearance of the arc. If the second point falls between the first and third poins, the arc has zero angle measure 219 but nonzero arc length 220; it is displayed as a segment 5 that starts at the first selected point, passes through the second point, and ends at the third point. If the second point is collinear with, but not between, the first and third points, the arc is not well defined: it disappears and any measurements that depend on it are undefined.

### ▼ Construct an Arc

Sketchpad provides two different ways to construct an arc.

| | |
|---|---|
| Use **Construct | Arc through 3 Points** 186 to construct an arc that passes through three selected points. | |
| Use **Construct | Arc on Circle** 185 to construct an arc that lies on a selected circle and is bounded by two selected points on the circle. | |

### ▼ Use an Arc

- Construct a point on an arc using the **Point** 112 tool or the **Construct | Point on Object** 176 command. Then animate 267 the point along the arc or use the point as the driver for a locus 10.

- Attach an independent point to the arc by using the **Edit** 144 | **Merge** 153 command.

- Use **Construct | Arc Interior | Arc Sector** 189 to construct the arc sector defined by an arc.

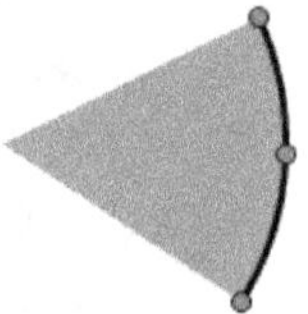

- Use **Construct | Arc Interior | Arc Segment** 189 to construct the arc segment defined by an arc.

- Use **Construct | Intersection** 177 to construct an arc's intersection with a straight object 90, a circle 6, another arc, a point locus 10, or a function plot 52.

### ▼ Measure an Arc

Use the Measure 216 menu to find:

- the **Radius** 221 of an arc
- the **Arc Length** 220 of an arc
- the **Arc Angle** 219 of an arc

## 1.5    Polygons and Other Interiors

Polygons and other interiors are Sketchpad objects that define a region of a plane. Interiors give your sketches substance and color, and allow you to measure the areas and perimeters of figures. Use the perimeter or circumference of an interior as a path on which to construct or animate points.

There are four kinds of interiors: polygons, circle interiors, arc sectors, and arc segments.

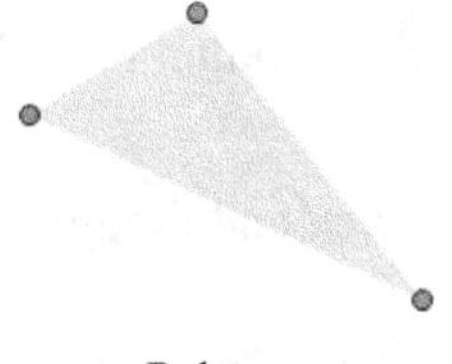

Polygon

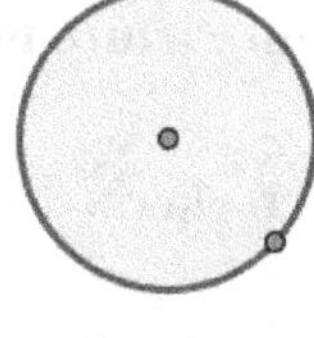

Circle Interior

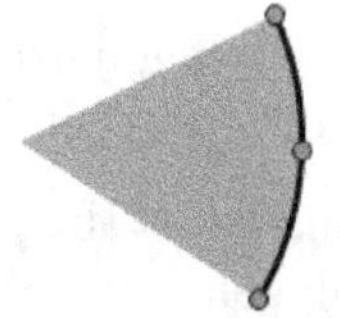

Arc Sector Interior

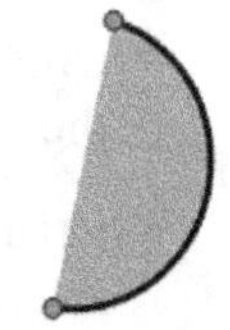

Arc Segment Interior

Polygons and interiors are path objects [89].

You can change the object properties [79], label properties [82], and display attributes [86] of interiors.

## ▼ Construct an Interior

| |
|---|
| Use **Construct \| Polygon Interior** [187] to construct a polygon defined by three or more selected vertex points. |
| Use the **Polygon** [118] tool to construct a polygon. |
| Use **Construct \| Circle Interior** [188] to construct the interior of each selected circle [6]. |
| Use **Construct \| Arc Interior \| Arc Sector** [189] to construct the sector interior of each selected arc [7]. An arc sector is bounded by the arc and by the radii to the two endpoints of the arc. |
| Use **Construct \| Arc Interior \| Arc Segment** [189] to construct the segment interior of each selected arc [7]. An arc segment is bounded by the arc and by the chord connecting the endpoints of the arc. |

## ▼ Use an Interior

- Construct a point on an interior using the **Point** [112] tool or the **Construct \| Point on Object** [176] command. Then animate [267] the point around the interior or use the point as the driver for a locus [10].

- Attach an independent point to the perimeter or circumference of the interior by using the **Edit \| Merge** [153] command.

- Change an interior's layer using **Bring to Front** [247] or **Send to Back** [247] from the Context [245] menu.

- Change the opacity of an interior using **Edit \| Properties \| Opacity** [92].

## ▼ Use a Polygon

Use a polygon in the same way as other interiors. In addition, there are some special ways in which you can use or change polygons.

- Construct all the segment edges of a selected polygon by choosing **Edit \| Select Parents** [152] and then choosing **Construct \| Segments** [178].

- Construct a new polygon and its segment edges at the same time by using the **Polygon and Edges** [118] tool.

- Construct a polygon's first intersection with a ray using **Construct \| First Intersection** [177].

- Crop a picture to a polygon using **Edit | Crop Picture to Polygon** 152, so that only the portion of the picture within the polygon is visible.

- Create an angle marker 60 by pressing the **Marker** 122 tool on a vertex and dragging into the interior of the polygon.

**Polygon Frame:** A polygon can be displayed with or without a frame that shows its perimeter. The frame is an optional display attribute of the polygon. It does not represent the segment edges of the polygon: You cannot select an individual segment edge of the frame or perform any construction or measurement on the frame, except for constructions or measurements that apply to the polygon itself. To construct the midpoint of a segment edge or measure the length of a segment edge of the polygon, you must explicitly construct that segment edge.

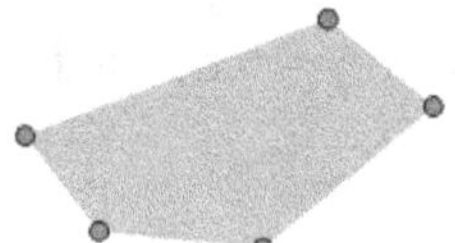

Polygon Displayed Without Frame

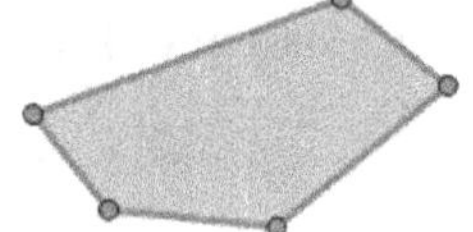

Polygon Displayed With Frame

- Show or hide the frame of the polygon using **Edit | Properties | Opacity** 92 and changing the setting for **Frame polygon perimeter.**

    To show only the frame, first show the frame and then set **Opacity** to 0%.

- Change the width of the polygon's frame using **Display | Line Style** 163.

- Determine whether new polygons display their frames by choosing **Edit | Preferences | Tools** 280 and changing the setting for **Frame new polygon perimeters.**

### ▼ Measure an Interior

Use the Measure 216 menu to find:

- the **Perimeter** 218 of a polygon or an arc interior.

- the **Circumference** 218 of a circle interior.

- the **Radius** 221 of a circle interior or an arc interior.

- the **Arc Angle** 219 or **Arc Length** 220 of an arc interior.

## 1.6    Loci

A mathematical locus is the set of all possible positions of an object that satisfy some specific condition.

A Sketchpad locus is a set of possible positions of an object as a point 2 moves along a path 89 or as a parameter 36 varies over a domain.

The point that moves along a path or the parameter that varies over a domain is called the *driver*.

The object that actually forms the locus is called the *driven object*. (The driven object must depend on the driver.)

The path along which the point moves, or the numeric domain within which the parameter varies, is called the *domain* of the driver.

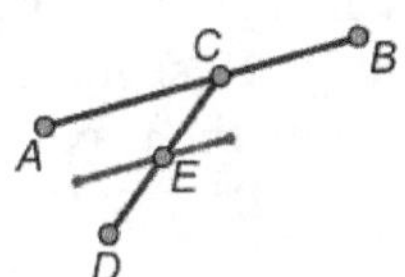

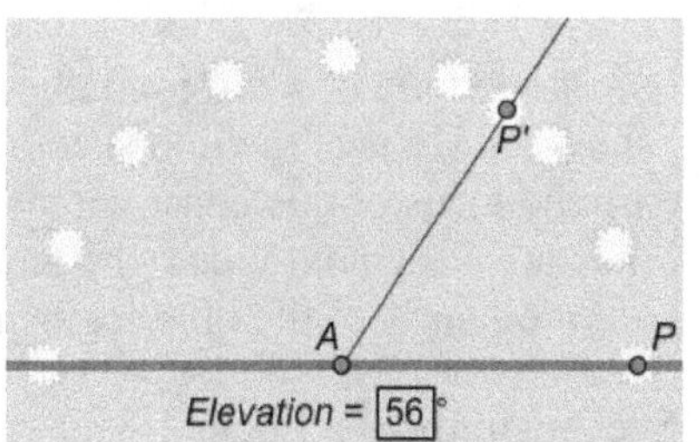



Driver: Point $C$
Domain: Segment $AB$
Driven object: Point $E$

Driver: Parameter *Elevation*
Domain: $0° \leq Elevation \leq 180°$
Driven object: Picture Attached to $P'$

The short definition of a Sketchpad locus, then, is this:

**A locus is a set of positions of a driven object as a driver varies over its domain.**

For a general mathematical locus, the number of possible positions is infinite. It would take a long time, and a lot of memory, to create such a locus in Sketchpad, so Sketchpad loci are approximations based on a finite (but potentially very large) set of positions. Each position in this set is called a *sample*.

There are several kinds of loci, named according to the type of the driven object.

| | |
|---|---|
| **Point Locus:** The driven object is a point. In this example, the parabola is the locus of point $C$ as point $A$ moves along the directrix.<br><br>A point locus is a path object [89]. | |
| **Non-point Locus:** The driven object is a geometric object, but not a point. In this example, the shape is the locus of circle $c_2$ as point $C$ moves around circle $c_1$. | |
| **Family of Functions:** The driven object is a function plot. This example shows the family of functions $f(x) = a \sin(x)$ as parameter $a$ (the driver) varies from -2 to 2. | |
| **Family of Curves:** The driven object is a point locus. This example shows the family of parabolas with the given directrix $d$ as the focus $F$ moves along a segment. | |
| $$\frac{dy}{dx} = \frac{x^2}{1-y^2}$$ | |
| **Family of Loci:** The driven object is a non-point locus. This example shows | |

> the slope field of a differential equation. The red segment is constructed to show the slope at the intersection of the two dashed lines. The blue segments are the locus of this constructed segment as point *A* moves along the *x*-axis; they are a segment locus (the locus of the red segment). The green segments are the family of the blue segment locus as point *B* moves along the *y*-axis.

See Anatomy of a Locus [14] for more details and examples.

You can change the object properties [79], label properties [82], and display attributes [86] of loci.

You can change the line style [163] of continuous loci, and you can change the point style [162] of discrete loci.

Change the opacity of a locus of an interior or a picture using **Edit | Properties | Opacity** [92].

Use **Edit | Properties | Plot** [95] to change the number of samples used to display a locus and to change whether it's displayed continuously or discretely.

### ▼ Construct a Locus

To construct a locus, you must first construct the driven object — the object whose locus you want to construct — in such a way that it depends on the driver. If the driver is a point on a path [176], the domain of the driver is the path (if it's of finite length) or a subset of the path (if it's infinite). If the driver is a parameter, you determine the domain when you create the locus. If the driver is an independent point, you must select a path [89] that doesn't depend on the driver to serve as the domain for the driver.

> The driven object can be a point, a straight object, a circle, an arc, an interior, a picture, a function plot, or a locus.

1. Select the driver and the driven object.

2. If the driver is an independent point [2], select the drive path — a path object that does not depend on [80] the driver.

3. Choose **Construct | Locus** [190].

### ▼ Use a Locus

- Use the **Point** [112] tool or the **Construct | Point on Object** [176] command to construct a point on a point locus. Then animate [267] the point along the locus or use the point as the driver point for a new locus [10].

- Use **Edit | Merge** [153] to merge an independent point to a point locus. Then animate [267] the point along the locus or use the point as the driver point for a new locus [10].

- Use **Construct | Intersection** [177] to construct the intersection of a point locus with a straight object, a circle, or an arc.

### ▼ Resize a Point Locus

If a point locus is based on a closed drive path (such as a circle) or a finite path (such as a segment or arc), the domain of the driver is fixed. But if the drive path is infinite and open (such as a ray or line), the domain of the driver—and therefore, the potential size of the locus — is unbounded. If possible, Sketchpad limits the domain based on the portion of the path that is visible on the screen. Such a point locus (on an infinite open domain) displays an arrowhead on the end of the locus.

To change the displayed length of such a point locus, use the **Arrow tool** 104 to drag the arrowhead at either end of the locus. Drag in the direction that the arrowhead points to increase the length of the locus; drag in the opposite direction to decrease the length of the locus.

### ▼ Change the Number of Samples

Mathematically, a locus may describe an infinite number of positions of the driven object. However, to display an infinite number of positions would require a computer to use an infinite amount of time, so Sketchpad instead displays a large (but not infinite) number of possible positions rather than all possible positions. Each position that Sketchpad does display is called a sample.

To change the number of samples for a selected locus:

- Press the + or − key. Each press of the key increases or decreases the number of samples.

- Choose **Increase Resolution** or **Decrease Resolution** from the Context 245 menu.

- Choose **Edit | Properties | Plot** 95 and type a new value for **Number of samples**.

You can also change the initial number of samples for newly constructed point loci. Choose **Edit | Advanced Preferences | Sampling** 283 and type a new value for **Number of samples for new point loci.**

> This setting has no effect on previously constructed loci.

### ▼ Display the Locus Continuously or Discretely

Change the display of a point locus between continuous and discrete using the Plot Properties for a Locus 95. To do so, select the locus, choose **Edit | Properties** 158, and go to the Plot 95 tab.

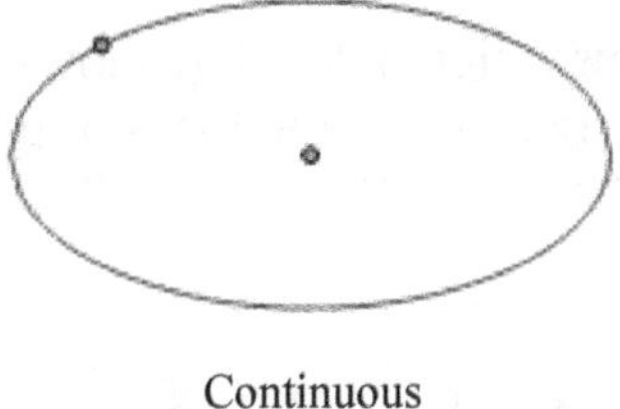

Continuous

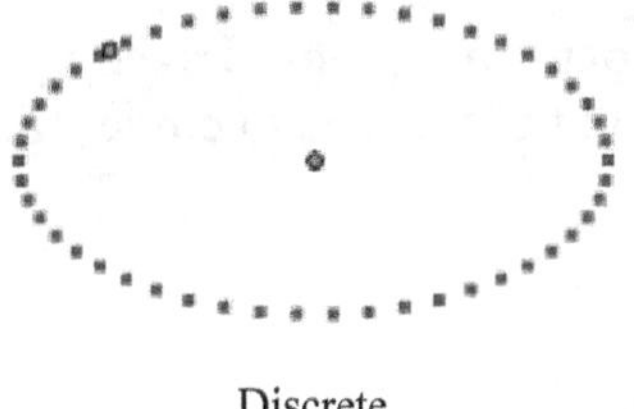

Discrete

### ▼ Display Arrowheads and Endpoints

If a point locus is based on an open domain, Sketchpad displays either a square endpoint or an arrowhead at each end of the locus. An arrowhead can be dragged in either direction to extend or contract the corresponding end of the domain. A square endpoint indicates that the corresponding end of the domain cannot be modified.

For instance, if the domain is a segment, the ends of the domain are fixed by the segment endpoints. If the domain is a ray, one end of the domain is fixed by the ray's initial point, but the other end can be extended or contracted.

For a parametric locus, both ends of the domain are always adjustable, either by dragging the arrowheads or by choosing **Edit | Properties | Plot** 95.

To determine whether arrowheads and endpoints are displayed on a point locus, choose **Edit | Properties | Plot** 95 and check or uncheck **Show arrowheads and endpoints.**

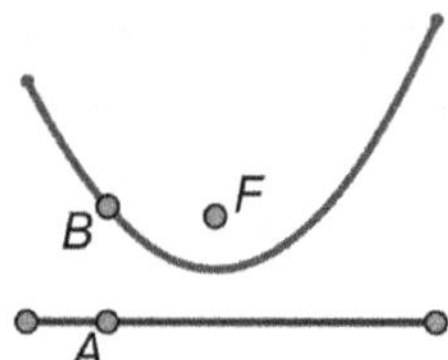

Driver: Point A
Domain: Segment
Two Endpoints

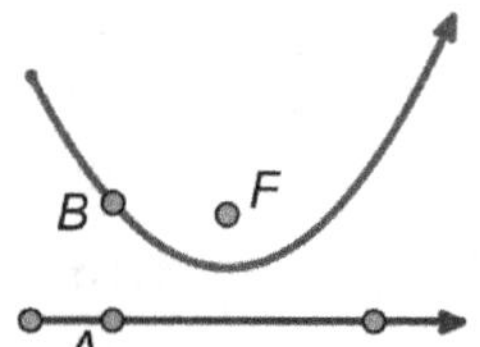

Driver: Point A
Domain: Ray
One Endpoint, One Arrow

## 1.6.1    Anatomy of a Locus

Recall the definition of a locus: **A locus is a set of positions of a *driven object* as a *driver* varies over its *domain*.**

Possible driven objects include points [2], straight objects [5], circles [6], arcs [7], interiors [8], pictures [18], function plots [52], and loci themselves.

Possible drivers are points on paths and parameters.

The domain for a point driver is normally its path, which might be a straight object, a circle, an arc, a polygon or other interior, a function plot, or another point locus. If the point driver is not constructed on a path, you must explicitly select a path to serve as the domain.

The domain for a parameter driver is a numeric domain that you specify when you construct the locus.

### Examples of Loci

**Locus of a point, driven by another point:** The illustration below left shows segment *CD* with end point *C* attached to circle *AB;* the illustration below right shows the locus of point *E* on the segment.

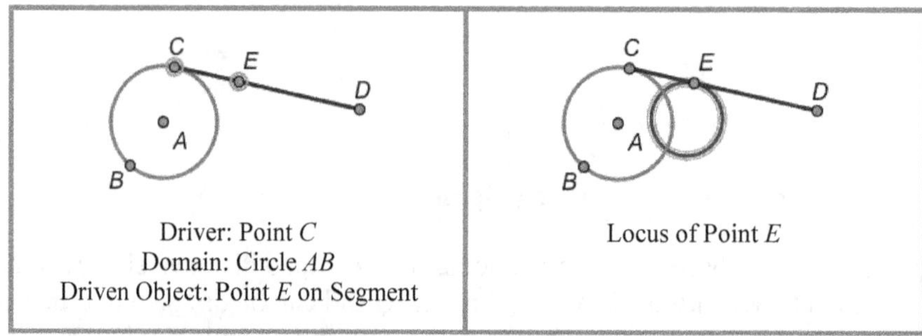

Driver: Point *C*
Domain: Circle *AB*
Driven Object: Point *E* on Segment

Locus of Point *E*

**Locus of a point, driven by another point:** In the illustration below left, point *P* is the intersection of line *j* (the perpendicular bisector of segment *CF*) and line *k* (the perpendicular to *d* through *C*). This construction guarantees that point *P* is equally distant from point *F* and segment *d*. The illustration below right shows a parabola: the locus of point *P* as point *C* moves along segment *d*.

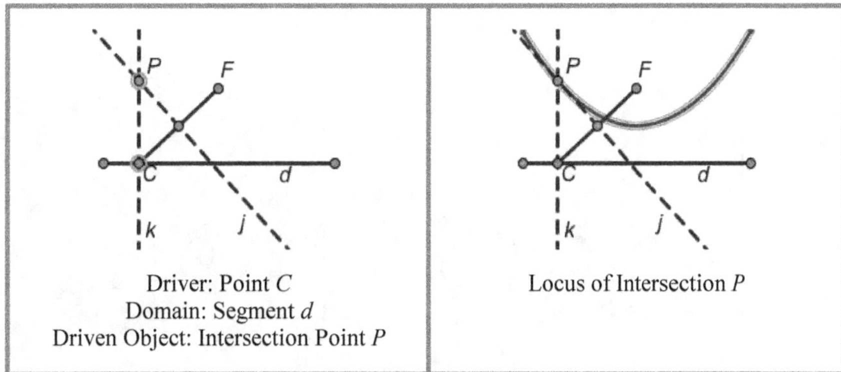

Driver: Point $C$
Domain: Segment $d$
Driven Object: Intersection Point $P$

Locus of Intersection $P$

**Locus of a picture, driven by a parameter:** In the illustration below left, parameter *Elevation* has been used to rotate point $P$ to $P'$, and a picture has been attached to $P'$. On the right is the locus of the picture as the parameter varies from $0°$ to $180°$.

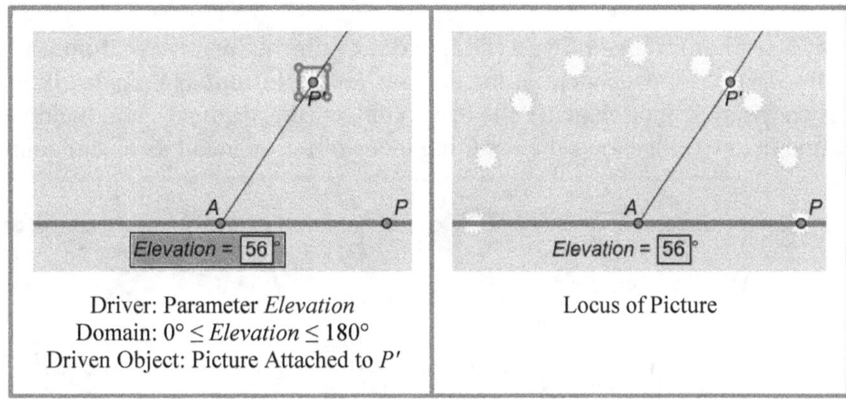


Driver: Parameter *Elevation*
Domain: $0° \leq Elevation \leq 180°$
Driven Object: Picture Attached to $P'$

Locus of Picture

**Locus of a segment, driven by a point:** In the illustration below left, segment $k$ is constructed to have a slope determined by a calculation involving its $x$ and $y$ values. As driver point $A$ moves along segment $CD$, the calculation's value changes, and so does the slope of segment $k$. The locus below right shows the locus of the slope segment for various positions of point $A$ on its domain.

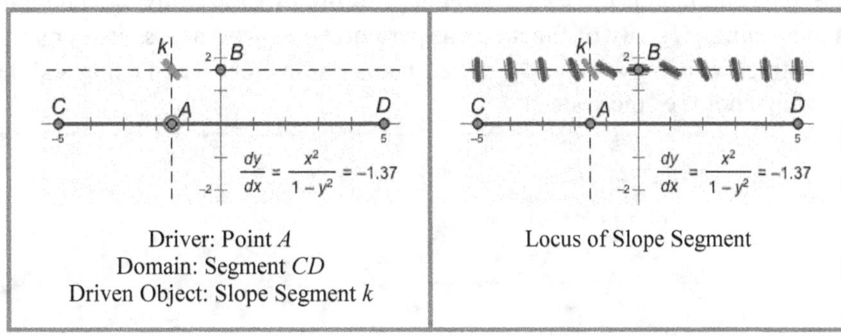


Driver: Point $A$
Domain: Segment $CD$
Driven Object: Slope Segment $k$

Locus of Slope Segment

**Locus of a circle, driven by a point:** In the illustration below left, circle $AB$ was constructed first, and then circle $CB$ was constructed with point $C$ defined on circle $AB$. Point $C$ is the driver, circle $CB$ is the driven object, and circle $AB$ is the domain. (As before, there's no need to select the domain, because point $C$ is constructed on it.) The illustration below right shows the resulting locus of circle $CB$.

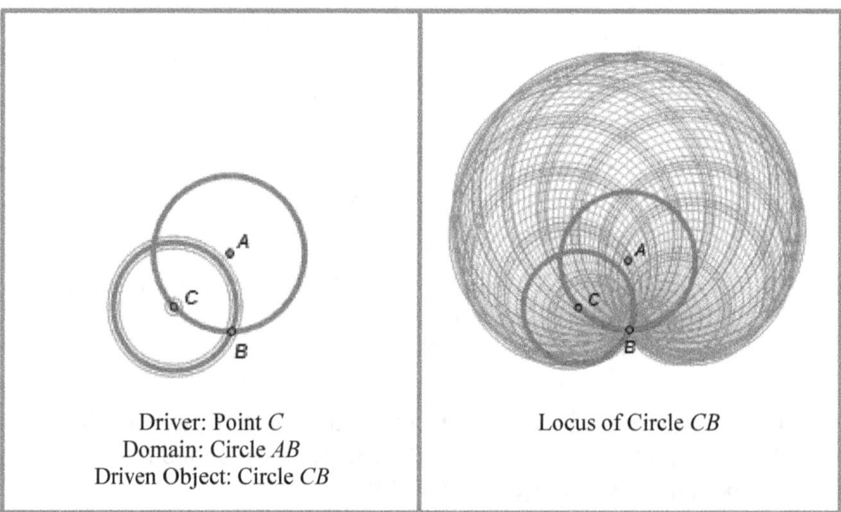

Driver: Point $C$
Domain: Circle $AB$
Driven Object: Circle $CB$

Locus of Circle $CB$

**Locus of a segment, driven by an independent point using an arc domain:** In the illustrations below, the driver is independent of the domain, so the selections include all three objects: the driver (the left endpoint of the segment), the driven object (the segment), and the domain (the arc). The illustration below right shows the resulting locus of the segment as its endpoint moves along the arc.

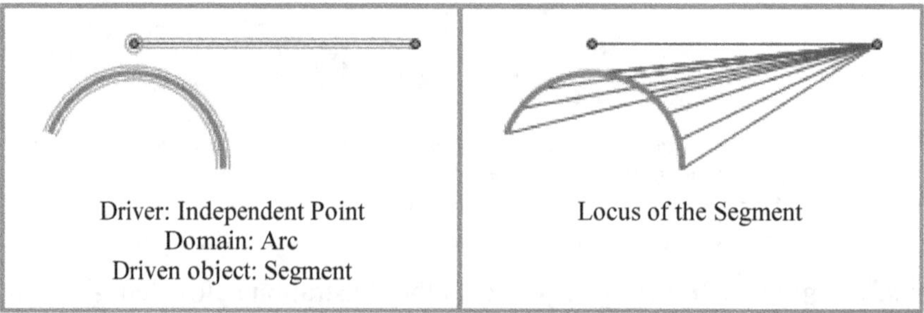

Driver: Independent Point
Domain: Arc
Driven object: Segment

Locus of the Segment

**Family of functions (locus of a function plot), driven by a parameter:** These illustrations show a family of functions. In this example the driver is parameter $a$. With the driver and the function plot selected, the command becomes **Construct | Family of Functions** 190. The illustration below right shows the resulting family of functions as parameter $a$ varies across its domain. (You can do this same construction using a slider 222 rather than a parameter. When using a slider, the driver would be the adjustable point of the slider.)

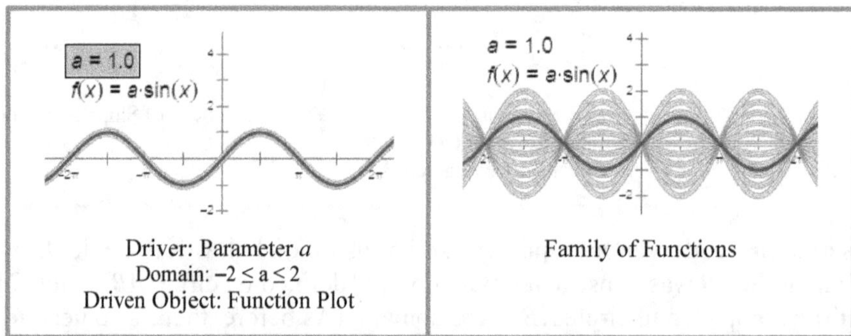


Driver: Parameter $a$
Domain: $-2 \le a \le 2$
Driven Object: Function Plot

Family of Functions

**Family of curves (locus of a point locus), driven by another point:** These illustrations show a family of point loci. In this example the driver is point $F$, the focus of the parabola locus. With the driver and the locus selected, the command becomes **Construct | Family of Curves** 190. The illustration below right shows the resulting family of parabolas as focus $F$ moves along its segment.

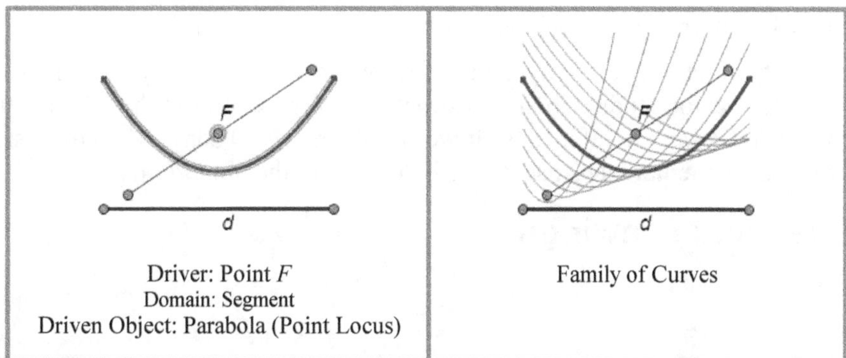

Driver: Point *F*
Domain: Segment
Driven Object: Parabola (Point Locus)

Family of Curves

**Family of loci (locus of a non-point locus), driven by a point:** These illustrations show a family of loci. Driver point *B* can move along the vertical segment, determining the vertical position of the red slope segment and of the locus of the slope segment. With the driver and the segment locus selected, the command becomes **Construct | Family of Loci** 190. The illustration below right shows the resulting family of loci: the slope field of the differential equation used to determine the slope of the red segment.

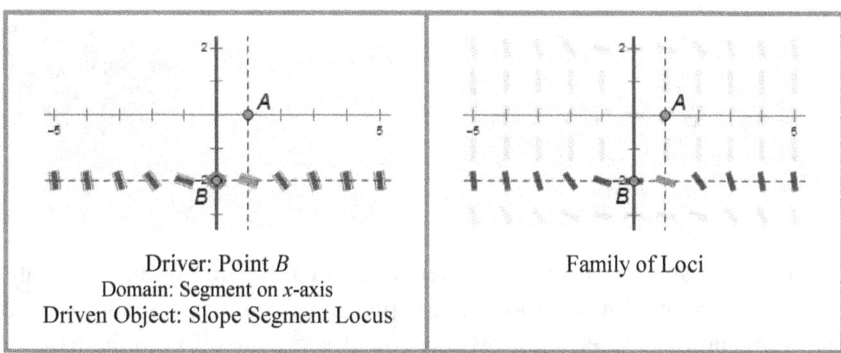

Driver: Point *B*
Domain: Segment on *x*-axis
Driven Object: Slope Segment Locus

Family of Loci

Two metaphors may help you better understand how the various parts of a locus relate to each other. One way of thinking about a locus with a point driver is as an abstract mathematical function, a function in which the elements of the ordered pairs are not necessarily numbers. For a locus driven by a parameter, the first element of the ordered pair is a specific value of the parameter (a number) and the second element is the corresponding position of the driven geometric object. A locus driven by a point is similar, but the first element is the position of the driver point. Sketchpad moves or varies the independent variable (the driver) along its domain (its path), while keeping track of the position of the dependent variable (the driven object). Each sample of the locus represents one value of the function, and the entire locus is an approximation of the range of the function. (It's an approximation because Sketchpad uses only a finite number of ordered pairs — or samples — in constructing the locus.) The abstract mathematical function in this analogy is the construction by which the independent variable (the driver) determines the position of the dependent variable (the driven object).

A second way of thinking about a locus is as a durable form of a traced 170 animation 171. An animating point or parameter (the driver) moves or varies along its domain and defines the position of some object (the driven object). If you trace that object, you eventually see *all* of its positions (its *locus*) for the animated positions or values of the driver. (Where animation and tracing require time and motion to trace out the locus of an object, a construction of its locus gives you the entire result instantaneously, allowing you to use time and motion to then further explore or manipulate that result.)

Choose **Edit | Properties | Plot** 95 to set the number of samples and to determine whether the locus is displayed in discrete or continuous form.

The difficulty of naming these dynamic concepts has a long history: When Johan De Witt and Sir Isaac Newton studied locus constructions of the conics in the 17th century, they used the term *directrix* to refer to what we call the *driver*. Today, when discussing the same type of locus, mathematicians use directrix to refer to the *domain* instead! 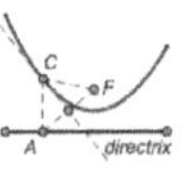

## 1.7     Pictures and Drawings

Pictures add color and interest to your sketches, and allow you to explore mathematical transformations[192] in dramatic ways. These pictures can be digital photos, scanned diagrams, images from the web, and even freehand drawings you create yourself with the **Marker**[122] tool. You can control their appearance by layering them, varying their opacity, and resizing them.

The term *drawing* refers to any picture you create with the **Marker** tool. Anything you can do to a picture in Sketchpad you can also do to a drawing.

You can use pictures in your geometric and mathematical explorations:

- Import a picture[19] from another program — or from the web — by copying and pasting or by dragging and dropping.

- Import a picture from Sketchpad's picture gallery by choosing **Help | Picture Gallery.**

- Attach a picture[19], as you import it, to one, two, or three points in order to control its location, size, and shape dynamically.

- Create a freehand drawing using the **Marker**[122] tool.

- Determine the location, size, and shape of an unattached picture by translating it or by rotating or dilating its resize handles[109].

- Determine the location, size, and shape of an attached picture by moving the point(s) to which it's attached.

- Determine the transparency of a picture using its Opacity Properties[92].

- Set the layer of a picture using its Context[245] menu.

- Transform pictures geometrically using the **Translate**[200], **Rotate**[203], **Dilate**[206], and **Reflect**[207] commands.

- Transform pictures using powerful custom transformations[211]. The resulting sampled transformed pictures[23] can involve a wide variety of special effects.

- Construct the locus[10] of an attached picture as a point or parameter varies. (The picture must depend on the point or parameter.)

- Construct the iterated image[24] of an attached picture by iterating a point or parameter upon which it depends.

- Crop[152] a picture to a polygon, so that the only visible portion of the picture is the portion that intersects the polygon.

- Define a function[232] based on a picture or drawing of its graph.

See Importing Pictures[19] for details on importing or creating a picture, attaching a picture to one or more points, replacing a picture, and determining the opacity of white portions of an imported picture.

See Transforming Pictures[22] for details on resizing a picture or drawing, reshaping a picture or drawing, and other ways of transforming a picture or drawing.

See Working with Pictures[23] for details on constructing the locus or iterated image of a picture drawing, changing the color of a drawing, and using a drawing or picture to define a function.

## 1.7.1  Importing Pictures

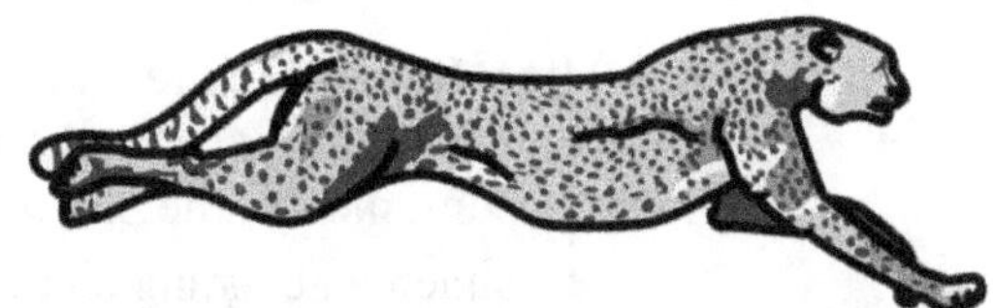

To bring a picture into Sketchpad from some other application, such as a web browser, use either of two methods: drag and drop or copy and paste. Drag and drop is often more convenient, but not all other applications support it. Copy and paste takes a few steps more, but is more widely supported, and can also allow you to define a picture's size and shape based on selected points.

Both methods allow you to import many kinds of pictures, including **.jpg, .gif, .png, .bmp,** and **.tif.**

### ▼ Drag and Drop a Picture

You can often drag either a picture or a picture file directly from some other application into your sketch window.. When you drag the picture or picture file over the sketch window, the pointer displays a + sign. Position the picture where you want it to appear and release the button to drop the picture.

You can drop the picture onto a point to attach the picture to the point[20].

You can drop the picture onto an existing picture to replace the existing picture[20].

Most browsers and many other applications support drag and drop, as do the desktops of both Mac and Windows systems.

To use a picture from an application that doesn't support drag and drop, copy the picture from that application and paste it into Sketchpad.

## ▼ Copy and Paste a Picture

You can copy a picture from another application and paste it into your sketch.

1. Select and copy the picture in the other application.

2. Switch to Sketchpad and choose **Edit | Paste Picture** [147].

    If there are no selected points in your sketch when you paste a picture, the result is an independent picture, not attached to any other sketch object.

    If there are one, two, or three selected points, the picture will be attached to the points.

    If there's a single selected picture (or a selected transformed or iterated image of a picture), the new picture will replace the selected one.

## ▼ Draw a Picture with the Marker Tool

To create a picture from scratch, choose the **Marker** [122] tool and begin drawing in an empty part of the sketch. Anything you draw will be included in the drawing until you choose a different tool or choose a menu command.

Use the resulting drawing like any other picture. You can even copy it or cut it and then paste it to attach it to one, two, or three points.

## ▼ Attach a Picture to Points

Control a picture's location dynamically by attaching it to a point. Control the size or shape of a picture by pasting it onto two or three selected points.

**One Point:** Paste a picture onto a single point, or drag a picture from another application and drop it onto a point. The new picture is centered on the point; when the point moves, the picture moves with it.

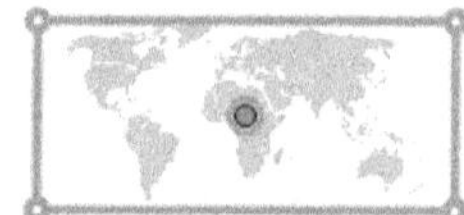

    To attach an existing picture to one point, select both the picture and the point and choose **Edit | Merge Picture to Point** [153].

**Two Points:** Paste a picture onto two points. The new picture is anchored between the two selected points; when the points move, the picture's location, size, and shape change to match the imaginary rectangle defined by the two points.

**Three Points:** Paste a picture onto three points. The new picture is stretched across the three points, with the bottom-left corner of the picture attached to the first point, the bottom-right corner to the second point, and the top-left corner to the third point. When the points move, the picture's location, size, and shape change to match the imaginary parallelogram defined by these three points.

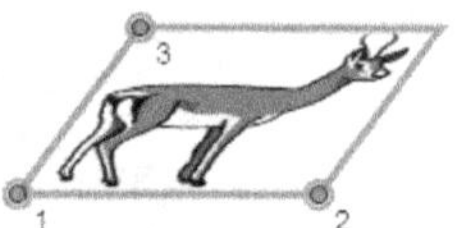

To attach an existing picture to two or three points, select the picture and choose **Edit | Cut** [145]. Then select the points and choose **Edit | Paste** [147].

## ▼ Replace an Existing Picture

Use either drag and drop or copy and paste to replace an existing picture.

### Drag and Drop:

1. Drag a picture from some other application.

2. In your sketch, drop it onto an existing picture or onto a transformed or iterated image of an existing picture.

### Copy and Paste:

1. Copy a picture from another application,

2. In Sketchpad, select a single picture or a transformed or iterated image of a picture.

3. Choose **Edit | Paste Replacement Picture** 147. The new picture replaces the existing one.

## ▼ Control the Transparency of White Portions When Importing a Picture

Sketchpad normally makes pure white portions of an imported picture completely transparent, so that you can see through it to other pictures or interiors underneath. (The transparency of these white portions of the picture is independent of the picture's opacity setting, 92 which applies to the rest of the picture.) If you hold the Shift key while you drop or paste a picture into Sketchpad, pure white portions of your picture will be treated like the rest of the picture, with their opacity determined by the picture's Opacity 92 properties.

Points, lines, circles, text, and all other non-layered objects are always drawn on top of pictures.

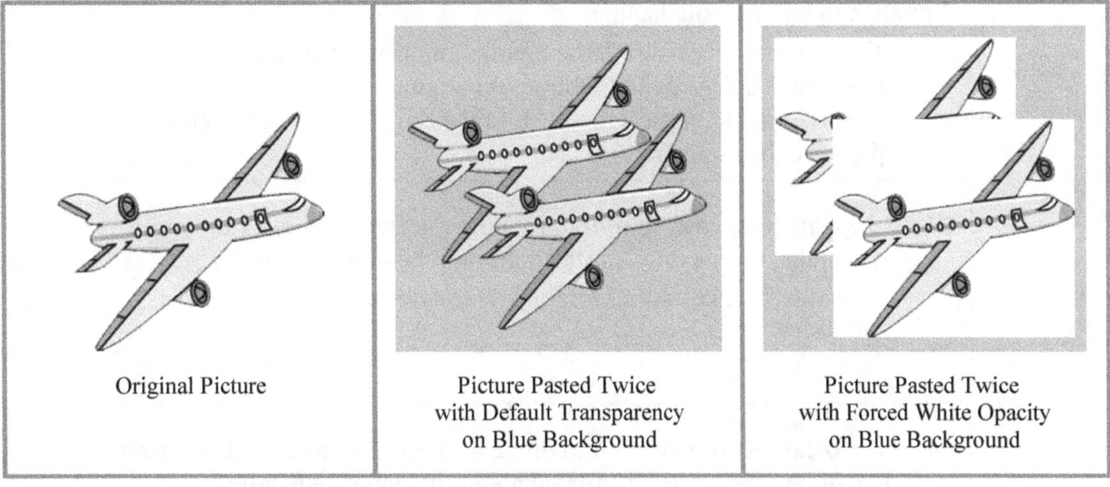

| Original Picture | Picture Pasted Twice with Default Transparency on Blue Background | Picture Pasted Twice with Forced White Opacity on Blue Background |

## ▼ Control the Image Size and Resolution When Importing a Picture

When you drop a picture that contains size information, it's normally imported at its original size, even though the picture may have a higher resolution than the screen. For instance, if you import a picture that's 0.5 inches square at 600 dpi, the original has 300 dots per side. When you import it to Windows, which normally has a screen resolution of 96 dpi, the picture will still be 0.5 inches square, so will have 48 dots per side.

To import the picture at its full original detail, hold the Alt key (Windows) or Option key (Mac) when you drop it. In the example, the picture will appear in Sketchpad with 300 dots per side, and will measure 300/96 = 3.125 inches on a side.

## 1.7.2   Transforming Pictures

Transformed images of pictures and drawings provide compelling and revealing ways to view and understand geometric and mathematical transformations. Sketchpad provides a number of ways of transforming pictures and creating objects based on pictures.

### ▼ Translate, Rotate, or Scale an Original Picture

If an original picture is independent or attached to a single point, you can transform the original in various ways.

 Note

If the picture is attached to two points or to three points, you must move the points to which it is attached in order to transform the original. You can still create transformed images of the original using either built-in or custom transformations.

- Use the **Translate Arrow** |104| tool to translate a selected picture to a different location in your sketch.

- Use the **Translate Arrow** |104| tool to scale a selected picture by dragging a resize handle vertically, horizontally, or both. The picture is resized relative to the opposite handle.

  You can stretch or shrink the picture independently in the horizontal and vertical directions, changing its aspect ratio (ratio of horizontal to vertical size). Hold the Shift key while resizing to preserve the aspect ratio.

- Use the **Rotate Arrow** |104| tool to rotate a selected picture by dragging a resize handle. The picture is rotated around the opposite handle.

  Hold the Shift key while rotating to constrain the rotation to multiples of 15°.

- Use the **Dilate Arrow** |104| tool to dilate a selected picture by dragging a resize handle. The picture is dilated about the opposite handle.

  The picture's aspect ratio (ratio of horizontal to vertical size) remains constant.

### ▼ Transform a Picture Using the Translate, Rotate, Dilate, and Reflect Commands

Use Sketchpad's built-in transformations to create a transformed image of a picture.

Select the picture and choose **Translate** |200|, **Rotate** |203|, **Reflect** |207|, or **Dilate** |206| from the Transform |192| menu. Depending on the transformation you want, you may need to mark various objects or values to serve as the center, distance, angle, ratio, or mirror.

These four transformations, and any transformation that combines them, are called *similarity transformations* because the transformed image is similar to the original: angles and ratios of distances are preserved.

## ▼ Transform a Picture Using a Custom Transformation

Use Sketchpad's custom transformations to create an arbitrary transformed image of a picture.

Use two points, one of which depends on the other, to define a custom transformation [211].

Select the picture you want to transform.

Choose the custom transformation you want from the bottom of the Transform menu.

If the two points that define the custom transformation are related using only Sketchpad's built-in transformations, the result is a normal transformed picture, just as if you had applied the various transformations directly to the original picture.

> Such a normal transformed picture can be displayed more quickly and with less distortion than a sampled picture.

If the two points that define the custom transformation are related in some other way, the result is a sampled transformed picture.

## ▼ Sampled Transformed Pictures

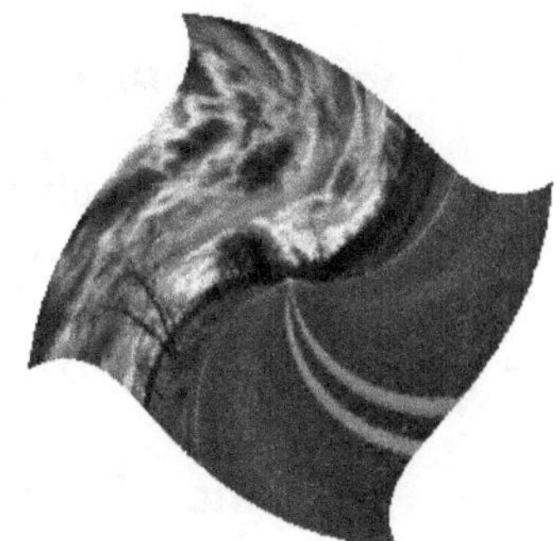

A *sampled transformed picture* is a custom transformed image of a picture in which the custom transformation doesn't rely strictly on Sketchpad's built-in transformations of translation, rotation, reflection, and dilation.

Transformed images of pictures constructed using the built-in transformations retain the essential shapes of their pre-images, though their size and orientation may change.

> A rectangular area in the pre-image always corresponds to a similar rectangular area in the image.

There is no guarantee of similarity for a transformed image of a picture constructed in other ways. The resulting image must be displayed using a sampling process: The transformed picture is displayed by sampling rectangles within the pre-image and then transforming each sample to create the transformed image. The smoothness and level of detail of the transformed path depend on the number of samples used. To change the number of samples and the allowable distortion, select the transformed picture and choose **Edit | Properties | Plot** [98].

## 1.7.3    Working with Pictures and Drawings

Transformed images of pictures and drawings provide compelling and revealing ways to view and understand geometric and mathematical transformations. Sketchpad provides a number of ways of transforming pictures and creating objects based on pictures.

### ▼ Create a Locus of a Picture or Drawing

To create the locus of a picture, the shape or position of the picture must depend on a point that can move along a path or on a parameter that can vary. Select the picture and the point or parameter and choose **Construct | Locus** 190. The locus of the picture appears.

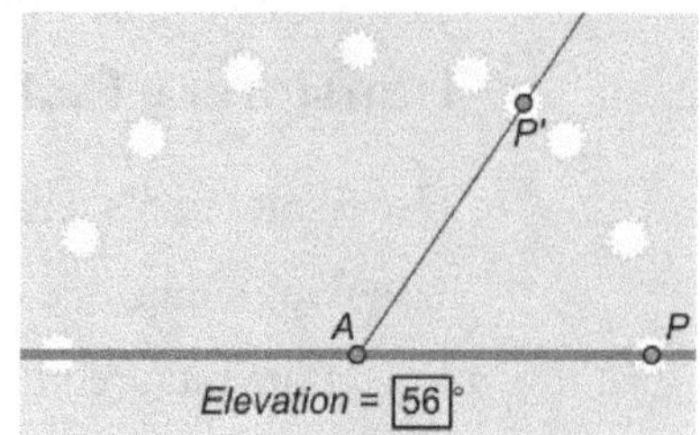


Driver: Parameter *Elevation*
Domain: $0° \leq Elevation \leq 180°$
Driven object: Picture Attached to $P'$

### ▼ Create an Iterated Image of a Picture or Drawing

To create an iterated image of a picture, the shape or position of the picture must depend on a pre-image point or parameter. Select the pre-image object(s) and choose **Transform | Iterate** 208. The picture is iterated along with the points it depends on.

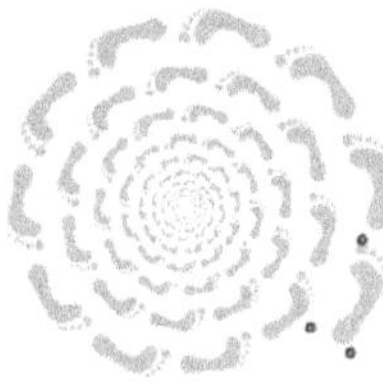

### ▼ Crop a Picture or Drawing to a Polygon

To display only the portion of a picture that overlaps a polygon, select the picture and polygon and choose **Edit | Crop Picture to Polygon** 152. The original picture is hidden, and the new cropped image appears in a layer on top of the polygon.

### ▼ Define a Function Based on a Drawing or Picture

To define a function based on the top edge of a picture or drawing, select it and choose **Number | Define Function from Drawing** 232. The value of the function is determined by the highest opaque pixel at each location in the function's domain.

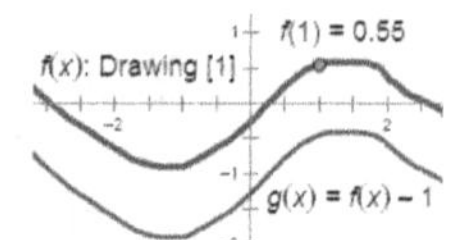


Use the Marker tool to draw the desired graph of your function, and then define a function based on your drawing. You can use it in the same way as other functions: evaluate it, transform it, and so forth.

You can even modify the function by creating a new drawing, copying it, and choosing **Edit | Paste Replacement Picture** 147.

### ▼ Change the Color of a Drawing

After you create a drawing using the **Marker** 122 tool, you can change its color using **Display | Color** 164.

## 1.8    Iterations and Iterated Images

An iteration is a rule that defines a repeated series of geometric or numeric operations. Each step of the iteration is performed on the results of the previous step, and the iteration rule defines how the results of one step are used to perform the next step.

An iterated image object is the set of iterated objects corresponding to a single constructed or calculated object from the original operation.

Use the **Transform | Iterate** [208] command to create an iteration.

An iteration can involve geometric constructions [26], numeric operations [28], or a combination of both.

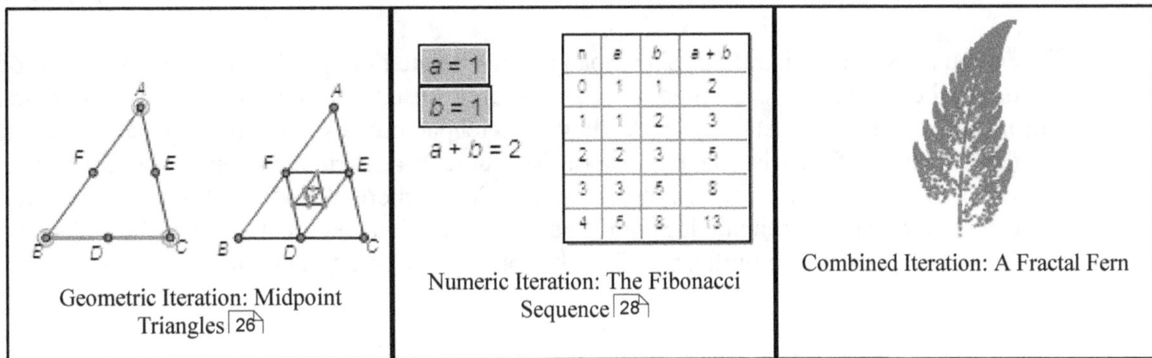

Geometric Iteration: Midpoint Triangles [26]

Numeric Iteration: The Fibonacci Sequence [28]

Combined Iteration: A Fractal Fern

The geometric constructions or numeric operations must define an output in terms of some input. The input can be independent points [2], points on a path [89], parameters [36] or independent calculations [39]. The iteration uses the output of one step as the input for the next step, and creates an iterated image object for each output object. The iterated image object shows the set of iterated positions or values of the original object on which it's based. This is called the *orbit* of that object.

**A geometric iteration** repeatedly applies a particular geometric construction to a set of points. At each step the construction is based on the point(s) produced by the previous step. For instance, you could construct a triangle from three points and then translate the original points to the right by 2 cm. When you iterate this construction, the first step constructs a new triangle from the translated points and produces three new translated points. Each additional iteration produces a new triangle and three new translated points. The sequence of triangles is the *orbit* of the original triangle you constructed.

**A numeric iteration** repeatedly applies a particular calculation. At each step the calculation uses the result(s) of the previous step. For instance, you can apply the calculation "add 2" repeatedly, starting with the *seed* value $a = 5$. The first step produces $5 + 2 = 7$, the second step produces $7 + 2 = 9$, and so forth. The sequence of output values (7, 9, 11, 13, ...) is the *orbit* of $a + 2$.

Sketchpad displays the orbit of iterated numeric or algebraic objects in a table. The first column shows the depth of iteration; additional columns show the orbits of the iterated numeric or algebraic objects.

In these examples, it's helpful to think of the iteration as having two distinct elements: the original calculation or construction (the seed or pre-image) and the mapping that takes each value or point to the next value or point in the sequence. Thus you might say that 5 maps to 7 in the numeric iteration, and that 47 maps to 49. Similarly, you might say that A maps to A' in the geometric iteration, and that A' maps to the next translated point. The entire iteration is defined by the pre-image (the seed) and the mapping. When you first apply the mapping to the pre-image, the result is the *first image*. As you continue to apply the mapping, you generate the second image, the third image, thes fourth image, and so on.

Most iterations use a single mapping to produce a single image of the original for each step of the iteration. However, some iterations (like this Sierpiński Gasket 33) produce multiple copies at each step 211.

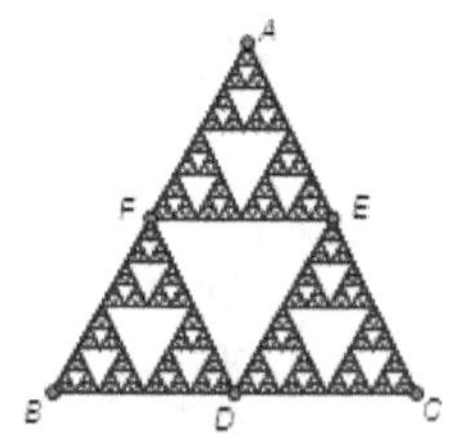

When you construct an iteration, Sketchpad creates iterated image objects of the objects produced by the original construction, even those objects that aren't used as input for the next step of the iteration. For instance, in the geometric example above, Sketchpad creates iterated image objects not only of the three vertices of the triangle, but also of the segments and triangle interior. Objects that can be iterated include geometric objects, measurements, loci, functions and function plots, pictures, and text attached to a point. (Sketchpad does not produce iterated image objects for families of functions, for families of loci, for action buttons, or for other previously-constructed iterated images.)

*Subtopics:*


*See Also:*


## 1.8.1    A Geometric Iteration

When you create an iteration, it can be a geometric iteration (based on independent points), a numeric iteration 28 (based on parameters), or a combination of the two. Here's an example of using a geometric iteration to create an iterated midpoint triangle.

To define a Sketchpad iteration, create the pre-image and then specify how its defining points or parameters should be mapped to produce the first iteration image. Sketchpad handles the iteration from there.

For instance, here are the pre-image and the first three images for the midpoint triangle iteration.

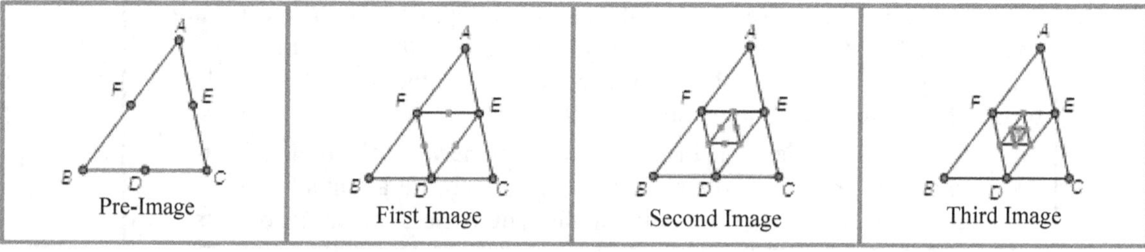

|  |  |  |  |
| --- | --- | --- | --- |
| Pre-Image | First Image | Second Image | Third Image |

The first image repeats the pre-image construction, but starting from midpoints *D*, *E,* and *F*. The map that defines this iteration is shown on the right.

Follow the steps below to construct this iterated midpoint triangle.

$$A \rightarrow D$$
$$B \rightarrow E$$
$$C \rightarrow F$$

1. Construct the pre-image triangle exactly as shown above.

 Note

Don't construct more objects than necessary in the pre-image. In this example, the pre-image construction includes three sides of the triangle and the three midpoints *D, E,* and *F.* Segments *DE, EF,* and *DF* will be constructed as part of the first image. If you also constructed these segments as part of the pre-image, they would be constructed again, resulting in unnecessary duplicated segments.

2. Select the independent seed points *A, B,* and *C* that define the pre-image.

3. Choose **Transform | Iterate** 208. The Iterate dialog box appears with the left column filled in.

 Note

You may need to drag the dialog box out of the way so that you can see your original triangle and its midpoints.

Use the Structure and Display pop-up menus 209 to control how the iteration is constructed and displayed.

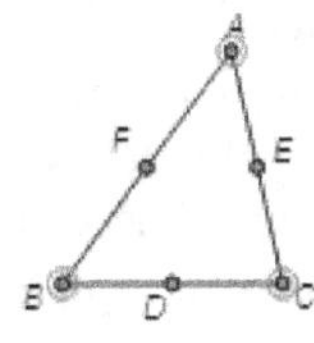

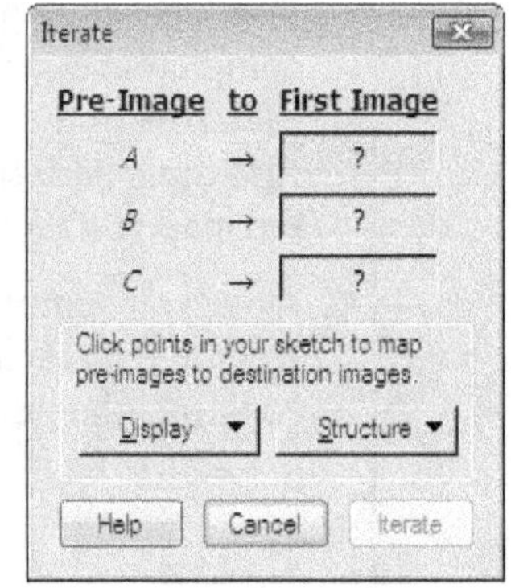

4. Click point *D* in the sketch to fill in the top box of the right column. This specifies that point *D* should play the same role in the first image that point *A* played in the pre-image,

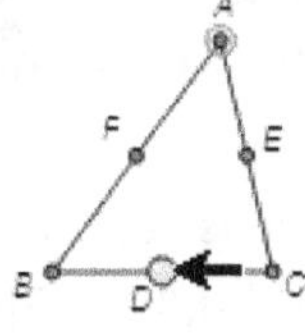

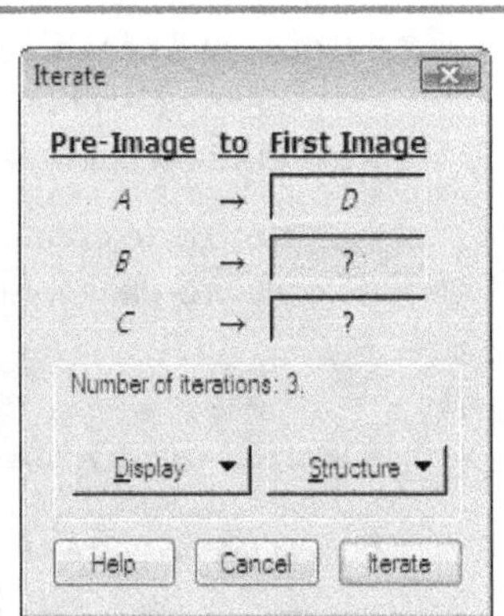

5. Similarly click points *E* and *F* in the sketch to complete the iteration map. As you complete each row of the map, Sketchpad displays the first three partial images based on the part of the map you've already completed.

6. After you complete the entire map, click the **Iterate** button to confirm the iteration rule and construct the iteration images.

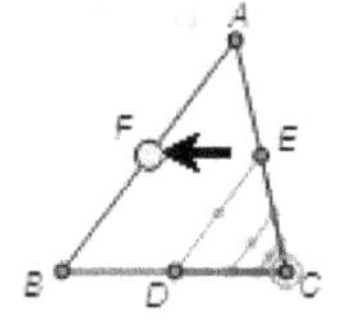
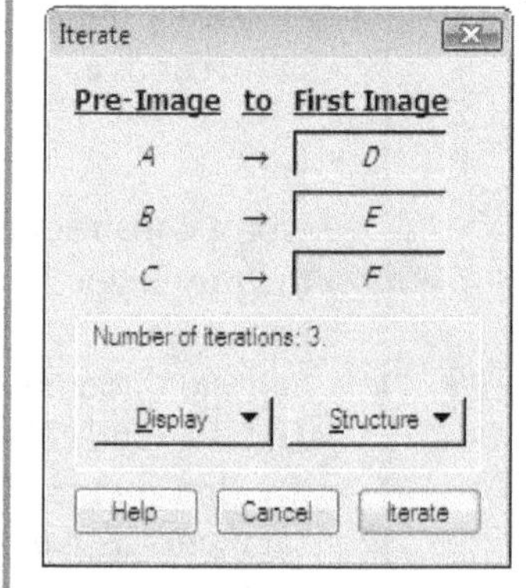

Sketchpad produces an iterated image object for each object affected by your mapping. In this example there are six iterated image objects, one for each vertex of your original triangle and one for each side. You can select and manipulate each iterated image object separately. For example, you could hide or clear[147] the three iterated image objects of your original triangle's vertices, or you could color each of the three iterated image objects of your original triangle's edges with a different color. 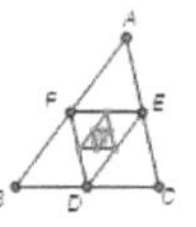

Sketchpad normally displays iterations to a depth of three, showing the first, second, and third iteration image. Use Iteration Properties[35] to change the depth of the iteration in order to show more or fewer iteration images, or select any of the iterated image objects and press + or − on the keyboard.

Note

You can also use a numeric value (a parameter or a calculation) to control the depth of an iteration. After selecting the seed points and/or values, select one additional parameter or calculation, and then hold Shift and choose **Transform | Iterate to Depth**[208]. (Notice the change in the command name.) The whole number part of the value you selected last will define the number of iterations. (When you choose **Iterate,** the minimum depth is one; when you choose **Iterate to Depth,** the minimum depth is zero.)

Use **Iterate** to create fractals such as the Sierpiński Gasket[33] by specifying more than one mapping of your pre-image. See Multiple Iteration Maps[211] for more information.

*See also:*
*A Numeric Iteration*[28]
*Working with Iterations*[30]
*How to Construct a Sierpiński Gasket*[33]
*Iteration Properties*[35]

## 1.8.2   A Numeric Iteration

When you create an iteration, it can be a geometric iteration[26] (based on independent points), a numeric iteration (based on parameters), or a combination of the two. Here's an example of using a numeric iteration to create a Fibonacci sequence.

To define a Sketchpad iteration, create the pre-image and then specify how its defining points or parameters should be mapped to produce the first iteration image. Sketchpad handles the iteration from there.

For instance, here are the pre-image and the first three images for a Fibonacci sequence iteration. The first image repeats the pre-image calculation, but instead of $a$ and $b$ uses the values of $b$ and $a + b$ from the previous step.

|              | a | b | a+b |
|--------------|---|---|-----|
| Pre-image    | 1 | 1 | 2   |
| First Image  | 1 | 2 | 3   |
| Second Image | 2 | 3 | 5   |
| Third Image  | 3 | 5 | 8   |

Here's the map that shows how to go from one step of the iteration to the next. It indicates that each new image will be created by starting with the values of $b$ and $a + b$ from the previous step.

$$a \rightarrow b$$
$$b \rightarrow a + b$$

Follow the steps below to construct this iterated Fibonacci sequence.

<table>
<tr>
<td>

1. Construct the pre-image by creating new parameters 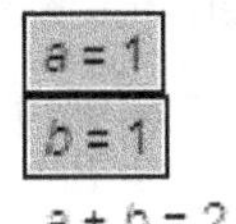 226 $a$ and $b$. Choose **Number | Calculate** 227 to compute $a + b$.

2. Select the parameters $a$ and $b$ that define the pre-image.

3. Choose **Transform | Iterate** 208. The Iterate dialog box appears with the left column filled in with the starting parameters from the pre-image.

>You may need to drag the dialog box out of the way so that you can see your parameters and calculation.

</td>
<td>

$a = 1$
$b = 1$
$a + b = 2$

</td>
<td>

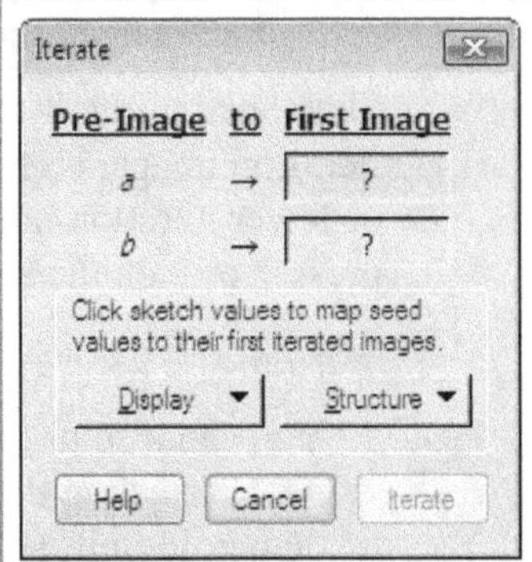


</td>
</tr>
<tr>
<td>

4. Click parameter $b$ in the sketch to fill in the top box of the right column. This specifies that parameter $b$ should play the same role in the first image that parameter $a$ played in the pre-image.

</td>
<td>

$a = 1$
$b = 1$ ◄
$a + b = 2$

</td>
<td>

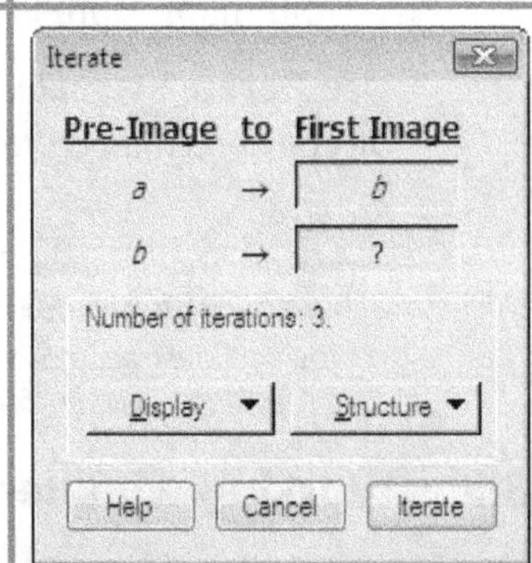


</td>
</tr>
<tr>
<td>

5. Similarly click $a + b$ in the sketch to complete the iteration map. This specifies that the result of $a + b$ should play the same role in the first image that parameter $b$ played in the pre-image.

</td>
<td>

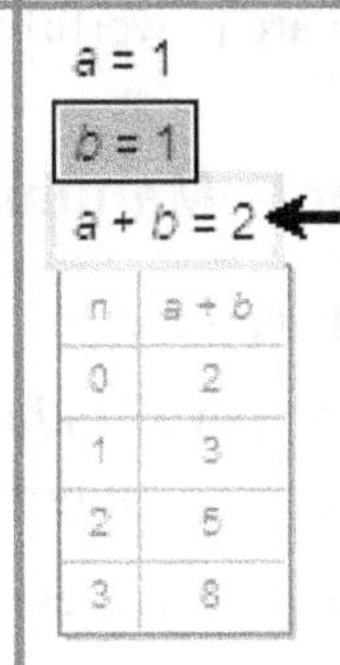

$a = 1$
$b = 1$
$a + b = 2$ ◄

| n | a + b |
|---|-------|
| 0 | 2     |
| 1 | 3     |
| 2 | 5     |
| 3 | 8     |

</td>
<td>

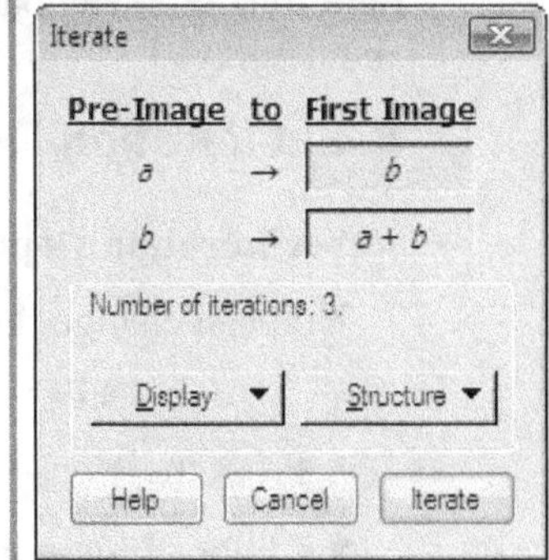


</td>
</tr>
</table>

6. After you complete the entire map, click the
   **Iterate** button to confirm the iteration rule.
   Sketchpad shows the result of this iteration as a
   table containing the results of $a + b$ for each step
   of the iteration.

| n | a + b |
|---|-------|
| 0 | 2 |
| 1 | 3 |
| 2 | 5 |
| 3 | 8 |

Sketchpad produces an iterated image object for each object affected by your mapping. In this example the pre-image has only a single object (the calculation $a + b$) that depends on the parameters, so the result is a single iterated image object: the table showing the value of the calculation.

If the pre-image contained other numeric or geometric objects that depended on the parameters, the iteration would generate an iterated image object for each such object.

Sketchpad normally displays iterations to a depth of three, showing the first, second, and third iteration image. Use Iteration Properties 35 to change the depth of the iteration in order to show more or fewer iteration images, or select any of the iterated image objects (the table in this example) and press + or − on the keyboard.

 Note

You can also use a numeric value (a parameter or a calculation) to control the depth of an iteration. After selecting the seed points and/or values, select one additional parameter or calculation, and then hold Shift and choose **Transform | Iterate to Depth** 208. (Notice the change in the command name.) The whole number part of the value you selected last will define the number of iterations. (When you choose **Iterate,** the minimum depth is one; when you choose **Iterate to Depth,** the minimum depth is zero.)

Use **Iterate** to create fractals such as the Sierpiński Gasket 33 by specifying more than one mapping of your pre-image. See Multiple Iteration Maps 211 for more information.

*See also:*
*A Geometric Iteration* 26
*Working with Iterations* 30
*How to Construct a Sierpiński Gasket* 33
*Iteration Properties* 35

## 1.8.3   Working with Iterations

Iterations in Sketchpad are powerful and flexible. Here are some of the ways they can be controlled and displayed.

### ▼ Set Iteration Depth Manually or Parametrically

**Set Iteration Depth Manually**

You can change the depth of an iteration in several ways.

To begin, select one or more iterated image objects.

- Press the + or − key. Each press of the key increases or decreases the depth of iteration by one.

- Choose **Increase Iterations** or **Decrease Iterations** from the Context 245 menu.

- Choose **Edit | Properties | Iteration** 35 and type a value for **Number of iterations.**

No matter which method you use, the minimum depth is one, and the maximum depth is limited by the number of iteration samples that can be displayed.

You can choose **Edit | Advanced Preferences | Sampling**[283] to change the maximum number of iteration samples allowed.

**Set Iteration Depth Parametrically**

When you define an iteration, use a measurement, parameter, or calculation[92] in your sketch to automatically determine the depth of the iteration. First select the points or parameters that define the pre-image, and then select the value that defines the iteration depth. (The depth value must be the last object selected.) Hold the Shift key while you pull down the Transform[192] menu, and the **Iterate**[208] command changes to **Iterate to Depth.** Choose this command and use the Iterate dialog box[209] to define the iteration normally. Once the iteration is defined, the integer part of the depth value determines the iteration depth.

Note

If the value is negative, the depth is set to zero.

If the value is too large to display all the iterated images, Sketchpad uses the maximum depth at which the images can be displayed.

When an iteration's depth is determined by the value of a parameter or calculation, you can't set the depth manually.

## ▼ Control the Appearance of Iterated Images

You can select, color, style, hide, or clear[147] individual iterated image objects. In this example, you might want to hide or clear the images of the iterated triangle's vertices, so that only the sides of the triangle are visible in your illustration. Click an image object to select the entire orbit (iterated image) of the corresponding pre-image object. Any change you make in color, point or line style, or visibility will affect the entire orbit of that pre-image object. The appearance of unselected orbits will not be affected.

You can display the full orbit of all objects in the iteration, or only the final iteration of all objects. This example shows the full orbit. If it showed the final iteration only, all inner triangles except the smallest would be hidden. To control this setting when you create the iteration, choose **Full orbit** or **Final iteration only** from the Display pop-up menu of the Iterate dialog box[209]. To change this setting after you've created the iteration, use the Iteration Properties[35] panel.

## ▼ Create or Suppress a Table of Iterated Values

If one or more visible measured values[92] change as a result of an iteration, Sketchpad creates a table[39] of iterated values when you create the iteration. The table contains one column for each visible value affected by the iteration, as well as an initial column — labeled $n$ — that indicates the level of iteration. Each row in the table describes the values of the measurements at the indicated level of iteration.

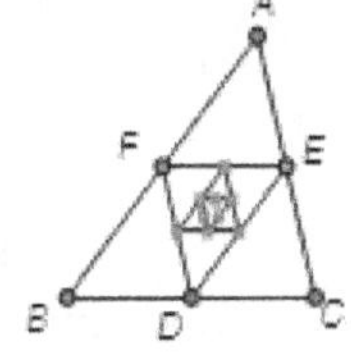

$$seed = 100.00$$

$$\frac{seed}{2} = 50.00$$

| $n$ | $\dfrac{seed}{2}$ |
|---|---|
| 0 | 50.00 |
| 1 | 25.00 |
| 2 | 12.50 |
| 3 | 6.25 |

For example, in a sketch containing a parameter *Seed,* with an initial value of 100, and a calculation *Seed / 2,* if you iterate the pre-image parameter *Seed* to the first image calculation *Seed / 2,* Sketchpad produces this table, containing the iterated images of *Seed / 2* as *Seed → Seed / 2.*

The number of rows in a table of iterated values changes automatically as you increase or decrease the depth of iteration. If you don't want to create the table of iterated values, uncheck **Tabulate Iterated Values** in the Structure pop-up menu of the Iterate dialog box [209]. If you create a table when you create the iteration and decide you no longer need it, select the table and clear [147] it from your sketch.

### ▼ Make Iterated Points on Paths Fixed or Random

Sometimes you may want to specify a point on an object [176] as the first image of an iterated pre-image. For instance, you might map a triangle's vertices to arbitrary points constructed on its three segment edges. In this case, the Structure pop-up menu of the Iterate dialog box [209] allows you to choose how Sketchpad will iterate these arbitrary points on objects.

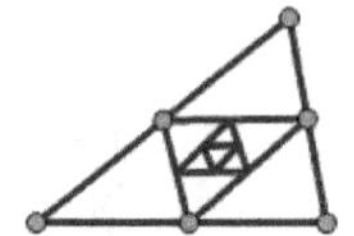

Same Relative Locations

**To Same Relative Locations:** Sketchpad displays each iterated image point at the same relative location as the first image you chose in the Iterate dialog box. If you drag the first image point to a new location on the path on which it's constructed, all of the iterated images adjust to the same relative new location.

**To New Random Locations:** Each iterated image of your initial point appears at a new, random location on its iterated path, independent of the location of the first image. This choice is useful if you're exploring geometric probability or other applications of randomness.

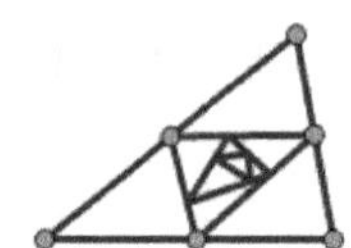

New Random Locations

> After you've constructed an iteration, use its Iteration Properties [35] panel to experiment with both choices for how to iterate points constructed on path objects.

When a selected iteration maps at least one point to a random location, you can randomize the locations either by choosing **Edit | Properties | Iteration** [35] and clicking the **Randomize Now** button or by pressing **!** (exclamation point) on the keyboard.

### ▼ Create the Terminal Point of an Iterated Point Image

On occasion you may want to use the very last point of an iterated point object. You may want to attach a construction to this point or measure some quantity that depends on the point.

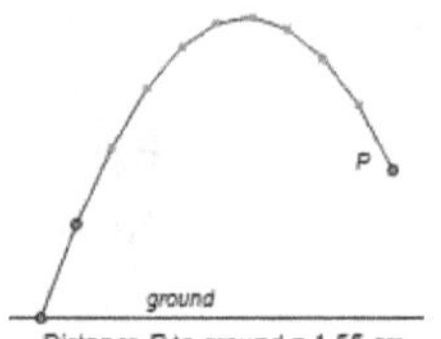

> You cannot construct a terminal point on an iteration that uses multiple iteration maps [211].

To construct the terminal point of an iterated point image, select the iterated image of the point and choose **Transform | Terminal Point** [208]. If the depth of the iteration changes, the terminal point moves accordingly.

> When you select an iterated point image, the **Iterate** [208] command changes to **Terminal Point.**

In this examples, an iteration was used to construct the flight path of a thrown ball. (The pre-image points, not shown here, defined the initial velocity of the ball and the strength of gravity.) To determine the height of the ball after some number of iterations, the terminal point was constructed and used to measure the distance [218] between the terminal point and the ground.

### ▼ Use Multiple Iteration Maps to Create Iterated Fractals and Tessellations

Use multiple iteration maps [211] to make each step of the iteration produce two or more copies of the original objects. Multiple maps allow you to create iterated fractals and tessellations such as the Sierpiński gasket [33], a binary tree, or a parallelogram tessellation.

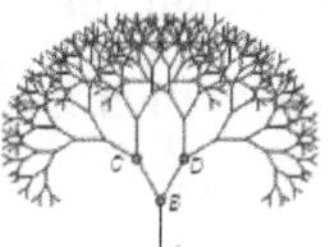 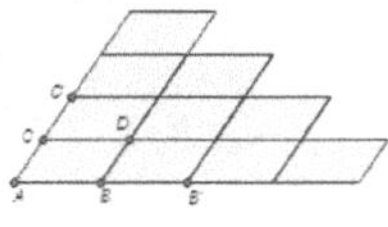

*See also:*


## 1.8.4    How to Construct a Sierpiński Gasket

Because iteration can be applied to any type of Sketchpad construction, the options that support it may at first seem complex. The best way to develop an understanding of iteration is to work through examples. In this example, you'll use iteration to define a fractal known as the Sierpiński gasket. This fractal is the limit of the process of replacing a triangle with three smaller interior triangles, one in each corner of the original triangle. Each of these three smaller triangles is replaced with three even smaller triangles; and so on. Since at each stage you are replacing a pre-image triangle with three different image triangles, you'll need three mappings to define the fractal.

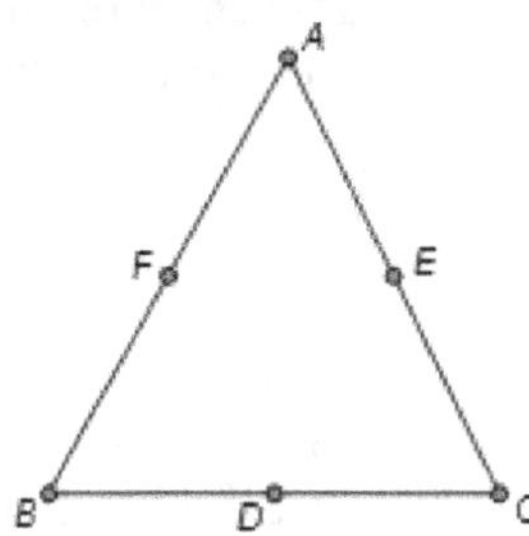

1. In a new sketch, use the **Segment** [115] tool to construct $\triangle ABC$.

2. Construct the midpoints [176] of your triangle's edges. Use the **Text** [120] tool to label the vertices $A$, $B$, and $C$, and the midpoints $D$, $E$, and $F$, as in the illustration.

   This is your pre-image triangle. Note that if you constructed the three corner triangles, $\triangle AFE$, $\triangle FBD$, and $\triangle EDC$, each of them would be similar to the pre-image and half its size.

3. Select the three points [106] $A$, $B$, and $C$, and choose **Transform | Iterate** [208].

4. In the Iterate dialog box, map $A \to F$, $B \to B$, and $C \to D$.

   This maps the original triangle to the lower-left corner $\triangle FBD$. You should see a series of triangles iterating into the lower-left corner of your original triangle.

   Note that in this step you map $B$ to itself, as this vertex is the same in both the original triangle and in the lower-left corner triangle.

5. Use the Structure pop-up menu to add a new mapping to your iteration rule. In the new mapping, map $A \to E$, $B \to D$, and $C \to C$.

   This iterates your pre-image triangle to the lower-right corner, while simultaneously — by the previous map — iterating each image to the lower-left corner.

6. Use the Structure pop-up menu again to add a third and final mapping to your iteration rule. In this

third mapping, map $A \to A$, $B \to F$, and $C \to E$.

This iterates your previous mappings toward the upper corner of the triangle.

7. Click **Iterate** to dismiss the dialog box.

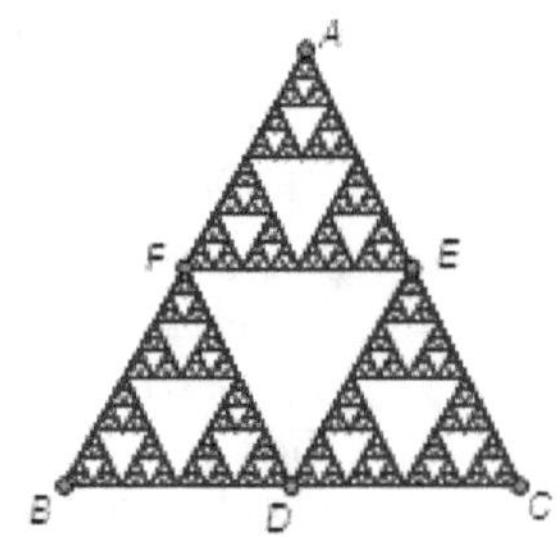

Be careful not to increase the number of iterations too quickly. Because each iteration adds three times as many new triangles as the previous iteration, the construction quickly becomes very complex! Sketchpad will start to slow down if your sketch contains iterations more complex than your computer can handle gracefully.

You can increase or decrease the number of displayed iterations by selecting any of the iterated images and pressing the + or – key on the keyboard.

If you could iterate an infinite number of times, this process would result in a Sierpiński gasket. If you imagine the area of your initial triangle as having been replaced by the area of three smaller triangles at each step, think for a moment about what happens to the total area of all of the smaller triangles as you increase the number of iterations. Because the three smaller triangles didn't cover the initial triangle, the area must be getting smaller. So with each iteration the area of the new triangles becomes smaller. What's the limit of the area? How do you know? What happens to the perimeter? Fractals frequently give rise to surprising properties.

You can visualize the areas by repeating the above steps in a new sketch, using a $\triangle ABC$ in which you have used **Construct | Triangle Interior** 187 to construct the triangle interior. When you're done specifying your three mappings, choose **Final iteration only** from the Display pop-up menu of the Iterate dialog box. After you close the dialog box, select and hide the original interior of $\triangle ABC$. Also hide the iterated images of the three sides of the triangle. (The iterated images of the three vertices are normally already hidden.)

You're left with only the iterated images of the triangle interior, so you can see how changing the depth of iteration affects the total area.

*See also:*

## 1.8.5 Iteration Properties

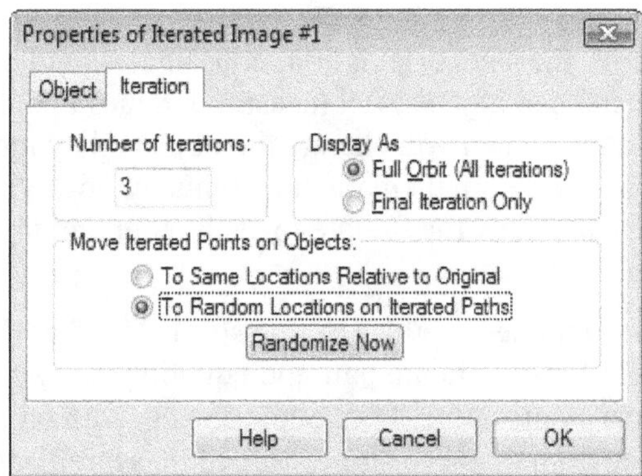

Iteration rules and iterated images 24 have an Iteration Properties 158 panel. Use this panel to set the number of iterations, to specify whether the iterations display all levels or only the final iteration, and to determine how random points in the iteration behave.

To open this dialog box, select any iterated image and choose **Edit | Properties** 158, or choose **Properties** 158 from the Context 245 menu.

**Number of iterations:** This number determines how many times the iteration is repeated. The minimum value you can use is 1; the maximum value depends on the iteration and is smaller for complex iterations that involve more than a single map. If the number of iterations — the iteration depth — was defined by a measurement or calculated value 30 when the iteration was first created, the current depth is displayed here but cannot be edited.

You can adjust the number of iterations without going to Properties by selecting an iterated image and pressing the + or – key while your sketch window is active.

Use the Sampling 283 panel of Advanced Preferences 160 to change the maximum depth allowed for iterations.

**Display as:** Set this to **Full orbit (all iterations)** to display all the iterated images (the images for every level of the iteration). Set this to **Final iteration only** to display only the images at the final level, as set by the **Number of iterations.**

**Move iterated points on objects:** This choice appears only for iterations in which one or more of the pre-image points is a point on a path 176. It determines how iterated images of such points behave. Set this to **Same locations relative to original** in order to have each iterated image appear in the same relative location on its path as the first image does on its original path. Set this to **Random locations on iterated paths** to have each iterated image appear at a random location on its path. When you've set this to **Random locations on iterated paths,** the **Randomize Now** button is enabled; you can click this button to assign each image point on the path a new random position.

You can randomize an iteration without going to Properties by selecting an iterated image and pressing the **!** key while your sketch window is active.

*See also:*

## 1.9  Measurements

Measurements quantify the size, orientation, coordinates, and other characteristics of Sketchpad objects. All measured values update dynamically in Sketchpad when you change the objects they measure. Often you can gain important mathematical insights by observing how measurements change and how they relate to each other and to other objects in the sketch.

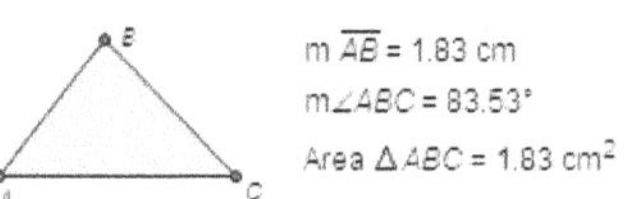


All measurements have numeric values of one sort or another. Most measurements (except for coordinate pair and equation measurements) have a single value. Like other objects with values 92 (parameters and calculations), these single-valued measurements can be serve many purposes 92 in your sketch. For instance, they can be used as distances 199, angles 195, and scale factors 197 in transformations 192; they can define calculations 39 and functions 45; they can be tabulated 39; and they can determine the scale of a coordinate system 41 or the position of a plotted point 239.

To create new measurements, first select the object(s) to measure, then choose a command from the Measure 216 menu.

## 1.10  Parameters

Parameters are simple numeric values 92. Unlike measurements and calculations, they do not depend on other objects for their value. A parameter is defined by a single number and an optional unit.

$t_1 = \boxed{1.0}$

$t_2 = 5°$

$t_3 = \boxed{3.00}$ cm

> Use parameters to define mathematical constructions when you want to explore the effects on the construction of varying a numeric quantity.

There are three kinds of parameters:

- Angle parameters have units of degrees or radians.

- Distance parameters have units of pixels, cm, or inches.

    A pixel is one dot on the screen, and is the smallest distance you can move the pointer.

- Dimensionless parameters have no units.

Once a parameter is defined, you can easily change its value by typing new values or by animating the parameter so that it changes value gradually over some numeric domain.

A parameter that displays an edit box can be edited directly in the sketch.

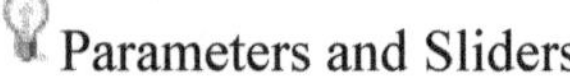

Parameters and Sliders

To create an easily modified numeric value, use either a parameter or a slider 222. A parameter is easier to create, and you can adjust it more precisely and easily animate it over a specific numeric range. A slider is harder to create, but you can manipulate it easily (by dragging the point at the tip) and control it in a continuous and satisfying way.

### ▼ Create a Parameter

- Choose **Number | New Parameter** 226 (or press Shift+Ctrl+P on Windows or Shift-⌘P on Mac), or

- Choose **New Parameter** from the Calculator's 257 Value pop-up menu (or press Shift+Ctrl+P on Windows or Shift-⌘P on Mac).

In the New Parameter dialog box that appears, type an initial value for the parameter and determine the units of the parameter.

 Note

The initial value you type determines both the value and the keyboard adjustment used when you press the + or − key. The keyboard adjustment is based on the least significant digit that you type for the value. For instance, if you type 5, the keyboard adjustment will be 1. If you type 5.0, the keyboard adjustment will be 0.1. And if you type 5.00, the keyboard adjustment will be 0.01. To change the keyboard adjustment later, choose **Edit | Properties | Parameter** 38.

### ▼ Change the Value of a Selected Parameter

$$t_1 = \boxed{2.0}$$

Use the parameter's Value Properties 94 to determine whether it has an edit box allowing its value to be edited directly in the sketch.

- Directly edit the value of a parameter that has an edit box. Click the **Arrow** 104 or **Text** 120 tool in the edit box to begin editing, or select the parameter and press Return or Tab. To finish editing, click outside the edit box or press Esc.

- Double-click the parameter with the **Arrow** 104 tool to use a dialog box to change the value or the label. (If the parameter has an edit box, you must double-click outside the edit box.)

- Choose **Edit | Properties** 158 and type a new number in the **Value** 94 panel.

- Select the parameter and press the + or − key on your keyboard 295.

- Choose **Increase Value** 247 or 247**Decrease Value** 247 from the Context 245 menu.

- Create a Movement button 71 to change the parameter to a particular value.

### ▼ Animate a Parameter

- Select it and choose **Display | Animate** 171, or

- Select it and click the Motion Controller's 263 Animate button, or

- Use an Animation action button 149. Select the parameter and choose **Edit | Action Buttons | Animation** 149.

### ▼ Modify a Parameter's Default Domain, Speed, or Other Properties

To determine how the parameter is animated and how its value changes when you press the + or − key, select the parameter and choose **Edit | Properties | Parameter** 38.

### ▼ Change a Parameter into a Calculation

Select the parameter, choose **Edit | Edit Parameter** 157, and use the Calculator 257 to redefine the parameter as a calculation based on other values.

## 1.10.1  Parameter Properties

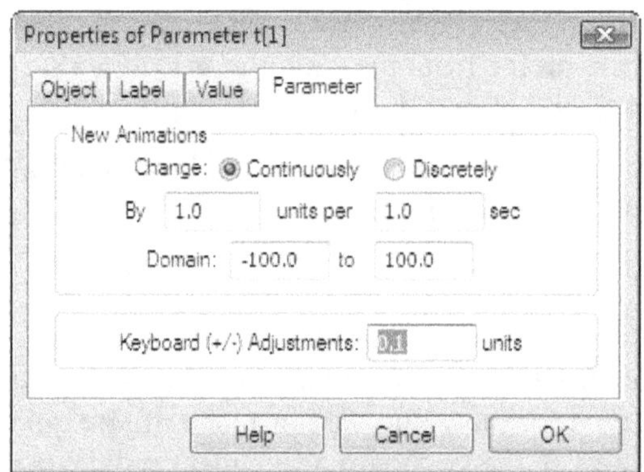

Parameters 36 have a Parameter Properties 158 panel. Use this panel to change the default animation 267 behavior of a parameter 36. The settings on this panel determine how the parameter's value changes when you animate 171 the parameter  or when you select the parameter and press the + or − key. These settings are also used as the initial settings when you create an Animation button 149 that animates the parameter.

To open this dialog box, select the parameter and choose **Edit | Properties** 158, or choose **Properties** 158 from the Context 245 menu.

**Change:** Choose **Discretely** to make the parameter's value jump by the amount in the units box each time it changes. Choose **Continuously** to make the value change gradually instead of by jumps.

**By units per sec:** These numbers determine how quickly the parameter's value changes. This rate is not exact; if your computer is busy with other tasks, the parameter may change more slowly than the rate you specify here.

**Domain:** These numbers determine the minimum and maximum values of the parameter during animation.

The domain affects animation only. You can still change the parameter manually to any value, regardless of the domain, by double-clicking it with the **Arrow** 104 tool, by choosing **Edit** 144 | **Edit Parameter** 157, or by pressing the + or − key.

**Keyboard (+/−) adjustments:** This number specifies the amount by which the parameter changes when you select it and press the + or − key on the keyboard 295.

 Note

Use the parameter's Value Properties 94 to determine whether it has an edit box allowing its value to be edited directly in the sketch.

## 1.11 Calculations

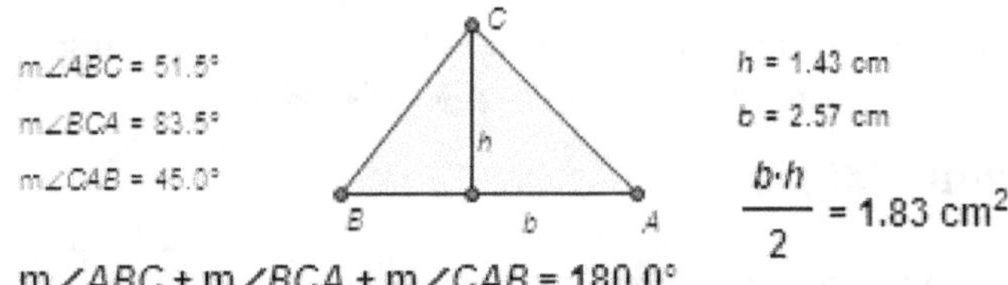

Calculations are mathematical expressions that relate one or more terms — such as measurement — by arithmetic. For example, after you measure the interior angles of a triangle, use Sketchpad's Calculator 257 to find the sum. Similarly, measure the base and height, and use the Calculator to find the area. You can define a calculation using measurements in your sketch as well as various mathematical operators, built-in functions, and even functions 45 that you've defined yourself.

As you change the measurements on which a calculation depends, the calculated result changes accordingly.

Like all value objects 92, calculations can be used to dynamically control other objects in your sketch.

By observing calculations you can gain valuable insights into mathematical relationships.

- To create a calculation, choose **Number | Calculate** 227.

- To edit an existing calculation, double-click it with the **Arrow** 104 tool or select it and choose **Edit | Edit Calculation** 157.

- To change a calculation into a parameter 36, edit the calculation to make it a simple numeric value with no operators or functions. (It can be dimensionless, or it can have angle or distance units.)

- To change a calculation's precision and how it's displayed, select the parameter and choose **Edit | Properties | Value** 94.

## 1.12 Tables

| Area $\odot CD$ | (Radius $\odot CD)^2$ | $\dfrac{\text{Area } \odot CD}{(\text{Radius } \odot CD)^2}$ |
|---|---|---|
| 2.54 cm$^2$ | 0.81 cm$^2$ | 3.1416 |
| 14.79 cm$^2$ | 4.71 cm$^2$ | 3.1416 |
| 36.46 cm$^2$ | 11.61 cm$^2$ | 3.1416 |
| 51.13 cm$^2$ | 16.27 cm$^2$ | 3.1416 |
| 92.80 cm$^2$ | 29.54 cm$^2$ | 3.1416 |

Use tables to examine how measurements 36, parameters 36, and calculations 39 change over time. Tables are organized in rows and columns, with each column describing a single object, and each row containing the object's value at the moment that the row was added to the table.

In addition to other single-valued measurements, you can tabulate functions 45, coordinates 223 and equations 225.

Once you create a table, you can add rows 228 to it one at a time, or you can instruct Sketchpad to collect rows automatically as you drag points or other objects that cause the tabulated values to change. You can also remove [229] one or all rows from an existing table.

The last row of a table normally tracks current values as they change in the sketch. You can turn this feature on or off by choosing **Edit | Properties | Table** [41].

### ▼ Working with Tables

- To create a table, select one or more measured values — measurements [36], calculations [39], coordinate pairs [223], equations [225], or text containing values — and choose **Number | Tabulate** [228].

- Sketchpad creates a table automatically when you create an iteration [24] that causes one or more values to change. Such a table depends on the depth of the iteration; you cannot add rows manually.

- To add a row to the table, capturing the current values of your tabulated measurements, double-click the table, or select the table and choose Number | Add Table Data [228]. In the dialog box that appears, choose **Add one entry now.**

- To capture changing values in a table automatically as you drag [104] or animate [171] measured objects in your sketch, choose **Number | Add Table Data** [228]. Sketchpad displays a dialog box that lets you specify how many rows to add, and how frequently to add them.

- To remove a table's most recently added row, or to remove all of the rows from a table, select the table and choose **Number | Remove Table Data** [229]. To remove a table's most recently added row, you can also double-click a table with the **Arrow** [104] tool while holding down the Shift key.

- To change the labels of a table's columns, change the name of the original measurement tabulated in that column.

- To change the precision or units of a table's data, change the precision [94] or units [276] of the original measurements tabulated by the table.

- To change the color of a table's borders, choose a color from the **Display | Color** [164] submenu.

- To change the color, font, size, or style of a table's text, use the Text Palette [270].

- To plot the data in a table, select the table and choose **Graph | Plot Table Data** [240].

- To export the data in a table to another application such as Fathom Dynamic Data Software or Microsoft Excel, select the table and choose **Edit | Copy** [146]. Switch to the other application and choose **Paste.** Sketchpad exports the table data to the other application as tab-delimited text.

## 1.12.1  Table Properties

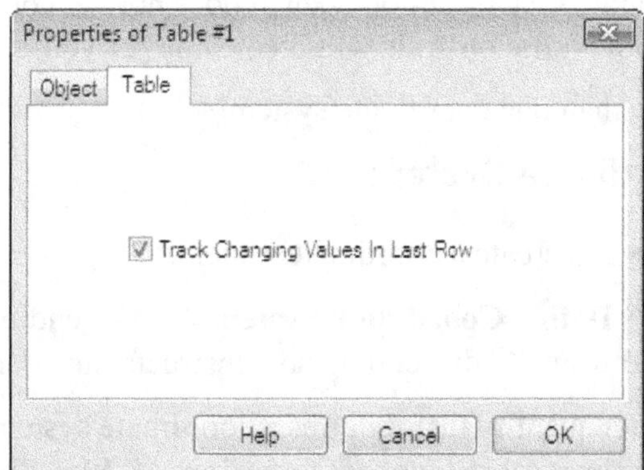

The Table Properties [158] panel appears only for tables [39] created using the **Number | Tabulate** [228] command. To open this panel, select the table and choose **Edit | Properties** [158] | Table, or choose **Properties** [158] from the Context [245] menu.

Use this panel to determine whether the last row of a table tracks the changing values in the sketch.

# 1.13    Coordinate Systems and Axes

A coordinate system allows you to measure coordinates, distances, slopes, and equations on the plane and to plot points and functions.

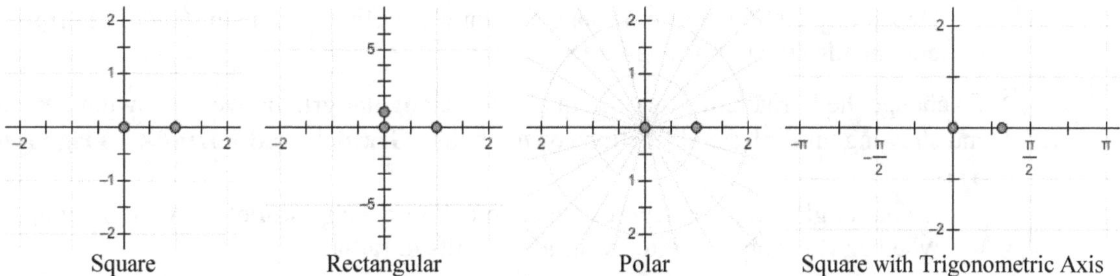



A coordinate system is defined by its origin, grid form, and the scale of its axes.

**Origin:** The origin is either a selected point or, if no point is selected when the coordinate system is constructed, a new point in the center of the screen.

**Grid Form:** Coordinate systems are either square (with the same scale on both axes) or rectangular (with different scales on the horizontal and vertical axes). The grid appears in either rectangular or polar form, and the axes can be measured either in decimal numbers or in multiples and fractions of $\pi$.

**Scale:** The scale of a square coordinate system is determined by a unit point or by a distance measurement in the sketch. The scale of a rectangular coordinate system is determined by two unit points (one for each axis) or by two distance measurements in the sketch.

A coordinate system has a horizontal and a vertical axis. The axes behave as lines [5] for most purposes.

Sketchpad's analytic measurements (**Coordinates** [223], **Coordinate Distance** [224], **Slope** [224], and **Equation** [225]) and plotting commands (**Plot Value on Axis** [237], **Plot Points** [239], **Plot as (x, y)** [239],

**Plot Table Data** [239], **Plot New Function** [241], and **Plot Parametric Curve** [241]) are defined in reference to a coordinate system. If your sketch does not yet contain a coordinate system, any of these commands except **Plot Value on Axis** will create one for you.

You can define more than one coordinate system [43].

### ▼ Create a Coordinate System

There are two ways to create a coordinate system.

- Choose **Graph | Define Coordinate System** [233]. Depending on the objects you've selected, there are several ways [233] this command constructs the coordinate system.

- Measure or plot a quantity that requires a coordinate system, such as **Coordinates** [223], **Coordinate Distance** [224], **Slope** [224], **Equation** [225], **Plot Points** [239], or **Plot New Function** [241]. If you don't already have a coordinate system, any of these commands will create one.

The default coordinate system is a square system with its origin at the center of your sketch window and a unit point at (1, 0).

You can define multiple coordinate systems [43], though there's seldom a need to do so.

### ▼ Modify a Coordinate System

- To change the grid form, choose **Graph** [233] **| Grid Form** [235] and choose a polar, square, or rectangular grid. On a polar grid, coordinates are measured and points are plotted in polar $(r, \theta)$ coordinates; on a square or rectangular grid, coordinates are measured and points are plotted in rectangular $(x, y)$ coordinates.

  Depending on how the coordinate system was defined, some of the Grid Form commands may not be available.

- To change the horizontal axis of a square or rectangular grid between showing decimal numbers and showing multiples and fractions of $\pi$, choose **Graph | Grid Form** [235] **| Trigonometric Axis.**

  If your angle units are degrees when you choose **Trigonometric Axis,** Sketchpad asks whether you want to switch your angle units to radians.

  You can also change an axis individually by selecting it and choosing **Edit** [144] **| Properties** [158] **| Axis** [44].

- To change both axes of a polar grid between showing decimal numbers and showing multiples and fractions of $\pi$, choose **Graph | Grid Form** [235] **| Trigonometric Axis.**

- To hide or show grid lines, choose **Graph | Hide Grid** [236] or **Graph | Show Grid** [236].

- To change the grid to show dot paper rather than grid lines, choose **Graph | Dotted Grid** [236]. (To use a coordinate system as a geoboard, see How to Construct a Geoboard [44].)

- To change the color of the grid lines, select the grid by clicking a grid intersection and choose a color from the **Display | Color** [164] submenu.

- To change the appearance of tick numbers, select the axis and choose the desired font, size, or style from the Text Palette [270].

- To make points snap to integer coordinate positions when dragged, choose **Graph** [233] **| Snap Points** [237].

## ▼ Use a Coordinate System

- To change the scale of a coordinate system, drag the unit point. Square coordinate systems have one unit point, and rectangular coordinate systems have two.

- To change the scale of a coordinate system, press and drag any axis tick number with the **Arrow** [104] tool.

  The scale of a coordinate system may be defined by an object in the sketch; such a coordinate system doesn't have a unit point and cannot be scaled by dragging axis tick numbers.

- If you have more than one coordinate system, make one of them active by choosing **Graph | Mark Coordinate System** [234].

- To plot a point on the coordinate system, choose **Graph | Plot Points** [239].

- To plot two measured values on the coordinate system, select them and choose **Graph | Plot as (x, y)** [239].

- To plot a point on an axis, choose **Graph | Plot Value on Axis** [237].

- To plot a new function [45] on the coordinate system, choose **Graph | Plot New Function** [241].

- To plot an existing function [45] on the coordinate system, select the function and choose **Graph | Plot Function** [241].



## 1.13.1 Multiple Coordinate Systems

Most sketches use no more than one coordinate system, but you can create additional coordinate systems if they're needed. (For instance, you might want more than one coordinate system to compare objects — the coordinates of a point or the plot of a function — in the two systems.)

When you have more than one coordinate system, only one of those systems is the *marked* or active coordinate system — the one that's used for measuring [216] and plotting [233].

To change which coordinate system is marked in your sketch:

1. Identify the coordinate system you want to mark by selecting its origin, unit point, either axis, or grid. (To select a grid, click the intersection of two grid lines.)

2. Choose **Graph | Mark Coordinate System** [234].

## 1.13.2  Axis Properties

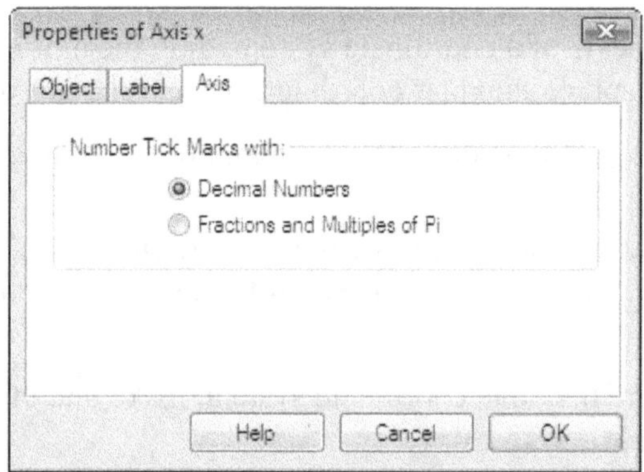

Axes have an Axis Properties |158| panel. Use this panel to determine whether an axis is numbered using decimal numbers or using fractions and multiples of $\pi$.

To open this panel, select the axis and choose **Edit | Properties** |158| **| Axis,** or choose **Properties** |158| from the Context |245| menu.

You can choose **Graph | Grid Form** |235| **| Trigonometric Axis** to change this setting for the horizontal axis of the marked coordinate system.

To change this setting for a vertical axis or in a sketch with multiple coordinate systems |43| , you must select the axis and choose **Edit | Properties** |158| **| Axis.**

## 1.13.3  How to Construct a Geoboard

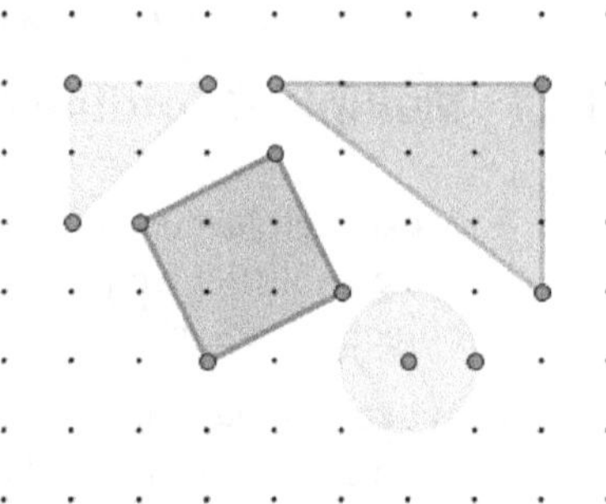

Use a coordinate system to make a Sketchpad geoboard. Objects you construct on this geoboard will always stick to the grid dots; segments look and behave just like rubber bands attached to the posts on a physical geoboard. And with a Sketchpad geoboard you can construct polygons and circles, and you can measure lengths and slopes of segments, perimeters and areas of polygons, and coordinates of points.

1. Choose **Edit | Preferences | Units** |276| and set **Distance Units** to **cm.**

2. Choose **Graph | Dotted Grid** |236|.

3. Select the grid (by clicking a grid intersection) to change the dot color (by choosing **Display | Color** |164|) or to change the dot size (by choosing **Display | Point Style** |162|).

4. Hide |167| the origin, unit point, and axes.

5. Choose **Graph | Snap Points** |237| to make points snap to the dots.

Your geoboard is ready for use.

### Scaling the Geoboard

When you turn on Snap Points, dragged and newly constructed points snap to integer coordinates. For proper operation of the geoboard, grid dots must also appear at integer coordinates; otherwise points won't snap properly to the grid.

If you adjust the scale by dragging the unit point before hiding it, take care that dots still appear at every integer, rather than at fractions or multiples of integers.

If you adjust the scale after drawing on the geoboard, the points you've constructed won't follow the dots. Drag the points to snap them to dots again.

### Measurements on the Geoboard

When you measure length [217], distance [218], area [219], perimeter [218], and circumference [218] on the geoboard, the results are in cm or cm$^2$. If you change the scale of the geoboard so it's is no longer exactly 1 cm, these measurements will no longer correspond to the dot positions.

Coordinate distance [224] measurements remain correct even with a changed scale.

# 1.14 Functions

$$f(x) = 2 \cdot x + 3$$
$$g(x) = a \cdot \sin(b \cdot x + c) + d$$
$$g'(x) = a \cdot b \cdot \cos(b \cdot x + c)$$
$$h(x)\text{: Drawing [1]}$$

A function is a rule that turns input values into output values. For example, the function $f(x) = 2 \cdot x + 3$ is a rule that says "To get an output value, start with the input value, multiply it by 2, and then add 3 to the result." To evaluate a function you follow the rule for a particular input value. For example, $f(5) = 2(5) + 3 = 13$.

There are three kinds of functions in Sketchpad:

- A symbolically defined function is defined by an algebraic rule.

- A derivative function is defined by finding the derivative of another function.

- A data-defined function is defined by the top stroke of a drawing or the top edge of a picture.

Once you've created a function, see Using Functions [47] for information on the many ways you can use a function.

### ▼ Create a Symbolically-Defined Function

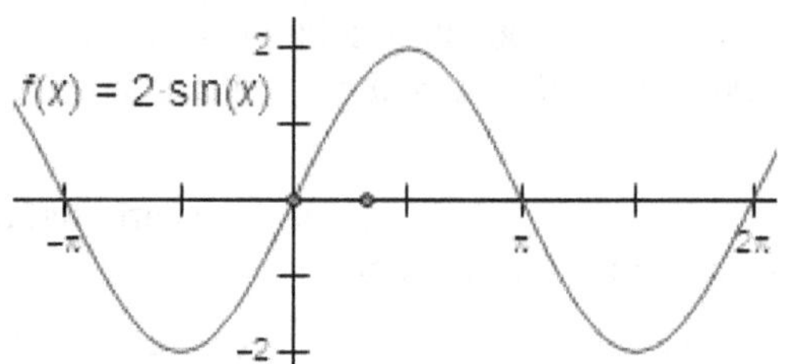

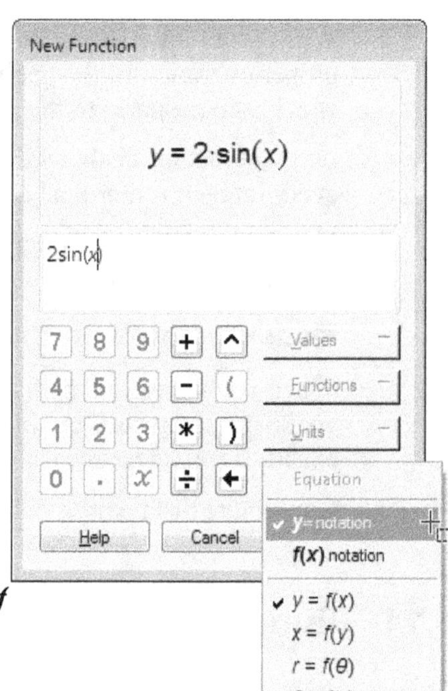


To define a new function symbolically, for example, $f(x) = 2 \cdot \sin(x)$, choose **Number | New Function** [230]. This command opens Sketchpad's Calculator [257] to allow you to define how the function is calculated.

To create a new function and plot it immediately, choose **Graph | Plot New Function** [241].

To determine whether the function appears in *y=* **notation** or *f(x)* **notation,** choose the desired notation from the Calculator's Equation pop-up menu [259].

*See also:*
*How to Use Signum to Construct a Piecewise Function* [262]

### ▼ Create a Derivative Function

To create a derivative function, select the function you want to differentiate and choose **Number | Define Derivative Function** [231].

The derivative of a symbolically-defined function is usually defined symbolically, as shown on the left. In rare cases of exceptionally complex symbolically-defined functions, the derivative may be defined approximately.

The derivative of a data-defined function (defined by a drawing or picture) is always defined approximately, as shown on the right.

| Symbolic Derivative | Approximate Derivative |
| --- | --- |
| $f(x) = -2 \cdot x^3 + 5 \cdot x^2 + x - 3$ <br> $f'(x) = -6 \cdot x^2 + 10 \cdot x + 1$ | $g(x)$: Drawing [1] <br> $g'(x) \approx \dfrac{g(x + 0.005) - g(x - 0.005)}{0.01}$ |

To determine whether the function appears in *y=* **notation** or *f(x)* **notation,** set the desired notation for the parent function.

Note

When Sketchpad creates a derivative function, it does not keep track of domain restrictions. For instance, $f(x) = \ln(x)$ has a domain restriction: it's defined for $x > 0$. Sketchpad shows the correct derivative, $f'(x) = 1/x$, but without the domain restriction: the derivative is defined for $x \neq 0$.

### ▼ Create a Data-Defined Function

f(x): Drawing [1]

Plot of f(x)

Smoothed Plot

To create a data-defined function based on a drawing [18] or picture [18], select the drawing or picture and choose **Number | Define Function from Drawing** [232]. The function is defined by the top edge of the drawing or picture.

To determine how much smoothing is applied to the function, choose **Edit | Properties | Function** [51]. An unsmoothed function follows every pixel in the top edge of the drawing or picture; a smoothed function reduces the small variations.

To determine whether the function appears in *y*= **notation** or *f*(*x*) **notation,** choose **Edit | Properties | Function** [51].

*Subtopics:*
  *Using Functions* [47]
  *Data-Defined Function Properties* [51]

## 1.14.1  Using Functions

Once you've defined a function, you can use it in many ways: you can plot it, use it in a calculation or to define another function, edit it, differentiate it, trace its family, construct its family, transform it symbolically, transform its plot with a custom transformation, use it to define a composite function or a parametric plot, or create a sound button.

### ▼ Plot a Function

To define a new function and plot it immediately on the marked coordinate system, choose **Graph | Plot New Function** [241] to open Sketchpad's Calculator [257]. Use it to define the function. When you close the Calculator, the function is plotted in the form that was set in the Calculator's Equation pop-up menu [259]. The function plot [52] can take any of four possible forms.

| | |
|---|---|
| $y = f(x)$ | This is the default form. |
| $x = f(y)$ | Use this form to plot the $y = f(x)$ function with the axes reversed. |
| $r = f(\theta)$ | You'll normally use the last two forms with a polar grid. [235] |
| $\theta = f(r)$ | Use this form to graph $\theta$ as a function of $r$ with a polar grid. [235] |

To plot one or more existing functions on the marked coordinate system, select every function you want to plot and choose **Graph** [233] **| Plot Function** [241]. Each function is plotted in the form that was set in the  Calculator's Equation pop-up menu [259] at the time you defined that function.

### ▼ Use a Function in a Calculation or Function Definition

Once you've defined a function, you can use it in later calculations and function definitions. To use an existing function in a calculation or in a new function definition, click the function in the sketch to enter it into the Calculator 257.

$f(x) = 2 \cdot x + 3 \quad g(x) = 3 \cdot f(x)$
$f(5) = 13.00 \quad g(5) = 39.00$

If the Calculator is hiding the function in the sketch, you may have to move the Calculator out of the way first.

You can also enter functions from the Function pop-up menu 259 in the Calculator. This menu includes many predefined functions, along with every selected user-defined function in your sketch.

### ▼ Edit a Symbolic Function

You can edit a symbolically defined function to change its definition or how it's plotted. Editing a function is useful, for example, if you've plotted the graph of $y = 2 \cdot \sin(x)$, and you want to change the function to $y = 3 \cdot \sin(x)$ in order to see how the two graphs differ. Instead of changing the constant to 3, you can insert a new parameter, allowing you to investigate the plots of the family of functions $y = a \cdot \sin(x)$.

To edit a function, select that function, choose **Edit | Edit Function** 157, and use the Calculator 257 to change the definition.

> You can also right-click (Win) or Ctrl-click (Mac) the function and choose **Edit Function** from the Context 245 menu.

When you edit a function, you can redefine it in any way you want — introducing new parameters and existing parameters, measurements and calculations — provided you don't create a circular definition. (A circular definition is one that uses a calculation that depends on the function you're editing. For instance, if you've defined a function $f$ and used it to calculate $f(3)$ in your sketch, you cannot later edit function $f$ so that it uses the calculation of $f(3)$ in its definition.)

> Choose a new equation form while editing a function to change how that function is plotted. For instance, to plot a function $f$ as a polar function, use the Equation pop-up menu to change the form from $y = f(x)$ to $r = f(\theta)$.

### ▼ Differentiate a Function

$$f(x) = -2 \cdot x^3 + 5 \cdot x^2 + x - 3$$
$$f'(x) = -6 \cdot x^2 + 10 \cdot x + 1$$

To create the derivative of a function with respect to its independent variable, select the function, then choose **Number | Define Derivative Function** 231. The result is a derivative function that can be plotted or evaluated like any other function.

### ▼ Trace a Family of Functions

Use parameters 36 or sliders 222 in a function definition to investigate families of functions.

> After you define the function and plot its graph, you can change or animate the parameter or slider, and observe or trace 170 the family of functions.

While you're using the function Calculator 257 to enter or edit a function, you can create a new parameter or use an existing parameter or other measurement from your sketch. When the value of this parameter or measurement changes, the definition of the function changes. For example, if you create parameter $a$ while you're specifying the function $f(x) = a \cdot \sin(x)$, you can investigate the behavior and plots of this entire family of functions, including functions such as $f(x) = -1 \cdot \sin(x)$ and $f(x) = 3 \cdot \sin(x)$ by varying the parameter $a$. Similarly, use three parameters as you define the function $f(x) = a \cdot x^2 + b \cdot x + c$ to investigate how changing each parameter's value affects the function plot.

With the Calculator 257 open add a new parameter to the function definition by choosing **Number | New Parameter** 226 from the Calculator's Values pop-up menu. Insert an existing parameter by clicking on it in the sketch.

### ▼ Construct a Family of Functions

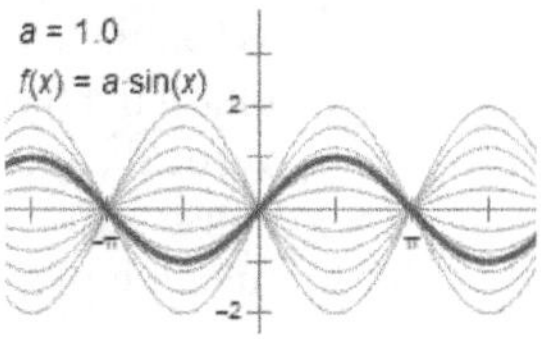


When you use a parameter 36 or slider 222 in a function definition, you can construct the family of functions.

You must first plot the function itself. Then select both the function plot and the parameter or slider used in the function definition and choose **Construct | Family of Functions** 190.

### ▼ Transform a Function Symbolically

Once you've defined a function $f(x)$ — for example, $f(x) = x^2$ — you can define and plot other functions that are transformations of $f(x)$. Transformations of $f(x)$ include such functions as these:

| | |
|---|---|
| $g(x) = 2 \cdot f(x)$ | vertical stretch |
| $h(x) = f(-x)$ | reflection across $y$-axis |
| $j(x) = -3 \cdot f(2x + 5) - 1$ | horizontal shrink (1/2) and translate left (5), followed by vertical reflection (–), stretch, (3), and translate down (–1) |
| $k(x) = f(x - h) + k$ | translate right by $h$ and up by $k$ |

To define a function that is a transformation of $f(x)$, choose **Number | New Function** 230. Define the new function as you normally would, inserting the original function $f$ wherever you want by clicking it in the sketch.

## ▼ Custom Transform a Function Plot

Once you've plotted a function, you can define a custom transformation 211 and apply the transformation to the function plot. For instance, if you define a custom transformation to do a vertical stretch, you can apply the transformation to a function plot.

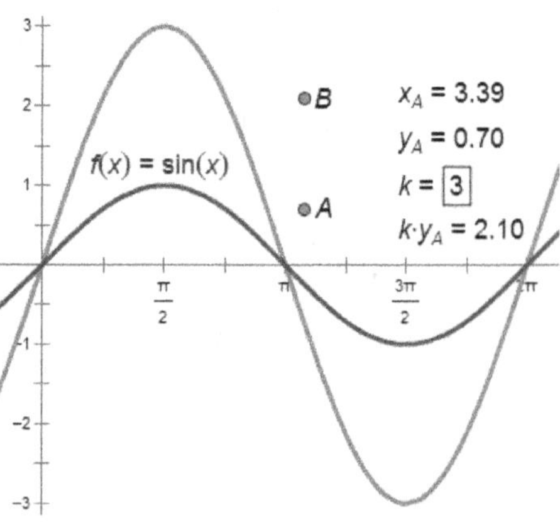


In this example, point $B$ is plotted at $(x_A, k{\cdot}y_A)$, and points $A$ and $B$ are used to define a custom transformation. This custom transformation is a vertical stretch by the factor $k$.

When the custom transformation is applied to the plot of $f(x)$ (in blue), the result is the vertically stretched red plot. Because the vertical stretch depends on parameter $k$, you can use this construction to explore a variety of vertical stretches of the blue function plot.

## ▼ Define a Composite Function

$$f(x) = 2{\cdot}x + 3$$
$$g(x) = x^2$$
$$h(x) = g(f(x))$$
$$h(5) = 169.00$$

*Composite functions* are functions of functions. For instance, $h(x) = g(f(x))$ is called the *composition* of functions $g$ and $f$, and can be thought of as follows: "First evaluate function $f$ at the input value. Then use this result as the input for function $g$. The result of evaluating $g$ is the output of the composite function $g(f(x))$."

To compose two functions $g$ and $f$, first define each function separately. Then choose **Number | New Function** 230 and use the function Calculator 257 to define the composite function $h(x)$. Click the functions $g$ and $f$ in the sketch to enter them into the Calculator.

> If the Calculator is hiding $f$ or $g$ in the sketch, you may have to move the Calculator out of the way first.

## ▼ Define a Parametric Plot

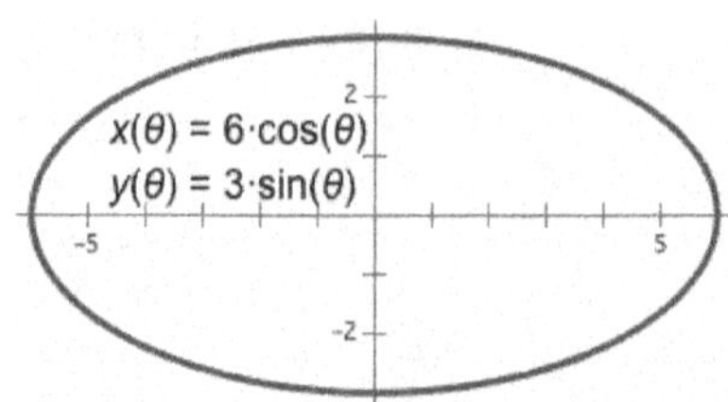


To create a parametric plot 54, define one function to represent the $x$ value of a point and a second function to represent the $y$ value. As the independent variable changes, the set of points determined by the values of the two functions defines the parametric plot.

In this example, given any value of $\theta$, the values of the two functions $x(\theta)$ and $y(\theta)$ determine a point.

To construct the parametric plot, select these two defining functions and choose **Graph | Plot Parametric Curve** 241. The resulting plot is in the shape of an ellipse.

### ▼ Create a Sound Button

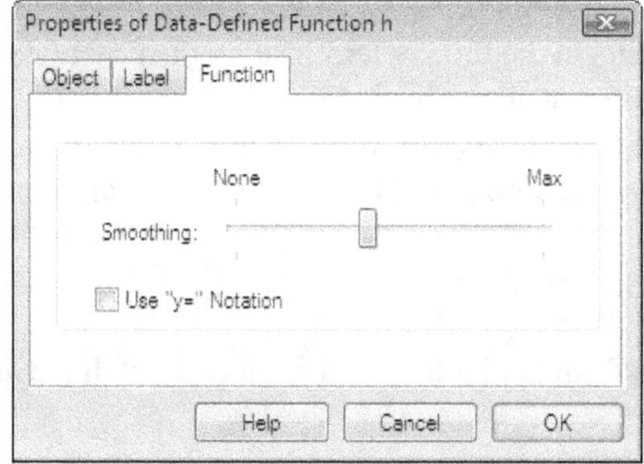


To create a Sound button [74], define a mathematical function to describes air pressure over time. (The volume is determined by the amplitude of the function, and the pitch is determined by its period.)

Then select the function and choose **Edit | Action Buttons | Sound** [150] to create the button. (You can also select two functions to play a sound in stereo.)

## 1.14.2  Data-Defined Function Properties

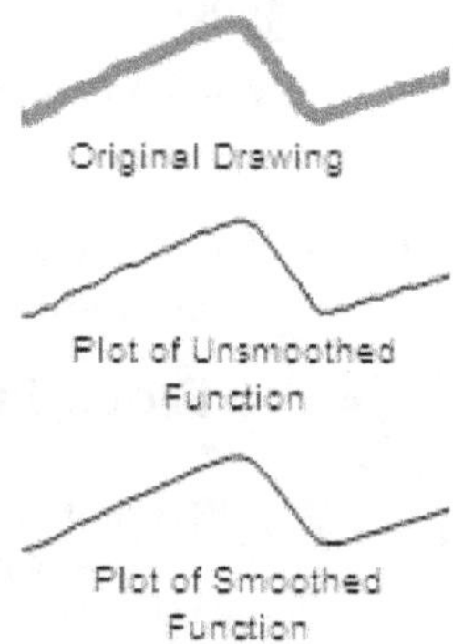

Data-defined functions (defined by the top edge of a drawing [18] or picture) [19] have a Functions Properties [158] panel.



**Smoothing:** Use this slider to determine the amount of smoothing to be applied to the function. An unsmoothed function follows every pixel in the top edge of the drawing or picture; a smoothed function reduces the small variations. The default setting, in the middle of the scale, applies a moderate amount of smoothing. (Smoothing can be particularly useful when investigating derivatives [231] of data-defined functions.)

 Note

Data-defined functions are based on a discrete set of elements, such as individual pixels in the picture whose image data defines the function. Thus at sufficiently high resolution, they are always step functions with derivatives of zero. The Smoothing slider allows you to introduce some simple curvature into the function, which may make it more suitable for various analytic purposes.

**Use "y=" notation:** Check this box to display the function in $y=$ notation rather than $f(x)$ notation.

## 1.15  Function Plots

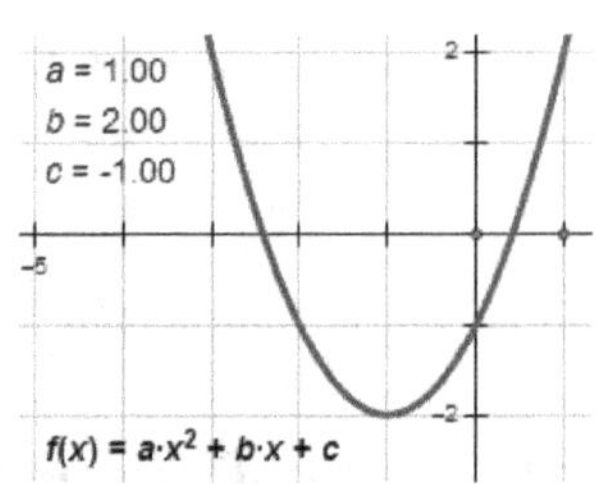



A function plot shows the graph of a particular function ⌐45⌐ on a coordinate system ⌐41⌐.

A function plot appears in the form determined by the Calculator's Equation pop-up menu ⌐259⌐ at the time the function was created or edited ⌐157⌐.

> To change the form of an existing function plot, double-click the function itself to open the Calculator. Then choose a new form from the Equation pop-up menu.

| | |
|---|---|
| $y = f(x)$ | This is the default form. |
| $x = f(y)$ | Use this form to plot the $y = f(x)$ function with the axes reversed. |
| $r = f(\theta)$ <br> $\theta = f(r)$ | You'll normally use these two forms with a polar grid ⌐235⌐. |

You can change the object properties ⌐79⌐, label properties ⌐82⌐, and display attributes ⌐86⌐ of function plots.

You can change the line style ⌐163⌐ of continuous function plots, and you can change the point style ⌐162⌐ of discrete function plots.

You can choose **Edit** ⌐144⌐ | **Properties** ⌐158⌐ | **Plot** ⌐95⌐ to change the number of samples used to display a function plot and to change whether it's displayed continuously or discretely.

### ▼ Plot a Function

To plot an existing function, select it and choose **Graph | Plot Function** ⌐241⌐.

To define a new function and plot it immediately, choose **Graph | Plot New Function** ⌐241⌐. Sketchpad's function Calculator ⌐257⌐ appears; use it to define the function.

> Functions are plotted on the marked coordinate system. If there is no coordinate system, Sketchpad creates one.

### ▼ Use a Function Plot

- Construct a point on a function plot using the **Point** ⌐112⌐ tool or the **Construct | Point on Object** ⌐176⌐ command. You can then animate ⌐267⌐ the point along the function plot or use the point in any other way you like.

- Use a function plot as a path along which to animate an independent point by using the **Edit | Merge** ⌐153⌐ command.

- Use **Construct | Intersection** 177 to construct the intersection of a function plot with a straight object, a circle, an arc, or another function plot using the same horizontal axis.

  Alternatively, click the intersection with the **Arrow** 104 tool or with any tool that can construct a point.

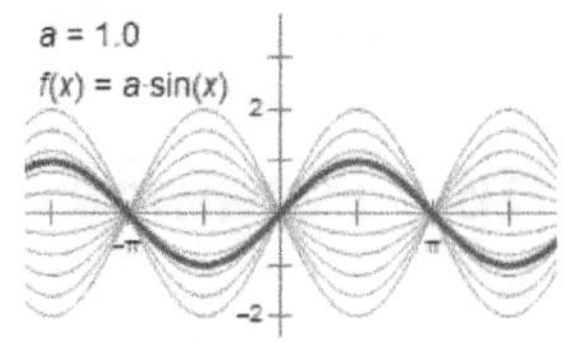


- Use **Construct | Family of Functions** 190 to construct a family of functions. The function definition must use a parameter or slider.

## Change the Domain of a Function Plot

When you plot a function, the domain is initially set to the width of the Sketchpad window.

To change the domain, you can:

- Select the function plot, choose **Edit | Properties | Plot** 97 , and enter new values for the ends of the domain.

- Use the **Arrow tool** 104 to drag the arrowheads that appears at the ends of the function plot.

  Drag in the direction that the arrowhead points to increase the length of the plot; drag in the opposite direction to decrease the length of the plot.

 Note

  To determine whether the arrowheads appear, choose **Edit | Properties | Plot** 97 .

## Change the Number of Samples

Mathematically, a function plot may include an infinite number of points. However, to display an infinite number of points would require a computer to use an infinite amount of time, so Sketchpad instead plots a large (but not infinite) number of points. Each point that Sketchpad does display is called a sample.

To change the number of samples for a selected function plot:

- Press the + or − key. Each press of the key increases or decreases the number of samples.

- Choose **Increase Resolution** or **Decrease Resolution** from the Context 245 menu.

- Choose **Edit | Properties | Plot** 97 and type a new value for the number of samples.

You can change the initial number of samples for newly constructed function plots. Choose **Edit | Advanced Preferences | Sampling** 283 and type a new value for **Number of samples for new function plots.** (This setting has no effect on already-existing function plots.)

## Display a Function Plot Continuously or Discretely

- Change between continuous and discrete display of a selected function plot by choosing **Edit | Properties | Plot** 97 .

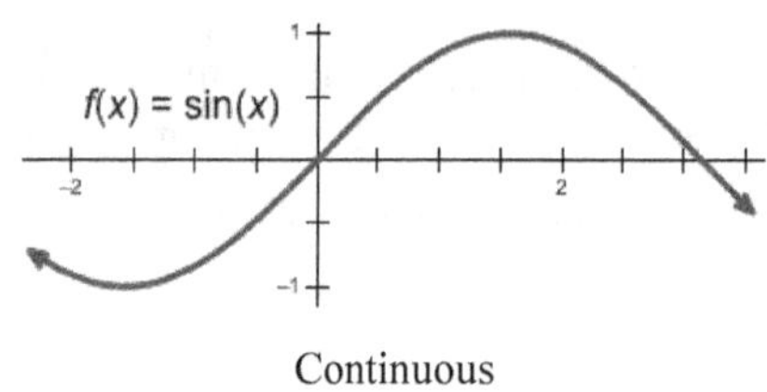

Continuous

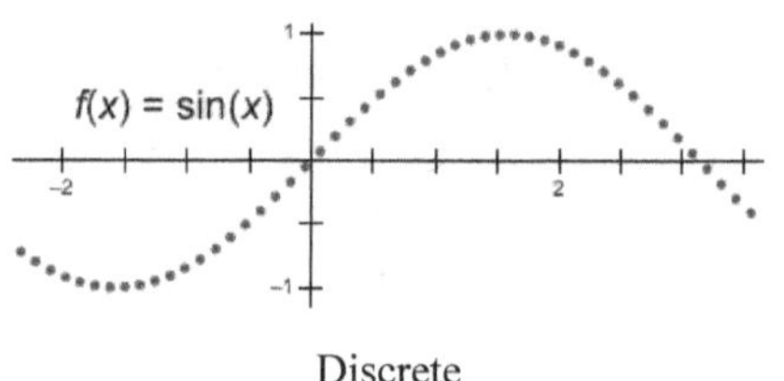

Discrete

# 1.16   Parametric Plots

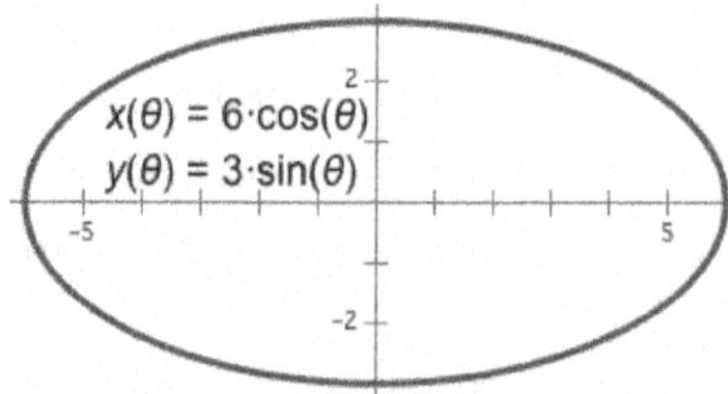


A parametric plot graphs two functions [45] on a coordinate system [41]. Each plotted point $(x, y)$ or $(r, \theta)$ corresponds to a particular value of an independent variable. This independent variable is used to evaluate each of the two functions. The value of the first function is used for the $x$ or $r$ value, and the value of the second function is used for the $y$ or $\theta$ value. (The independent variable is usually called a *parameter,* although it's not a separate object like a Sketchpad parameter [36].)

When you define a parametric plot, you choose whether to plot it using rectangular or polar coordinates, and you choose the domain for the independent variable.

You can change the object properties [79], label properties [82], and display attributes [86] of a parametric plot.

You can change the line style [163] of a continuous parametric plot, and you can change the point style [162] of a discrete parametric plot.

You can choose **Edit | Properties | Plot** [97] to change the number of samples used to display a parametric plot and to change whether it's displayed continuously or discretely.

### ▼ Construct a Parametric Plot

To construct a parametric plot, select two functions and choose **Graph | Plot Parametric Curve** [241]. Alternatively, you can choose the command and, with the Plot Curve dialog box open, click the functions in the sketch to enter them into the dialog box.

The parametric plot appears on the marked coordinate system. If there is no coordinate system, Sketchpad creates one.

### ▼ Use a Parametric Plot

- Construct a point on a parametric plot using the **Point** [112] tool or the **Construct | Point on Object** [176] command. You can then animate [267] the point along the parametric plot or use the point in any other way you like.

- Use a parametric plot as a path along which to animate an independent point by using the **Edit | Merge** [153] command.

- Use **Construct** [174] **| Intersection** [177] to construct the intersection of a parametric plot with a straight object, a circle, or an arc.

Alternatively, click the intersection with the **Arrow** 104 tool or with any tool that can construct a point.

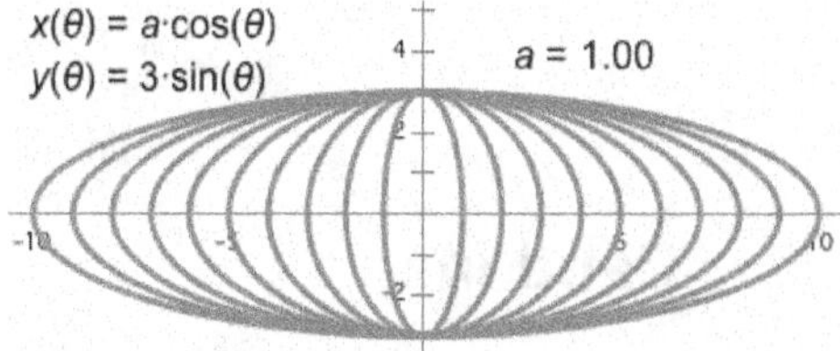


- If one or both defining functions depend on a parameter or slider, you can construct a family of parametric plots. Select both the parametric plot and the parameter or slider and choose **Construct | Family of Curves** 190.

### ▼ Change the Domain of a Parametric Plot

When you construct a parametric plot, enter the upper and lower limits of the domain in the Plot Curve 241 dialog box.

To change the domain, you can:

- Select the parametric plot, choose **Edit | Properties | Plot** 97, and enter new values for the limits of the domain.

- Use the **Arrow** 104 tool to drag the arrowheads that appears at the ends of the parametric plot. Drag in the direction that the arrowhead points to increase the length of the plot; drag in the opposite direction to decrease the length of the plot.

 Note

To determine whether the arrowheads appear, choose **Edit | Properties | Plot** 97.

### ▼ Change the Number of Samples

Mathematically, a parametric plot may include an infinite number of points. However, to display an infinite number of points would require a computer to use an infinite amount of time, so Sketchpad instead plots a large (but not infinite) number points. Each point that Sketchpad does display is called a sample.

To change the number of samples for a selected parametric plot:

- Press the + or – key. Each press of the key increases or decreases the number of samples.

- Choose **Increase Resolution** or **Decrease Resolution** from the Context 245 menu.

- Choose **Edit | Properties | Plot** 97 and type a new value for the number of samples.

You can change the initial number of samples for newly constructed parametric plot. Choose **Edit | Advanced Preferences | Sampling** 283 and type a new value for **Number of samples for new function plots.** (This setting applies to new function plots and parametric plots, but has no effect on already-existing plots.)

### ▼ Display a Parametric Plot Continuously or Discretely

- Change between continuous and discrete display of a selected parametric plot by choosing **Edit | Properties | Plot** 97.

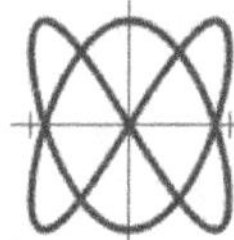

Continuous

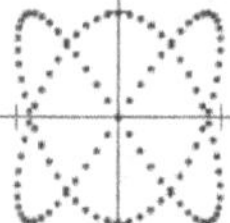

Discrete

# 1.17 Captions

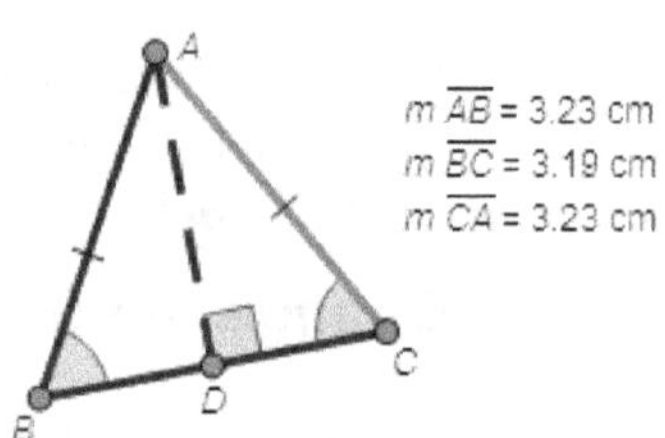



Use captions in Sketchpad to display text that identifies or explains your sketch. Sketchpad captions enable you to express mathematical ideas by using geometric and algebraic notation and by directly including in your caption links to the geometric and algebraic objects in your sketch.

- Include Hot Text™ |57| links to objects and values in your sketch. Hot Text values change dynamically as the value in the sketch changes, and Hot Text labels change if the label of the object is changed. Use Hot Text to see an updated form of an equation or conjecture, to highlight or flash linked objects in a sketch, or to press an action button. (In the illustration, the pointer is over the Hot Text link to segment CA, causing the segment to be highlighted in the construction.)

- Use the Text Palette |270| to set the font, size, style, and color of your text.

- Use mathematical formatting |272| to include segment overbars, grouping symbols, fractions, and other mathematical notation.

- Use Unicode |314| to to insert mathematical symbols (such as $\nabla f = \langle \partial f / \partial x, \partial f / \partial y \rangle$ ) and to write in multiple alphabets (such as живая математика, in Russian or 动感几何, in Chinese).

As with other text objects |99|, you can position a caption, format an entire caption, and merge a caption to a point.

### ▼ Create a New Caption

Click and drag the **Text** |120| tool in empty space in your sketch to create a caption |120|. Then begin typing.

You can also double-click the **Text** tool to create a caption.

### ▼ Edit a Caption

To begin editing, create a new caption or double-click an existing caption with either the **Text** |120| tool or the **Arrow** |104| tool.

While editing a caption, you can use a number of caption-editing keyboard shortcuts |297| in addition to the methods listed here.

- To move the insertion point, click at the desired location or use the arrow keys on your keyboard.

- To select a single word, double-click it.

- To select text, press and drag, or hold the Shift key while clicking in a new location.

- To select text word-by-word, double-click and hold on the first word you want to select. Drag in any direction to select more words.

- To change the font, size, style, or color of the selected text, use the Text Palette 270.

- To turn italic, bold, or underline style on or off for the selected text, press Ctrl+I, Ctrl+B, or Ctrl+U on Windows, or ⌘I, ⌘B, or ⌘U on Mac.

- To insert a Hot Text 57 link to an object, click in the sketch on the object to which you want to link. (The pointer shows a + sign when it's over an object you can insert as Hot Text.)

- To resize the caption 120 and reflow the contents, drag the caption's resize handle.

### ▼ Set the Alignment of a Caption

Hold the Alt key (Windows) or Option key (Macintosh) while pressing one of the arrow keys to set the text alignment.

- Use the left arrow for left alignment.

- Use the right arrow for right alignment.

- Use either the up or down arrow for centered alignment.

*Subtopics:*
  *Hot Text* 57

## 1.17.1 Hot Text™

Captions can contain Hot Text™ links to object labels and values in your sketch.

A Hot Text value changes dynamically as the value in the sketch changes, and a Hot Text label changes if the label of the object is changed. Use Hot Text to see an updated form of an equation or conjecture, to highlight or flash linked objects in a sketch, or to perform an action associated with an action button.

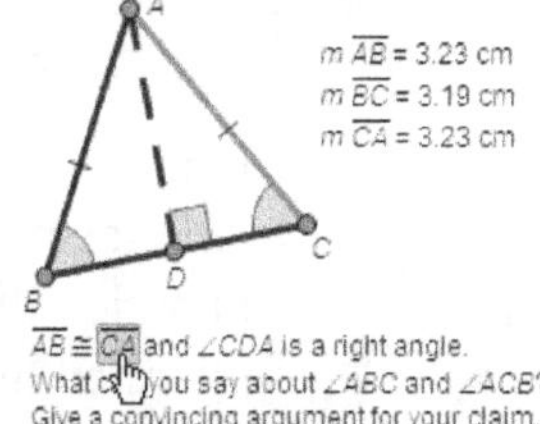


### ▼ Use Hot Text to Write and Read Mathematics

As a mathematical writer, use Hot Text to:

- Conveniently include proper mathematical notation for objects in your sketch.

- Build explanations that refer to current names and values of objects in your sketch.

- Document a construction, including links to hidden construction lines.

- Share your writing so that others can read along, seeing the connections to your construction and performing actions on that construction.

- Create equations or other mathematical expressions that incorporate values of sliders, parameters, or measurements.

- Write an annotated proof of a mathematical theorem.

As a mathematical reader, you can use Hot Text in sketches created by others to:

- See text and equations change in response to your manipulation of a sketch.

- Use links to follow an argument or explanation described in a caption.

- Press buttons within a caption to perform an action in the sketch.

## ▼ Insert Hot Text Labels and Values

While you're editing a caption, insert Hot Text labels and values into the caption.

> The pointer shows a + sign when it's over an object you can insert.

> You cannot create a Hot Text link to an object that depends on the caption in which you want to insert the Hot Text.

- To insert a link to an object, click the object in the sketch. (The Default Link column of the table below lists the form of the link inserted for various objects.)

- To insert a variant form of link to an object, hold the Shift key while you click the object, and choose a form from the pop-up menu that appears. (Available Variant Link forms are shown in the table below.)

- To insert the label of an angle, create an angle marker by either pressing on one leg of the angle and dragging to the other leg, or by pressing on the vertex and dragging into the interior of the angle.

> You can do this while you're editing the caption, without switching to the **Marker** [122] tool. When you drag, the pointer appears as the **Marker** [122] tool pointer. When you finish, Sketchpad creates the angle marker and inserts a Hot Text link into your caption.

| Object | Default Link | Variant Link(s) (Requires Shift Key) |
|---|---|---|
| Point [2], Straight Object [90], Circle [6], Arc [7], Polygon or Other Interior [8], Locus [10], Function Plot [52], Parametric Plot [54], Picture [18], Iterated Image [24], Table [39] | Label | None |
| Parameter [36] | Value | Label, Value as Addend |
| Measurement [36] | Value | Label, Value as Addend, Name |
| Calculation [39] | Value | Label, Value as Addend, Calculation |
| Function [45] | Full Equation | Label, Definition, Independent Variable, Dependent Variable |
| Angle Marker [60] (drag in the sketch to create a marker and insert its name) | Name | Label |
| Action Button [66] (pressing the link activates the | Label | None |

| | | |
|---|---|---|
| button) | | |
| Caption 56 (hold the Shift key to insert a caption) | None | Full Text |

**Label:** The label 81 of the object

**Value:** The value 92 of a parameter, measurement, or calculation

**Value as Addend:** The value preceded by a sign for addition or subtraction, depending on whether the value is positive or negative (This choice allows you to build nicely formatted expressions within a caption.)

**Name:** The original name of a measurement or angle marker, showing what the object measures or marks (for instance, m$\angle ABC$ for an angle measurement )

**Calculation:** The expression that defines the calculation (for instance, $\pi \cdot r^2$)

**Full Equation:** The equation of the function as it's displayed in the sketch (for instance, $f(x) = 2 \cdot x + 3$)

**Definition:** The definition of the function, without the label (For instance, $2 \cdot x + 3$)

**Independent Variable:** The label of the independent variable used in the function (for instance, $x$ )

**Dependent Variable:** The label of the dependent variable (for instance, $f$)

**Full Text:** The full text of the caption (An included caption doesn't flash like other Hot Text links, but is updated automatically.)

## ▼ Use a Caption Containing Hot Text

After you've created a caption containing Hot Text, you can interact with the caption.

- Manipulate the sketch so that a value changes, and observe the changing value in the caption.

- Change the label of an object in the sketch. The caption is updated automatically.

- Highlight a linked sketch object by moving the pointer over the Hot Text link to the object.
  Both the Hot Text and the object to which it links are highlighted (provided the object is visible), and Sketchpad displays an appropriate status message.

- Flash a linked sketch object by pressing and holding the pointer on the Hot Text link to the object.
  The object flashes even if it's hidden.

- Press a linked action button by pressing and releasing the pointer on the Hot Text link to the action button.

- Click a Hot Text link from various dialog boxes, including the Calculator, Plot Points, Iteration, and Transform dialog boxes.
  The linked object will be used as though you had clicked on it directly.

## ▼ Create a Mathematical Expression Containing Values

When you graph a function (like $y = ax^2 + bx + c$) or calculate the value of an expression (like $mx + b$), you may want to create a caption that shows the function or expression with the numeric values inserted in place of the names of the parameters or measurements.

For instance, if you've used two parameters ($m = 2$ and $b = 3$) to calculate the value of $mx + b$, you might want to create a caption that shows the expression as $2x + 3$.

This is easy as long as $b$ is positive. While editing your caption, you can follow these steps:

1. Click parameter $m$ in the sketch to insert "2".

2. Type "$x +$ ".

3. Click parameter $b$ in the sketch to insert "3".

The result appears as "$2x + 3$."

But if you change the value of b to make it negative (for instance, $b = -1$), the caption shows "$2x + -1$" instead of "$2x - 1$."

To create an improved display, follow these steps instead:

1. Click parameter $m$ in the sketch to insert "2".

2. Type "$x$ ".

3. Hold the Shift key while you click parameter $b$ in the sketch. From the menu that appears, choose **Value as addend.**

The result of choosing **Value as addend** is a value that always includes the sign (either $+$ or $-$). Depending on the value of $b$, the expression will appear correctly, as $2x + 3$ or $2x - 1$.

## 1.18    Angle Markers and Tick Marks

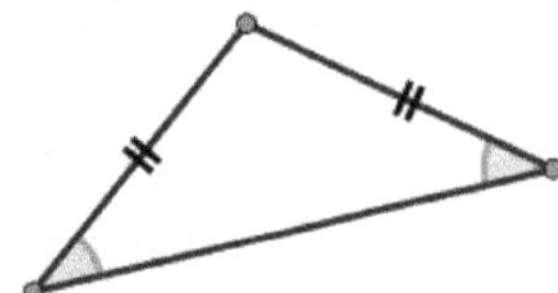

Angle markers 60 and tick marks 63 are display ornaments created by the **Marker** 122 tool. You can place an angle marker where two straight objects 90 meet to form an angle or at the vertex of a polygon 8 , and you can place a tick mark on a segment, ray, line, 5 or other path 89 .

Angle markers and tick marks cannot be transformed.

*Subtopics:*
 *Angle Markers* 60
 *Tick Marks* 63
 *Marker Properties* 65

### 1.18.1    Angle Markers

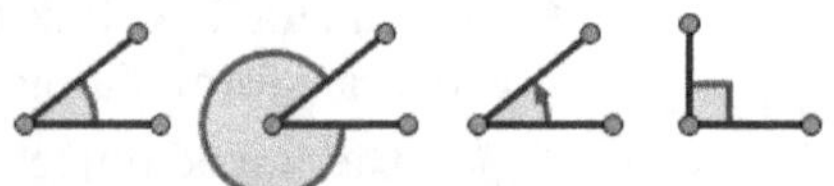

Use angle markers to draw attention to an angle, to measure an angle, or to denote special relationships such as congruent angles. Angle markers are normally displayed as filled arcs. When you drag or construct an angle so that it's a right angle, the marker is displayed as a right-angle marker.

You cannot trace, transform, or iterate angle markers.

Note

Sketchpad doesn't detect congruent angles automatically, but leaves it up to you to form and test conjectures by making appropriate observations and measurements. You can use angle markers to indicate confirmed conjectures, untested conjectures, or in any other way that you find useful.

### ▼ Create an Angle Marker

There are two ways to create an angle marker:

- Drag the **Marker** 122 tool from one side of the angle to the other.

- Drag the **Marker** 122 tool from the vertex to the interior of the angle.

    If there are several possible angles with the same vertex, the position to which you drag determines which angle is marked.

Note

While you are editing a caption, use either method to create a new angle marker and insert a Hot Text 57 link into the caption.

### ▼ Determine an Angle Marker's Definition

Angle markers can be defined in four ways.

| Definition | Example | Magnitude (degrees) | Magnitude (directed degrees) | Magnitude (radians) |
|---|---|---|---|---|
| **Simple:** A simple angle is always less than or equal to a straight angle. It represents the smallest turn at the vertex from the initial ray (or first point) to the terminal ray (or third point). Its direction can be counter-clockwise or clockwise. | $m\angle ABC = 38°$ <br> $m\angle DEF = -40°$ | $0° \leq \theta \leq 180°$ | $-180° < \theta \leq 180°$ | $-\pi < \theta \leq -\pi$ |
| **Reflex:** A reflex angle is always greater than or equal to a straight angle. It represents the greatest turn at the vertex from the initial ray (or first point) to the terminal ray (or third point). Its direction can be counter-clockwise or clockwise. | $m\angle ABC = -322°$ <br> $m\angle DEF = 320°$ | $180° \leq \theta \leq 360°$ | $-360° \leq \theta \leq -180°$ or $180° < \theta < 360°$ | $-2\pi \leq \theta \leq -\pi$ or $\pi < \theta < 2\pi$ |
| **Counter-clockwise:** A counter-clockwise angle represents the counter-clockwise turn at the vertex from the initial ray (or first point) to the terminal ray (or third point). It can be either simple or reflex. | $m\angle ABC = 38°$ <br> $m\angle DEF = 320°$ | $0° \leq \theta < 360°$ | $0° \leq \theta < 360°$ | $0 \leq \theta < 2\pi$ |
| **Clockwise:** A clockwise angle represents the clockwise turn at the vertex from the initial ray (or first point) to the terminal ray (or third point). It can be either simple or reflex. | $m\angle ABC = -322°$ <br> $m\angle DEF = -40°$ | $0° \leq \theta < 360°$ | $-360° < \theta \leq 0°$ | $-2\pi < \theta \leq 0$ |

When you use the **Marker** 122 tool to create an angle marker, the resulting angle marker is either simple or reflex depending on the direction in which you drag.

To change an angle marker to use a different definition, select it and choose **Edit | Properties | Marker** 65.

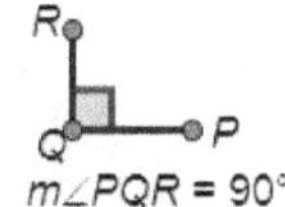


If a simple, counter-clockwise, or clockwise angle marker has a magnitude of exactly 90°, it appears as a right-angle marker.

### ▼ Change the Appearance of an Angle Marker

Angle markers display one or more closely spaced strokes, the region defined by the innermost stroke, and an optional arrowhead to indicate direction.

When the marked angle is a right angle, the stroke(s) form a customary right-angle marker. Otherwise the strokes appear as arcs.

- To change the number of strokes, click the angle marker with the **Marker** 122 tool.

- To set the color of the marker, choose **Display | Color** 164.

- To set the thickness of the marker's stroke(s), choose **Display | Line Style** 163.

- To set the radius of the stroke(s), choose **Display | Point Style** 162.

- To change the definition of the marker, the number of strokes, and whether or not an arrowhead appears, choose **Edit | Properties | Marker** 65.

- To determine the transparency of the region, choose **Edit | Properties | Opacity** 92. You can make the region completely transparent, so that only the stroke(s) are visible.

### ▼ Use an Angle Marker

To measure an angle, select an angle marker and choose **Measure | Angle** 218.

To use an angle marker as a marked angle for rotation or polar translation, select it and choose **Transform | Mark Angle** 195, or click the angle marker in the sketch while using the Rotation or Translation dialog box.

To create an angle marker while you're editing a caption, press and drag in the sketch to designate the angle, just as you would using the **Marker** 122 tool. Sketchpad creates the angle marker and inserts a Hot Text link into the caption in one step.

*See also:*
*Construct an Angle of Fixed Measure* 204
*Construct Congruent Angles* 204

## 1.18.2  Tick Marks

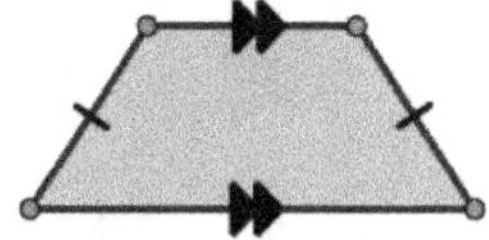

Use tick marks to denote special relationships (such as congruence or parallelism) that relate objects to each other.

Tick marks do not have labels of their own, and you cannot trace, transform, or iterate them.

 Note

Sketchpad doesn't detect congruence or parallelism automatically, but leaves it up to you to form and test conjectures by making appropriate observations and measurements. You can use tick marks to indicate confirmed conjectures, untested conjectures, or in any other way that you find useful.

### ▼ Create a Tick Mark

To create a tick mark, click the **Marker** 122 tool on a path.

The path can be a segment, ray, line 5, or other path object 89.

### ▼ Determine a Tick Mark's Type

There are four types of tick marks.

- Crossbar

- Open arrow

- Hollow arrow

- Solid arrow

Crossbars are often used to indicate congruent segments, and arrows are often used to indicate parallel straight objects.

A newly created tick mark is a crossbar. To change a tick mark to a different type, select it and choose **Edit | Properties | Marker** 65.

### ▼ Change the Appearance of a Tick Mark

Tick marks display one or more closely spaced decorations (either bars or arrows) on the path.

- To move a tick mark to a different position on its path, drag the marker with the **Marker** 122 tool.

- To change the number of bars or arrows, click it with the **Marker** 122 tool.

- To set the color of the bars or arrows, choose **Display** 162 | **Color** 164.

- To set the thickness of the bars or arrows, choose **Display** 162 | **Line Style** 163.

- To determine the type of marker, the number of bars or arrows, and the direction of the arrows, choose **Edit | Properties | Marker** 65.

You can also use the Context 245 menu to change the thickness, color, type, and direction of the bars or arrows.

## 1.18.3  Marker Properties

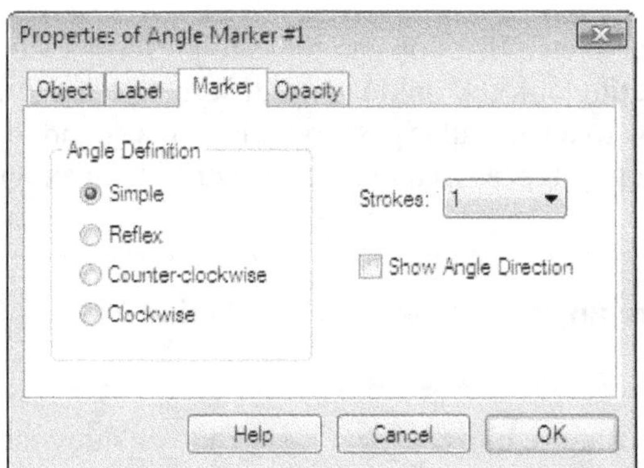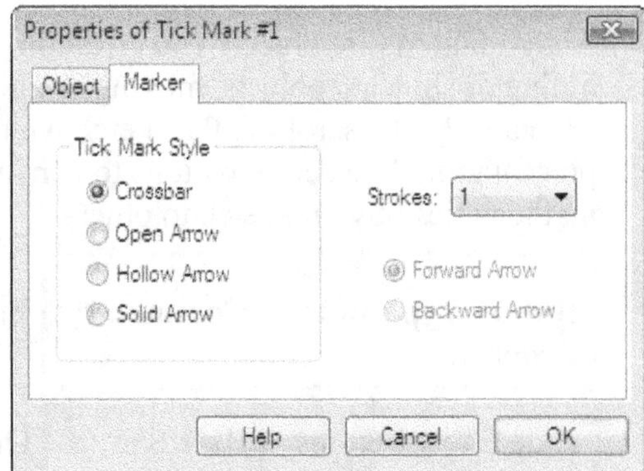

### ▾ Angle Marker Properties

Use the Marker Properties 158 panel to determine the definition of the angle being marked, the number of ticks used to mark it, and whether the direction of the angle is shown by an arrowhead.

**Simple**: A simple angle is less than a straight angle. No matter how the angle changes, the marker shows the smaller angle. A simple angle marker will always show the interior angle in a triangle.

**Reflex:** A reflex angle is greater than a straight angle. No matter how the angle changes, the marker shows the larger angle. A reflex angle marker will always show the angle outside a triangle.

**Counter-clockwise:** A counter-clockwise angle goes counter-clockwise from the initial side of the angle to the terminal side. Depending on the orientation of the sides, the angle may be less than or greater than a straight angle.

**Clockwise:** A clockwise angle goes clockwise from the initial side of the angle to the terminal side. Depending on the orientation of the sides, the angle may be less than or greater than a straight angle.

 Note

In a polygon that might become concave, use counter-clockwise or clockwise angle markers, rather than simple or reflex angle markers, so that the angle marker stays in its original location inside our outside the polygon.

**Strokes:** Determine the number of strokes used to display the angle marker.

**Show angle direction:** Check this box to display an arrowhead that shows the direction of the angle.

### ▾ Tick Mark Properties

Use this Properties 158 panel to determine the type of tick mark to display, the number of ticks to show, and whether to display arrows forward or backward.

**Tick Mark Style:** Set the type of marker to **Crossbar, Open arrow, Hollow arrow,** or **Solid arrow.**

**Strokes:** Determine the number of bars or arrows used to display the tick mark.

**Forward/Backward arrow:** Determine which way the arrow points.

## 1.19   Action Buttons

Action buttons are objects you create in your sketch that you can press to perform a variety of actions: hiding or showing objects, moving or animating objects, linking to a different page in your document or to a web site, scrolling the sketch window to a particular position, playing a sound, or making a presentation. Use action buttons to conveniently repeat frequent actions or to help explain the mathematics of your sketch to others.

| Button Type and Properties | Command | Prerequisites | Action |
|---|---|---|---|
| Hide/Show Buttons and Properties 67 | Hide/Show 149 | One or more objects | Hide or show the selected objects |
| Animation Buttons and Properties 69 | Animation 149 | One or more geometric objects or parameters | Animate the selected objects |
| Movement Buttons and Properties 71 | Movement 150 | One or more pairs of points or values. The first object of each pair must be free to move or vary. (In other words, the first object can be a point that's free to move or a parameter.) | Move or vary the first object of each pair toward the second object of the pair |
| Presentation Buttons and Properties 72 | Presentation 150 | One or more action buttons | Present the action buttons simultaneously or sequentially |
| Sound Buttons 74 | Sound 150 | One or two functions | Make a sound defined by the function(s) |
| Link Buttons and Properties 75 | Link 151 | None | Link to another page, another sketch, or a location on the Internet |
| Scroll Buttons and Properties 76 | Scroll 151 | One point | Scroll the window based on the location of the point |

### ▾ Using Action Buttons

To create an action button, choose a command from the **Edit | Action Buttons** 148 submenu. Once you've created an action button, you can do several things with it.

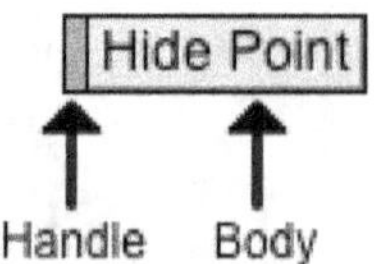


- Start the button's action by pressing the button body (not the handle) with the **Arrow** 104 tool. (If a button stays down after you press it, this indicates that its action is still continuing. Click the button a second time to stop its action.)

- Select the button by clicking the handle (not the body) with the **Arrow** 104 tool. Once the button is selected, you can hide it, clear 147 it, and perform other actions on it.

- Change the button's font, size, style, and color by first clicking the handle to select it and then using the Text Palette 270.

- Move the button to a different position by using the **Arrow** 104 tool to drag the button's handle.

- Change the button's label by double-clicking it with the **Text** 120 tool.

- Change the button's properties by choosing **Edit | Properties** 158 or by choosing **Properties** from the Context 245 menu.

#### ▼ Action Button Keyboard Shortcuts

You can create a keyboard shortcut for an action button, so that it performs its action when you press a key on the keyboard. In the button's label, insert an ampersand symbol (&) before the letter that you want to use as the shortcut. That letter will then be underlined to indicate the keyboard shortcut. For example if you label an animation button *Animate &Point,* the *P* will be underlined, so that the button appears as *Animate Point.* On the keyboard, *P* will be the shortcut key.

To perform the button's action, press the shortcut key on your keyboard. To stop the action, press the shortcut key a second time.

> The keyboard shortcut works only when the action button is visible.

> If several action buttons have the same keyboard shortcut, then pressing the key starts all of their actions.

 Note

Keyboard shortcuts are not case-sensitive unless there are separate buttons with uppercase and lowercase shortcuts for the same letter. For instance, if a sketch has a single button labeled *Animate Point,* you can start the action by pressing either "P" or "p." If a sketch has buttons labeled *Animate Point* and *Hide all points,* pressing "P" activates the former, and pressing "p" activates the latter.

## 1.19.1 Hide/Show Buttons and Properties

A Hide/Show button hides or shows a group of objects.

#### ▼ Creating and Using a Hide/Show Button

Select one or more objects and choose **Edit | Action Buttons | Hide/Show** 149 to create a Hide/Show button that toggles between Hide and Show, depending on whether the objects it controls are visible or not.

To create two buttons, one that hides and one that shows, hold the Shift key when you choose the command.

Use Hide and Show buttons when there are details in a sketch which you sometimes want visible and sometimes want hidden. For example, your sketch might use a single triangle to show the construction of the circumcenter, centroid, and orthocenter. If you show all the construction lines at the same time, the sketch will be very confusing. Use Hide/Show buttons to show or hide the construction lines for each of the three different constructions.

Normally the button's label changes from *Hide* to *Show* depending on whether the objects it controls are visible |86| or not. (When all the objects are hidden, the button is labeled *Show;* otherwise it's labeled *Hide.*)

Choose **Edit | Properties** |158| **| Hide/Show** to:

- Set a button to always hide, always show, or toggle between hiding and showing.

- Determine whether a button selects its objects after showing them.

- Fade objects in or out when the button shows or hides them.

Note

When you change a button's selection and fading behavior, Sketchpad remembers your new settings and uses them for future Hide/Show buttons.

## ▼ Hide/Show Properties

This Properties |158| panel appears only for Hide/Show |149| action buttons |66|. Use it to determine the type of action the Hide/Show button performs as well as various aspects of how it works.

To open this dialog box, select the button and choose **Edit | Properties** |158|, or choose **Properties** |158| from the Context |245| menu.

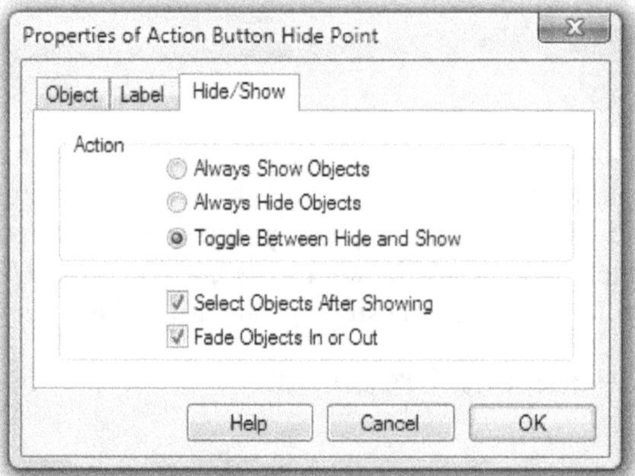

**Action:**

- Choose **Always show objects** to make this button into a Show button, which always shows the objects to which it applies.

- Choose **Always hide objects** to make this button into a Hide button, which always hides the objects to which it applies.

- Choose **Toggle between hide and show** to make this a toggling Hide/Show button, which hides objects when clicked on if one or more of its objects are showing and shows its objects when clicked on if they are all hidden.

  If you don't change the default label of a toggling Hide/Show button, the label changes between *Hide* and *Show* to describe the action it will perform next.

**Effects:**

- If **Select objects after showing** is checked, when you click on a Show button all its parental objects [80] are selected. (Even if some of the button's objects are already showing, they are still selected.) If unchecked, the objects are left unselected after they're shown.

- If **Fade objects in or out** is checked, objects fade in or out of visibility gradually when you click on the Hide/Show button. If unchecked, objects appear or disappear immediately.

## 1.19.2  Animation Buttons and Properties

An Animation button [149] animates one or more geometric objects or parameters [36].

### ▼ Creating and Using an Animation Button

`Animate Points`

Use Animation buttons to automate motion in your sketch. Use an Animation button to move a point along its path, to move an independent point randomly in the plane, or to vary a parameter.

Select one or more objects and choose **Edit | Action Buttons | Animation** [149] to create an Animation button.

How the button animates each selected object depends on the nature of the object:

- Independent points animate freely in the plane.

- Points on path objects [89] animate along their paths.

- Other geometric objects animate the points on which they depend.

- Parameters animate by changing their numeric value.

You may want to familiarize yourself with the **Display | Animate** [171] command before creating action buttons to perform animations for you.

Press an Animation button once to start the animation. The button remains pressed until the animation is finished. You can press the button a second time, while the button is still pressed, to stop the animation.

> You can also press the Esc [294] key to stop the animation. (You may have to press the Esc key more than once.)

Choose **Edit | Properties** [158] **| Animate** to set the animation speed and direction and to set the domain for parameter animation.

### ▼ Animate Properties

This Properties [158] panel appears only for Animation [149] action buttons [66]. Use it to determine how each animated object moves.

The panel appears automatically when you choose **Edit | Action Buttons | Animation** [149] to create a button. After you've created the button, choose **Edit | Properties** [158], or choose **Properties** [158] from the Context [245] menu, to display the panel again to make further adjustments.

Animate Properties for a Point

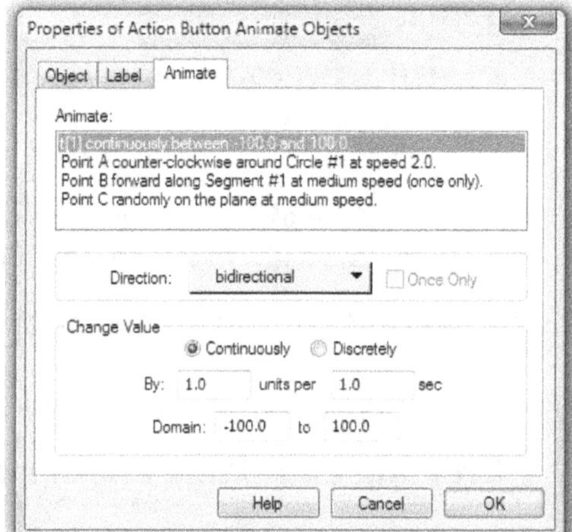

Animate Properties for a Parameter

The list at the top describes the motion of each animated object. If the button animates an object that can move only by moving its parents 80, the list shows the parents rather than the object itself. (For instance, if you animate the segment connecting point $A$ and point $B$, the list shows $A$ and $B$ rather than the segment.) When you highlight an object in the list, the details of that object's animation settings appear in the lower part of the panel, and you can modify the settings to change how the animation works.

**Direction:** For parameters 36 and for points that are animated on paths 89, you can set the direction of the motion. The available choices depend on the kind of path and the kind of objects. For instance, if the path is a circle or circle interior, the first two choices are **counter-clockwise** and **clockwise.** If the path is a straight object, arc, or polygon, the first two choices are **forward** and **backward**. If the animated object is a parameter, the first two choices are **increasing** and **decreasing**.

> Animating points or parameters randomly is particularly useful for investigations in probability, statistics, and chaos.

If you choose **random** direction for a point on a path, the point moves to a new randomly chosen position on its path each time it moves. Similarly, if you choose **random** direction for a parameter, the parameter takes on a new random value within its domain each time it moves.

**Once Only:** For a parameter or a point on a path which is moving randomly, check **Once only** to stop the animation after it generates a single new random position or value. For points on paths that aren't moving randomly, check **Once only** to stop the animation when the point returns to its starting position. For parameters which aren't moving randomly, check **Once only** to stop the animation when the parameter returns to its starting value.

> The **Once only** check box is not available when a point or parameter is moving bidirectionally.

**Speed:** This portion of the dialog box appears only for points, and allows you to set the speed of an animated point to slow, medium, fast, or some other desired value.

**Change Value:** This portion of the dialog box appears only for parameters, and allows you to determine how a parameter's value changes. You can determine whether the value changes continuously or discretely (jumping by an increment you set each time the value changes). You can determine how quickly the value changes, and the domain within which the value can vary.

> The rate at which a parameter changes its value is not exact. Sketchpad tries to change

the value at the designated rate, but if your computer is busy with other tasks, the
parameter may change more slowly than the rate set here.

*See also:*
Principles of Animation 267

## 1.19.3 Movement Buttons and Properties

A Movement button moves one or more points or parameters 36 toward defined destinations.

### ▼ Creating and Using a Movement Button

| Move A → B |

To create a Movement button, you must select one or more pairs of objects:

- A pair of points must consist of a point to move and a destination point toward which the first
  point can move.

- A pair of values must consist of a value to change (which must be a parameter) and a destination
  value toward which the parameter can change.

Select one or more such pairs of objects and choose **Edit | Action Buttons | Movement** 150 to
create a Movement button.

Press a Movement button once to start the movement. The button remains pressed until all moving
points and parameters reach their destinations. You can press the button a second time, while the
button is still down, to stop the movement.

> You can also press the Esc 294 key to stop the movement. (You may have to press the
> Esc key more than once.)

After creating a Movement button, you may want to hide the destination points or values so only
the moving points or parameters are visible.

Choose **Edit | Properties** 158 | **Move** to set the speed and behavior of the movement.

### ▼ Move Properties

This Properties 158 panel appears only for Movement 150 action buttons 66. Use it to determine the
speed and behavior of the movement.

> The panel appears automatically when you choose **Edit | Action Buttons | Movement**
> 150 to create a button. After you've created the button, choose **Edit | Properties** 158, or
> choose **Properties** 158 from the Context 245 menu, to display the panel again to make
> further adjustments.

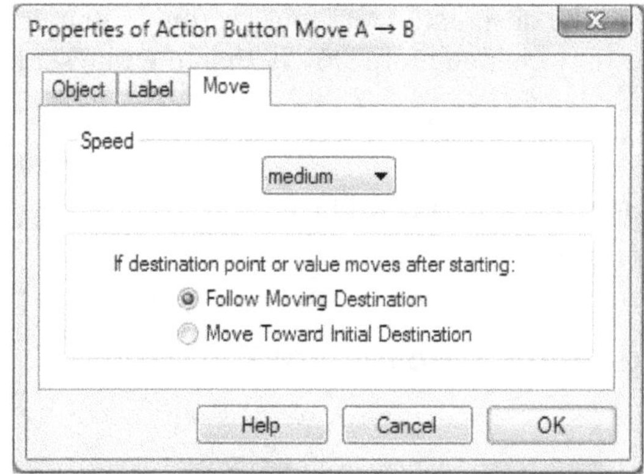

You can set the movement speed to slow, medium, fast, or instant.

If the destination point or value moves while the Movement button is active, you can decide how the moving point will travel. Choose **Follow moving destination** to have the moving point or parameter alter direction as the destination point or value moves, always continuing to move toward that destination. Choose **Move toward initial destination** to have the moving object travel directly to the location or value of the destination at the instant the Movement button was clicked, stopping when it reaches that initial destination.

> If the destination is moving, it's possible that the moving object will never reach the destination and will keep moving forever. You can take advantage of this "perpetual motion" to model kinematic systems.

### 1.19.4  Presentation Buttons and Properties

A Presentation button|150| automatically activates a group of other action buttons. The buttons can be activated either simultaneously or in sequence. Use a Presentation button to choreograph a complex set of motions or to present a Sketchpad slide show.

#### ▼ Creating and Using a Presentation Button

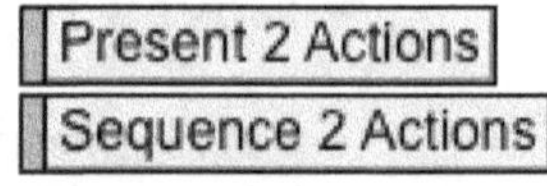


Select one or more action buttons and choose **Edit | Action Buttons | Presentation**|150| to create a Presentation button.

Pressing a Presentation button has the same effect as pressing each of its selected parent|80| actions either all at once or one after the other. Use Presentation buttons when you want to combine several related actions into a single button for ease of use.

Press a Presentation button once to start the presentation. The button remains pressed until all its actions complete. You can press Esc|294| or press the button a second time, while the button is still pressed, to stop a simultaneous presentation or to move a sequential presentation on to the next action in the sequence.

> You can also press the Esc |294| key to advance or end the presentation. (You may have to press the Esc key more than once.)

Choose **Edit | Properties**|158| **| Presentation** to determine whether the actions occur simultaneously or in sequence, and to set a variety of other options.

## ▼ Presentation Properties

This Properties|158| panel appears only for Presentation|150| action buttons|66|. Use it to determine how the presentation takes place.

> The panel appears automatically when you choose **Edit | Action Buttons | Presentation**|150| to create a button. After you've created the button, choose **Edit | Properties**|158|, or choose **Properties**|158| from the Context|245| menu, to display the panel again to make further adjustments.

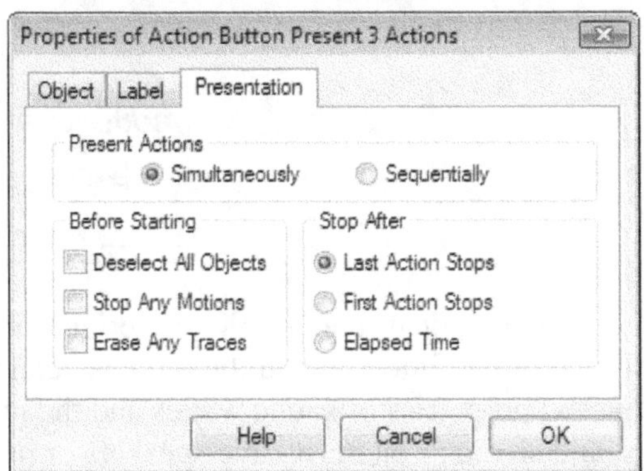
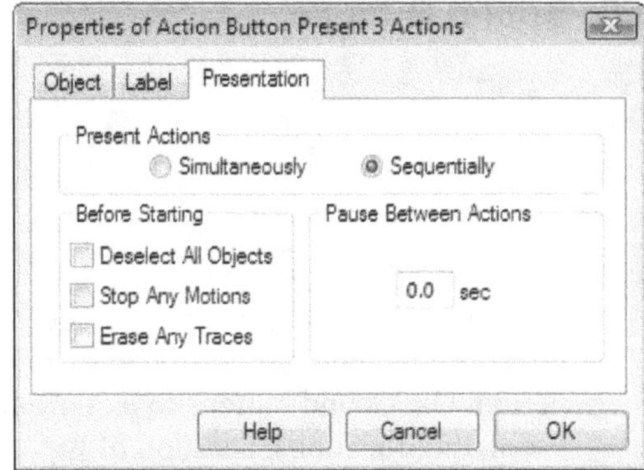

This panel appears only for Presentation action buttons|150|. A Presentation button presents the actions of a list of other parental|80| buttons|66|.

**Present Actions:** This choice is available only when the Presentation button presents more than one action button.

Choose **Simultaneously** to activate all actions of the presentation at the same time. Pressing a simultaneous Presentation button has the same effect as activating all of its parental action buttons at once. If you choose **Simultaneously,** you can also set stopping conditions for the Presentation (see **Stop after,** below).

Choose **Sequentially** to activate the actions of the presentation one after another in the order in which they were selected when the button was created. Pressing a sequential Presentation button has the same effect as pressing the parental action buttons one at a time, waiting for each activated action to be completed before proceeding to the next parental action. If you choose **Sequentially,** you can also specify a pause between presented actions (see **Pause between actions,** below).

> When you present a parental Animation button|149| sequentially, Sketchpad waits for the animation to complete before continuing the sequenced presentation. If the animation doesn't complete on its own, you can stop it yourself either by releasing the pressed Animation button, by choosing Stop Animation|172| from the Display|162| menu while the animation is continuing, or by pressing Esc|295|. Once the animation stops, the sequenced presentation resumes.

**Before Starting:** Check these options to specify any additional effects you'd like to occur at the moment the Presentation button is pressed. Based on your choices, Sketchpad will deselect any previously selected objects|106|, stop any previously started animations, and erase any previously displayed traces|170| before commencing the presentation.

**Stop After:** This choice is available only when the presentation presents actions simultaneously. Choose **Stop after last action stops** to allow each of the presented actions to proceed independently, and the presentation to complete only when the last presented action has finished. Choose **Stop after first action stops** to stop all of the presented actions stop as soon as the first one stops. (This choice can be useful for coordinating two or more animation or movement  actions.) Choose **Stop after elapsed time** to enter an overall duration in seconds for the presented actions. (This choice is useful when you want to stop an animation after a fixed amount of time.)

**Pause Between Actions:** This choice is available only when the presentation presents actions sequentially. Enter the amount of time (in seconds) you want to pause between each step of the sequence. If you enter zero, each step of the presentation commences as soon as the previous step completes. If you enter a nonzero pause, Sketchpad waits between steps.

## 1.19.5  Sound Buttons

$frequency = \boxed{440}$

$f(x) = \sin(2 \cdot \pi \cdot frequency \cdot x)$

$\boxed{\text{Hear Function } f}$

A Sound button 150 plays a sound defined by a mathematical function that describes air pressure over time. The volume is determined by the amplitude of the function, and the pitch is determined by its period. Use Sound buttons to explore the physical properties of sound waves and the mathematics of oscillating functions, addition of waveforms, and so on. (You must have speakers or headphones to hear the sound created by a Sound button.)

Note

The amplitude of the function should be between −1 and 1. Larger amplitudes may result in clipping and distortion of the sound.

When one function is selected, the resulting Sound button plays a monophonic sound. When two functions are selected, the resulting Sound button plays stereo sounds, with one function controlling the right channel and the other controlling the left channel.

Select one or two functions and choose **Edit | Action Buttons | Sound** 150 to create a Sound button.

The independent variable of the sound function represents time in seconds, so to play a sound at 440 hertz (A above middle C) you could use this function: $f(x) = \sin(440 \cdot 2\pi \cdot x)$. Sketchpad will warn you if you try to play a sound with a frequency or amplitude below the audible range.

Press the button to listen to the function. To stop playing the sound, press the button a second time or press the Esc key. (You may need to press Esc more than once.)

For example:

1. Use **Edit | Preferences | Units** 276 to set your sketch's preferred angle unit to radians.

2. Choose **Number | New Function** 230 and define $f(x) = \sin(440 \cdot 2\pi \cdot x)$.

3. With the function selected, choose **Edit | Action Buttons | Sound** 150. Sketchpad creates a Sound button.

4. Press the button. Sketchpad plays a sound that repeats 440 times per second. This frequency corresponds to A above middle C.

5. Press the button again to release it and stop listening to the function.

Use parameters in your function definition to create more interesting sounds. For instance, if you use the function $f(x) = a \cdot \sin(freq \cdot 2\pi \cdot x)$, you can vary parameter $a$ to change the volume and parameter $freq$ to change the frequency while you're playing the sound.

## 1.19.6   Link Buttons and Properties

A Link button [151] links to a different page in the current document, or to a web site or other location defined by a URL.

### ▼ Creating and Using a Link Button

Use a Link button to make it easy to navigate among pages in a multiple-page document, to open a web site related to the topic of your sketch, or to open a local Sketchpad or Help system document.

Use **File | Document Options** [142] to add multiple pages to your document [250].

Choose **Edit | Action Buttons | Link** [151] to create a Link button.

Choose **Edit | Properties** [158] **| Link** to open the Properties panel and set the destination page or URL for the button.

Use URLs to connect your sketch to related mathematical, historic, or reference material available on the Internet. Use special codes in the URL to link to local Sketchpad documents, Help system documents, or other local documents.

### ▼ Link Properties

This Properties [158] panel appears only for Link [151] action buttons [66]. Use it to determine the destination of the Link button.

The panel appears automatically when you choose **Edit | Action Buttons | Link** [151] to create a button. After you've created the button, choose **Edit | Properties** [158], or choose **Properties** [158] from the Context [245] menu, to display the panel again to make further adjustments.

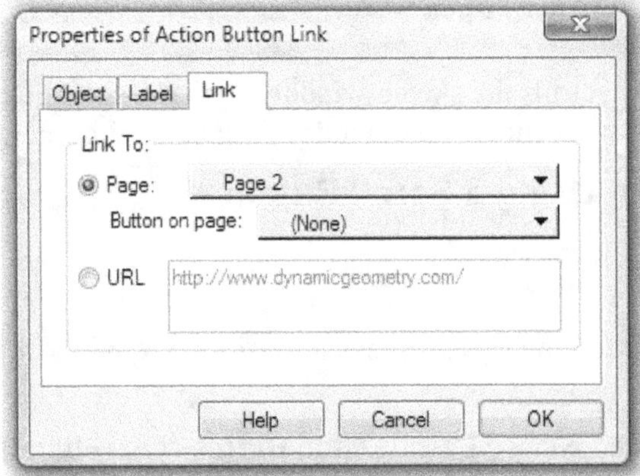


Use the **Link to** radio buttons to choose whether the button links to another page in your document [250] or to an Internet URL such as a web page.

**Page:** When this is chosen, the Link button will link to a different page of the current document. Use the Page pop-up menu to decide which page to link to. If that page contains any action buttons of its own (such as an Animation or Movement button), you can choose one of those buttons in the **Button on page** pop-up menu to cause the Link button to automatically activate the specified button on the linked page.

Use the **File | Document Options** [142] command to add or copy new pages into your document.

**URL:** When this is chosen, the Link button will link to a URL using your web browser. You must enter the actual URL, which can be a web site (if the URL starts with "http://") or a local file or program (if the URL starts with "file://").

The default URL points to the Sketchpad Resource Center, a web site with additional resources and information about using Sketchpad. You can copy web page URLs from your browser and paste them here.

URL links can also use these special URL forms:

**sketchdoc://**: Start the URL with "sketchdoc://" to link to a document in the same folder as the current document. For example, if you want to link from a document named Demo1.gsp to a document called Demo2.gsp in the same folder, you could use the URL "sketchdoc://Demo2.gsp".

**sketchapp://**: Start the URL with "sketchapp://" to link to a document in the same folder as the Sketchpad application.

**help://**: Start the URL with "help://" to link to a document in the Sketchpad Help folder.

Internet URLs such as http:// web sites are only accessible if you're connected to the Internet and have a web browser installed on your computer. Local file:// URLs are useful for accessing documents or other resources on your specific computer, but these buttons may not function if you open your document on another computer (which may well be missing the linked-to documents). The relative URLs — sketchdoc://, sketchapp://, and help:// — let you refer to resources in relation to "known locations" on any computer running Sketchpad. They can be handy if you're creating a folder of linked documents you'd like to share with others. As long as you use sketchdoc:// URLs, the documents within that folder will stay linked together no matter where you move or copy that folder.

## 1.19.7 Scroll Buttons and Properties

A Scroll button [151] scrolls the sketch window so that a specific point in the sketch becomes located either at the window's center or the window's upper-left corner.

### ▼ Creating and Using a Scroll Button

| Scroll |

Use a Scroll button in large sketches to position the window to show a particular part of your sketch.

Select a point and choose **Edit | Action Buttons | Scroll** [151] to create a Scroll button.

Choose **Edit | Properties** [158] **| Scroll** to determine whether the scroll target will be at the window's center or upper-left corner.

top left corner of the window or in the center of the window..

### ▼ Scroll Properties

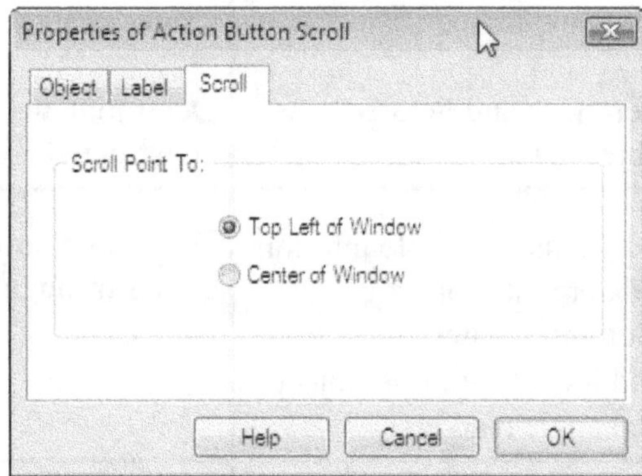

This Properties[158] panel appears only for Scroll[151] action buttons[66]. Use it to scroll the window in which it's located to show a specific portion of that window. The scrolling action is based on a point[2], and works in one of two ways: so that the point is at the top-left corner of the window, or so that the point is centered in the window.

The panel appears automatically when you choose **Edit | Action Buttons | Scroll**[151] to create a button. After you've created the button, choose **Edit | Properties**[158], or choose **Properties**[158] from the Context[245] menu, to display the panel again if you want to change the scrolling action.

# 1.20 Common Object Attributes

There are two main ways to control the appearance and properties of Sketchpad objects.

- Modify display attributes[86] such as color, object visibility, label visibility, display style, and tracing by using the Display[162] menu, the Text Palette[270], or the **Text**[120] tool.

- Inspect and modify the properties of an object by selecting it and choosing **Edit | Properties**[158] or choosing **Properties**[158] from the Context[245] menu.

Different properties apply to different kinds of objects:

| Properties | Applies to | Purpose |
|---|---|---|
| Object[79] | All objects | Inspect and navigate an object's relationship to other objects[80] in the sketch and control whether the object is visible and whether it can be selected by the **Arrow**[106] tool. |
| Label[82] | All objects that can show labels | Change an object's label, show or hide the label, and determine whether the label is used in custom tools. |
| Value[94] | Measurements[36], calculations[39], and parameters[36] | Determine how a parameter, measurement, or calculation is displayed. |
| Parameter[38] | Parameters[36] | Determine how a parameter is animated and how it's controlled by keyboard adjustments. |

| | | |
|---|---|---|
| Table 41 | Tables 39 | Determine whether the last row of a table tracks values as they change in the sketch. |
| Iteration 35 | Iterations and iterated objects 24 | Determine the depth of an iteration and how it's constructed. |
| Opacity 92 | Polygons 8, circle interiors 8, arc interiors 8, pictures 18, iterations 24 and loci 10 of these objects, and angle markers 60 | Determine how much the background of an object shows through it |
| Axis 44 | Axes 41 | Determine whether an axis is numbered using decimal numbers or using fractions and multiples of $\pi$. |
| Function 51 | Data-defined functions 51 | Determine the amount of smoothing to be applied to the function. |
| Plot 95 | Loci 10 and sampled transformed paths 89 | Determine the number of samples and whether the locus or sampled transformed path is displayed continuously or discretely. |
| Plot 97 | Function plots 52 and parametric plots 54 | Determine the domain of the plot, the number of samples, and whether the plot is displayed continuously or discretely. |
| Plot 98 | Sampled transformed picture 23 | Determine the number of samples and allowable distortion for a sampled transformed picture. |
| Marker 65 | Angle markers 60 and tick marks 63 | Determine the type of tick mark or angle marker, the number of strokes used to mark it, and the form in which it's displayed. |
| Hide/Show 67 | Hide/Show buttons 67 | Determine whether the button hides or shows and whether it selects the objects when it shows them. |
| Animate 69 | Animation buttons 69 | Determine the type and speed of an animation. |
| Move 71 | Movement buttons 71 | Determine the speed of a movement and how it handles a moving destination. |
| Presentation 72 | Presentation buttons 72 | Determine whether the presentation is simultaneous or sequential and control various aspects of the presentation. |
| Link 75 | Link buttons 75 | Determine the destination of the link. |

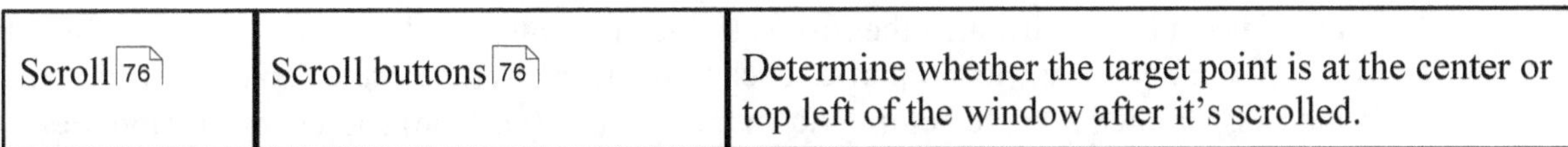


| Scroll | Scroll buttons | Determine whether the target point is at the center or top left of the window after it's scrolled. |
| --- | --- | --- |

## 1.20.1 Object Properties

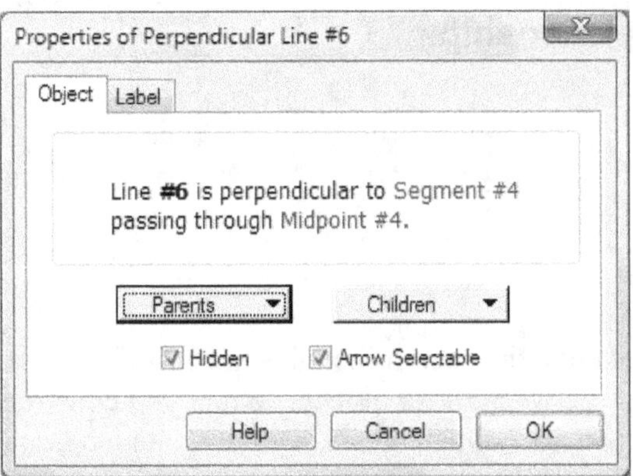


All Sketchpad objects have an Object Properties[158] panel. When you select an object and choose **Edit | Properties**[158], the left-hand tab shows the Object panel.

The Object panel has four parts:

**Object Description:** This is the object's geometric definition, described in terms of its relationship to its parents (the objects that geometrically define the object). Click any parent object's blue text to show the properties of that parent.

**Parents, Children:** Click on either the Parents or Children pop-up menu to see the parents or children[80] of the current object. As you move through either menu, the corresponding object in the sketch is highlighted. If you choose a parent or child from either the Parents or Children pop-up menu, the Properties dialog box switches to show the properties of this related object. (When you switch to a different object in this way, any changes you've made in the properties of the original object become permanent.)

Note

Use the Parents and Children menus to navigate the family tree — that is, to find an object's ancestors or descendants. This is a good way to learn how an object was constructed, to locate an object in a complicated sketch, or to display one particular hidden object.

**Hidden:** Use this checkbox to determine whether the object is hidden or visible. (You can also hide objects by choosing **Display | Hide**[167] and show them by choosing **Display | Show All Hidden**[167].)

**Arrow Selectable:** Use this checkbox to determine whether the object can be selected by clicking it with the **Arrow**[104] tool. Normally, you'll leave this property checked. If you clear the check-box, the object will no longer be selected when clicked by the Arrow tool or when you use the selection rectangle. This can be handy when you're working with something like a pasted picture that you want to use as the backdrop of a geometric measurement activity. If you don't want to accidentally select and drag the picture while working "on top" of it, you may want to make it not arrow-selectable.

Note

Use the **Arrow Selectable** setting judiciously; objects that are not arrow-selectable are difficult to manipulate and may frustrate other users. If you encounter an object that's not arrow-selectable, you may want to change its properties back to being arrow-selectable. To do so, right-click it (in Windows), or click it while holding down the Ctrl key (on Mac) and choose **Properties** 158 from the Context 245 menu.

While you're viewing one object's properties, you can switch to a different object by clicking that object in the sketch.

## 1.20.2 Parent-Child Relationships

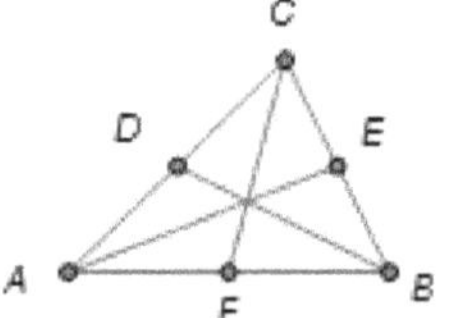


When you create a sketch, the sketch includes not just the geometric objects you've constructed, but also the relationships between those objects. When you construct the triangle shown here, your sketch includes more than just six points and six segments; it also includes the relationships between those 12 objects.

> Without relationships, this figure could be dragged apart into six disconnected points and six disconnected segments.

For instance, midpoint E depends on segment BC. When you choose **Construct | Midpoint** 176 you create not just a point but also a relationship, so that point E will always be the midpoint of segment BC. You can describe that relationship by saying that point E is the *child* of segment BC and that segment BC is the *parent* of point E. Similarly, segment AB is the child of its two endpoints A and B, and those endpoints are the parents of segment AB.

An object that has no parents is an *independent object,* and an object that does have parents is a *dependent object.* An independent object does not depend on any other object for its position or value. The position or value of a dependent object is determined by its parents.

These parent-child relationships define the mathematics of your sketch and are crucial to the way your sketches behave when you explore them by dragging. Object relationships keep the triangle together as a triangle, and relationships make the midpoints stay where they belong when you drag a vertex.

Think of a sketch as a family tree, defined both by the objects in the sketch and by their parent-child relationships.

These parent and child relationships are implicit in everything you do in Sketchpad. For instance, to use a command from the Construct 174 menu, you must first select certain objects (*prerequisites*) in your sketch. These prerequisites become the parents of the newly constructed child.

There are several ways to explore an object's mathematical definition — its family tree:

- Click it with the **Information** 124 tool. A balloon appears listing the object's parents, children, or both, and you can click the links in the balloon to explore those parents or children in turn.

  > To determine whether balloons show parents, children, or both, choose **Edit | Preferences | Tools** 280 and change the setting for the **Information** Tool.

- Select the object and choose **Edit | Select Parents** 152 or **Edit | Select Children** 152.

- Select the object and choose **Edit | Properties** 158. On the **Object** 81 panel, use the Parents and Children pop-up menus.

You can even rearrange your sketch's family tree using the **Edit | Split** 153 and **Edit | Merge** 153 commands.

## 1.20.3  Object Labels

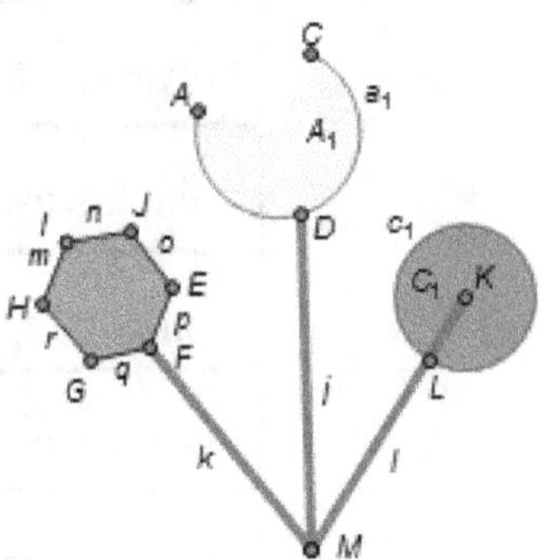

All points 2 , straight objects 5 , circles 6 , arcs 7 , interiors 8 , point loci 10 , measurements 36 , parameters 36 , calculations 39 , functions 45 , function plots 52 , angle markers 60 , and action buttons 66 have labels. A label is text used to identify the object.

- Show or hide labels using the **Text** 120 tool or **Display | Show/Hide Labels** 168 .

- Reposition labels by dragging with the **Text** tool.

- Edit labels by double-clicking with the **Text** tool, by choosing **Display | Label Objects** 168 , or by choosing **Edit | Properties | Label** 82 .

   When you edit a label, you can insert subscripts, symbols, Greek letters, and Unicode 82 using support built into Sketchpad or using Unicode input methods 314 available from your operating system.

- Change how labels are used in custom tools by choosing **Edit | Properties | Label** 81 .

**Italics:** Sketchpad normally uses italics for most objects, including geometric objects, measurements, parameters, and user-defined function names. Calculations and function definitions appear using mathematical italics in a way that is consistent with the most common standards for mathematical typesetting.

   To turn mathematical italicization on or off, choose Edit | Preferences | Text 279 and use the **Italicize all mathematical labels** checkbox.

**Default Labels:** Here are the default labels Sketchpad uses for various kinds of objects:

| Object | Default Label |
|---|---|
| Point 2 | $A, B, C, \ldots$ |
| Straight Object 5 | $j, k, l, \ldots$ |
| Circle 6 | $c_1, c_2, c_3, \ldots$ |
| Circle Interior 8 | $C_1, C_2, C_3, \ldots$ |

| | |
|---|---|
| Arc [7] | $a_1, a_2, a_3, \ldots$ |
| Arc Interiors [8] | $A_1, A_2, A_3, \ldots$ |
| Polygon [8] * | $P_1, P_2, P_3, \ldots$ |
| Point Locus [10] | $L_1, L_2, L_3, \ldots$ |
| Non-point Locus [10] | Cannot show a label |
| Function [45] | $f, g, h, \ldots$ |
| Function Plot [52] | $y=f(x), \ldots$ |
| Measurement [92] | $m_1, m_2, m_3, \ldots$ |
| Parameter [92] | $t_1, t_2, t_3, \ldots$ |
| Caption [56] | Cannot show a label |
| Pictures [18] | Cannot show a label |
| Action Buttons [66] | Label depends on action |

* When possible, a polygon with six or fewer vertices is labeled according to its vertices.

If Sketchpad runs out of letters with which to label points or straight objects, it starts over with $A_1$ and $j_1$.

## 1.20.4  Label Properties

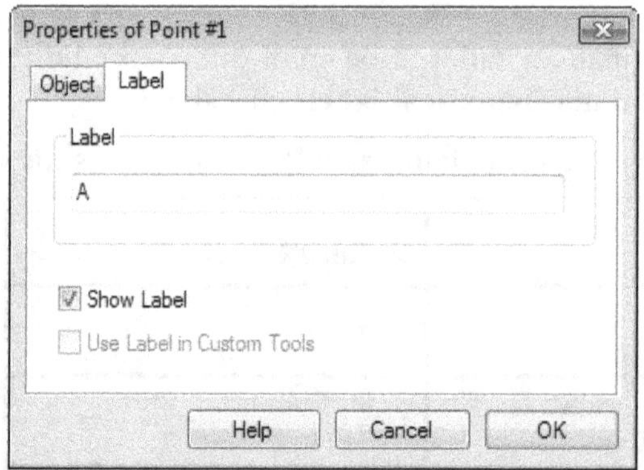

All points [2], straight objects [5], circles [6], arcs [7], interiors [8], point loci [10], measurements [36], parameters [36], calculations [39], functions [45], function plots [52], angle markers [60], and action buttons [66] have labels.

Click the **Text** [120] tool on an object to show or hide its labels, and use the Text Palette [270] to change the label's style.

Every object that can show a label has a Label Properties [158] panel. Use the Label panel to change an object's label, to change whether and how the label is displayed, and to change how the label is used in custom tools.

To show the Label panel, double-click an object's label with the **Text** [120] tool, or select a single object and choose either **Display | Label Object** [168] or **Edit | Properties** [158] | **Label.**

You can also choose **Properties** [158] from the Context [245] menu.

### ▼ Label

This box displays the label of the selected object. It will be blank if a label has not yet been assigned to the object. Type a new label to change the object's label.

If you enter a label that contains something in square brackets — such as A[1] — the part inside the brackets will be displayed as a subscript in your sketch. In other words, type A[1] in Label Properties to display $A_1$ in your sketch.

### ▼ Show Label

Use this checkbox to hide or show the object's label. (This checkbox is unavailable for objects that cannot display separate labels, such as measures.)

The final checkbox in Label Properties differs depending on the context. It appears as **Use label in custom tools** for a sketch object; but for a custom tool [125] object it appears as either **Automatically match sketch object** (for a given) or **Use label in sketches** (for a step).

### ▼ Use Label in Custom Tools (Sketch objects only)

This checkbox appears only when viewing properties for a sketch object.

Use this checkbox to determine whether the label of a sketch object will be used in custom tools. [125] Normally, custom tools assign new and unique labels to the objects they create. So, if you are defining a custom tool that includes an object with a special label — such as *hypotenuse* or *orthocenter* — and you want to duplicate that label whenever the tool is used, you should check this option. This box is enabled only for objects that can be created by a tool. It is disabled for independent points and other independent objects.

### ▼ Use Label in Sketches (Custom tool step objects only)

This checkbox appears only when using the Script View [286] to view properties of a custom tool [125] step object [288].

Use this checkbox to determine whether or not the label of a custom tool [125] step object [288] will be used in sketches. With this box unchecked, the custom tool assigns a new and unique label when it constructs this step. When the box is checked, the step's label is used whenever a sketch object is constructed from this step.

### ▼ Automatically Match Sketch Object (Custom tool given objects only)

This checkbox appears only when using the Script View [286] to view properties of a custom tool [125] given object [288].

Use this checkbox to determine whether or not a custom tool given object will be automatically matched to an object with the same label in the sketch. When this box is checked, the object appears in Script View [286] as an assumed given. When not checked, the object appears in Script View as a normal given. See Automatically Match a Given Object [318] for more information.

## ▼ Unicode

Labels can include any Unicode character or symbol supported by your operating system. Some operating systems allow you to enter Unicode characters directly from the keyboard; alternatively you can generate them in another application (such as the **Character Map** 314 on Windows or the **Special Characters** 314 pane on Macintosh) and insert them into Sketchpad.

You can enter some Unicode characters into labels by typing their names in curly brackets. See the section on Subscripts, Symbols, and Greek Letters.

## ▼ Subscripts, Symbols, Greek Letters, and Unicode

**Subscripts:** To include a subscript in a label, type the subscript in square brackets. For instance, if you type A[1] the label will appear as $A_1$ in your sketch.

Labels can include any Unicode character or symbol supported by your operating system. Sketchpad provides a convenient way to enter the symbols and Greek letters listed below.To enter other Unicode symbols and letters, use the Unicode input methods 314 available from your operating system.

**Symbols:** To enter in a label any of the symbols here, type the code shown in the table. The code will be converted to the symbol in the label.

| Symbol | Code |
| --- | --- |
| → | {->} |
| ⇒ | {=>} |
| ∠ | {<} or {angle} |
| ∟ | {rightangle} |
| ° | {degree} |
| ≤ | {lte} |
| ≥ | {gte} |
| ½ | {1/2} |
| ¹ (superscript) | {^1} |
| ² (superscript) | {^2} |
| ³ (superscript) | {^3} |

**Greek Letters:** To enter a Greek letter in a label, type the code shown in the table. The code will be converted to the Greek letter.

No code is listed for uppercase Greek letters that have equivalent Latin forms. For these letters, use the Latin form instead.

| Letter | Code | Letter | Code |
|---|---|---|---|
| α | {alpha} | A | |
| β | {beta} | B | |
| γ | {gamma} | Γ | {Gamma} |
| δ | {delta} | Δ | {Delta} |
| ε | {epsilon} | E | |
| ζ | {zeta} | Z | |
| η | {eta} | H | |
| θ | {theta} | Θ | {Theta} |
| ι | {iota} | I | |
| κ | {kappa} | K | |
| λ | {lambda} | Λ | {Lambda} |
| μ | {mu} | M | |
| ν | {nu} | N | |
| ξ | {xi} | Ξ | {Xi} |
| ο | {omicron} | O | |
| π | {pi} | Π | {Pi} |
| ρ | {rho} | P | |
| σ | {sigma} | Σ | {Sigma} |
| τ | {tau} | T | |
| υ | {upsilon} | Y | {Upsilon} |
| φ | {phi} | Φ | {Phi} |
| χ | {chi} | X | |
| ψ | {psi} | Ψ | {Psi} |
| ω | {omega} | Ω | {Omega} |

## 1.20.5 Display Attributes

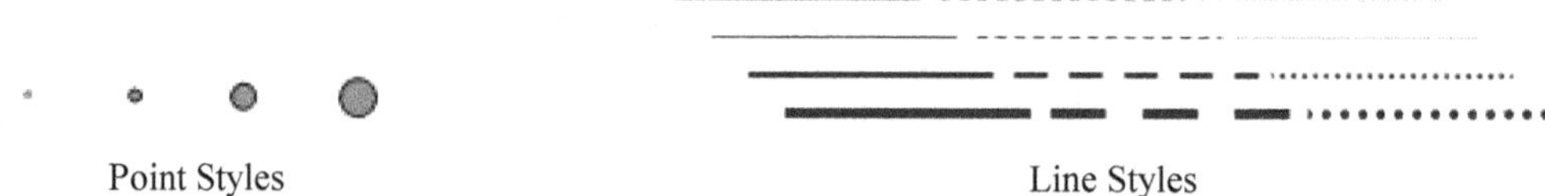

Point Styles            Line Styles

There are several display attributes that many objects have in common. You can set most of these attributes using the Display [162] menu, the Text Palette [270], or the **Text** [120] tool.

You can also use the Context [245] menu to change an object's display attributes.

To set other properties of an object, select the object and choose **Edit | Properties** [158], or choose **Properties** [158] from the Context [245] menu.

### ▼ Color

Every object in Sketchpad can be colored.

- Set an object's color by selecting the object and choosing **Display | Color** [164].

- Set the label color of an object by selecting it and using the Text Palette [270].

- Set the text color of an object that shows only text by selecting the object and using either **Display | Color** [164] or the Text Palette [270].

 Note

When you apply a new color to selected objects, you also set the default color in which new objects will be created. If you hold the Shift key while applying a color, you'll affect only the selected objects, without affecting the default color for new objects you create later.

### ▼ Label

Most objects in Sketchpad can be labeled.

- Show or hide an object's label by clicking the object with the **Text** [120] tool or by selecting the object and choosing **Display | Show/Hide Label** [168].

- Change an object's label by double-clicking the label with the **Text** [120] tool or by selecting the object and choosing either **Display | Label** [168] or **Edit | Properties | Label** [82].

- Change the font, size, style or color of a label by selecting the object and using the Text Palette [270].

### ▼ Visibility

Objects can be hidden from view, although such objects remain present in the sketch and continue to control or influence other objects.

- Hide one or more objects by selecting the object(s) and choosing **Display | Hide** [167].

- Show all hidden objects by choosing **Display | Show All Hidden** [167].

- Show a single hidden object by selecting a related object, using **Edit | Properties | Object** [79] to navigate to the desired object, and using its Object Properties [79] to show it.

- Make it easy to show or hide particular objects by creating a Hide/Show [149] action button.

## ▼ Point Style

Points  and geometric objects based on points (including iterated images [24] of points, discrete function plots [52], discrete point loci [10], and angle markers [60]) can be displayed in four sizes: Dot, Small, Medium, and Large.

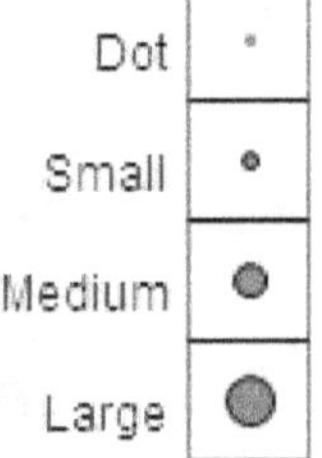

To set the point style for selected objects, use the **Display | Point Style** [162] submenu.

> When all selected objects share a common point style, a checkmark appears in the submenu next to that style.

**Note**

When you change a selected object's point style, Sketchpad remembers your chosen point style for future similar objects in the current sketch. To change an object's point style without changing the setting for future objects, hold the Shift key while choosing the command.

## ▼ Line Style

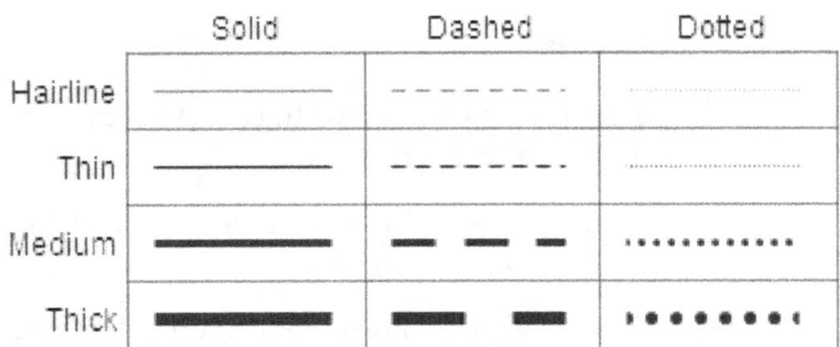

Many geometric objects (including segments, rays, lines [5], circles [6], arcs [7], continuous point loci [10], continuous function plots [52], polygon frames [8], angle markers [60], and path tick marks [63]) are displayed with straight or curved lines. The appearance of these lines is determined by the object's line style.

To set the line style for selected objects, use the **Display | Line Style** [163] submenu and choose either a width (**Hairline, Thin, Medium, Thick**) or pattern (**Solid, Dashed, Dotted**).

> When all selected objects share a common line style, a checkmark appears in the submenu next to that style.

If you construct a path coinciding with, or collinear to, a longer path (for example, when you construct an arc on the circumference of a circle, or when you construct a segment along a line), the longer path is automatically dashed to better display the shorter path coinciding with it.

**Note**

When you change a selected object's line style, Sketchpad remembers your chosen style for future similar objects in the current sketch. To change an object's line style without changing the setting for future objects, hold down the Shift key while choosing the command.

## ▼ Opacity

Most objects that occupy a two-dimensional area (polygons [8], circle interiors [8], arc interiors [8]

, pictures[18], iterations[24] and loci[10] of these objects, and angle markers[60]) can be made translucent[91], allowing similar objects below them to show through.

To set the opacity of a translucent object on a scale from 10% (nearly transparent) to 100% (completely opaque), select the object and choose **Edit | Properties | Opacity**[92]. (A completely transparent object would be invisible, so the only objects that can be set to 0% opacity are objects that have a visible frame or stroke such as framed polygons and angle markers.)

### ▼ Animation

Geometric objects and parameters can be animated[267] so that they move or change of their own accord.

- Start an animation by selecting one or more objects and either choosing **Display | Animate**[171] or pressing the Motion Controller's[263] Start button.

- Make it easy to move or animate particular objects by selecting them and creating an Animation[149] or Movement[150] action button.

*See also:*
*Principles of Animation*[267]

### ▼ Tracing

Geometric objects can be traced. As they move they leave behind on the screen a trace showing where they've been.

- Trace an object by selecting it and choosing **Display | Trace**[170]. (The check mark next to the command will turn on.)

- Stop tracing an already traced object by selecting it and choosing **Display | Trace**[170]. (The check mark next to the command will turn off.)

- Erase the collected traces from the screen by choosing **Display | Erase Traces**[171].

- Determine whether — and how quickly — traces fade from the screen by choosing **Edit | Preferences | Color**[277].

## 1.21  Object Categories

Several kinds of Sketchpad objects share important characteristics with each other. The most important of these are path objects[89], straight objects[90], layered objects[91], translucent objects[91], values[92], plots and sampled objects[95], and text objects[99].

Path objects[89] can be used to construct points that stay on the path and as animation paths. You can also measure the position of a point on a path and plot a point at a particular position on a path.

Straight objects[90] include segments, rays, lines, and axes.

Layered objects[91] occupy a two-dimensional area and appear in a layered order when they overlap.

Translucent objects[91] have adjustable opacity to allow objects in layers below them to show through.

Values[92] can be used in calculations[39], function[45] definitions, and to determine distances, angles, and scale factors for transformations[192].

Plots and sampled objects[95] can be used to show the locus[10] of a geometric object or to show the graph [52] of a function.

Text objects [99] can be aligned with each other.

## 1.21.1  Path Objects

There are certain geometric objects in Sketchpad on which you can construct and animate points. These objects are collectively referred to as path objects. Path objects include:

- straight objects [5] (segments, rays, lines, and axes)
- circles [6]
- arcs [7]
- polygons and other interiors [8] (the perimeter of an interior forms that interior's path)
- point loci [10], function plots [52], parametric plots [54], and sampled transformed paths

There are several commands that apply to path objects:

- Use **Construct | Point on Object** [176] to construct a point on a path.
- Use **Display | Line Style** [163] to set the style of most paths.
- Use **Measure | Value of Point** [221] to measure the value of a point on a path.
- Use **Graph | Plot Value on Path** [237] to plot a point at a particular position on a path.

### ▼ Animate a Point on a Path

When a point is constructed on a path, you can animate [267] the point along the path, either by choosing **Display | Animate** [171] or by creating an Animation button [69].

To animate an independent point along a path, you must first merge [153] the point to the path.

### ▼ Construct a Locus Using a Path as the Domain

If the intended locus driver [10] is a point on a path, the path is automatically used as the domain [10] for the driver.

Select the driver and the driven object. Then choose **Construct | Locus** [190].

If the intended locus driver is an independent point, you must select the path to use it as the domain.

Select the driver, the path, and the driven object. Then choose **Construct | Locus** [190].

### ▼ Auto-Dashing of Coincident or Collinear Paths

When you construct a path coinciding with, or collinear to, a longer path (for example, when you construct an arc on the circumference of a circle, or when you construct a segment along a line), the longer path becomes dashed to better display the shorter path coinciding with it.

Since a dashed coordinate axis is seldom desirable, a straight object constructed on an axis is drawn one step thicker than the default line thickness.

### ▼ Transform a Path Using Translation, Rotation, Dilation, and Reflection

Use Sketchpad's built-in transformations to create a transformed image of a path.

Select the picture and choose **Translate** [200], **Rotate** [203], **Reflect** [207], or **Dilate** [206] from the Transform [192] menu. Depending on the transformation you want, you may need to mark various objects or values to serve as center, distance, angle, ratio, or mirror.

> These four transformations, and any transformation that combines them, are called similarity transformations because the transformed image is similar to the original: angles and ratios of distances are preserved.

### ▼ Transform a Path Using a Custom Transformation

Use Sketchpad's custom transformations to create an arbitrary transformed image of a path.

Use two points, one of which depends on the other, to define a custom transformation. [211]

Select the path you want to transform.

Choose the custom transformation from the bottom of the Transform [192] menu.

If the two points that define the custom transformation are related using only Sketchpad's built-in transformations, the result is a normal transformed path, just as if you had applied the various transformations directly to the original path.

> If the pre-image path is a segment, the image remains a segment; if the pre-image is a circle, so is the image.

If the two points that define the custom transformation are related in some other way, the result is a sampled transformed path.

### ▼ Sampled Transformed Paths

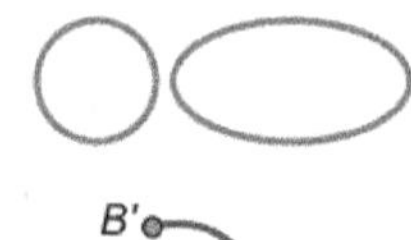

A *sampled transformed path* is a custom transformed image of a path object in which the custom transformation [211] doesn't rely strictly on Sketchpad's built-in transformations of translation, rotation, reflection, and dilation.

Transformed images constructed using the built-in transformations retain their essential shapes as straight objects, circles, arcs, polygons, and so forth.

Transformed images constructed in other ways may not retain the shapes of their pre-images. For instance, horizontal scaling transforms a circle into an ellipse, and swirling transforms a segment into a curve.

Such transformed images must be displayed using a sampling process: The transformed path is displayed by sampling locations along the pre-image and then transforming each sample to create the transformed image. The smoothness and level of detail of the transformed path depend on the number of samples used. To change the number of samples and whether the transformed path is displayed as continuous or discrete, select the transformed path and choose **Edit | Properties | Plot** [95].

When polygons and other interiors are transformed in this way, only the border of the interior is transformed.

## 1.21.2  Straight Objects

There are several kinds of straight objects in Sketchpad: segments, rays, lines [5], and axes [41].

Straight objects are path objects, [89] so you can use them in the same way you use other path objects.

In addition, there are several construction commands and measurement commands that require a straight object as a prerequisite:

- **Construct | Parallel Line** [179]
- **Construct | Perpendicular Line** [180]
- **Measure | Slope** [224]
- **Measure | Distance** [218] (to measure the distance from a point to a straight object)

### 1.21.3 Layered Objects

Most objects that occupy a two-dimensional area (polygons [8], circle interiors [8], arc interiors [8], pictures [18], and iterations [24] and loci [10] of these objects) appear in a layered order when they overlap.

To change the layer of a selected object, choose **Bring to Front** [247] or **Send to Back** [247] from the object's Context menu [245].

Layered objects are also translucent [91]: you can set their opacity to allow similar objects below them to show through. To change the opacity of a layered object, select it and choose **Edit | Properties | Opacity** [92].

### 1.21.4 Translucent Objects

Most objects that occupy a two-dimensional area (polygons [8], circle interiors [8], arc interiors [8], pictures [18], iterations [24] and loci [10] of these objects, and angle markers [60]) have adjustable opacity, so that they can allow similar objects in layers below them to show through.

The opacity scale goes from 0% (completely transparent) to 100% (completely opaque).

When you create pictures, iterated pictures, and loci they are initially opaque, with an opacity of 100%.

When you create polygons, circle interiors and arc interiors they are initially translucent, with an opacity of 50%.

To change the opacity of a translucent object, select it and choose **Edit | Properties | Opacity** [92].

For most objects, the minimum opacity value is 10%, so that you cannot set them to be completely transparent. (A completely transparent object would be invisible.)

Framed polygons and angle markers are the only objects that can be made completely transparent.

- Polygons have an optional frame in addition to their interiors. If a polygon's frame is showing, you can set its opacity to 0% to show only its frame.

- Angle markers have one or more strokes in addition to their filled areas. You can set an angle marker's opacity to 0% to show only its strokes(s).

Most translucent objects are also layered objects [91], and can be arranged above or below other similar objects by using **Bring to Front** [247] or **Send to Back** [247] from the object's Context menu [245]. The only exceptions are angle markers, which are always visible above other layered objects.

### 1.21.4.1 Opacity Properties

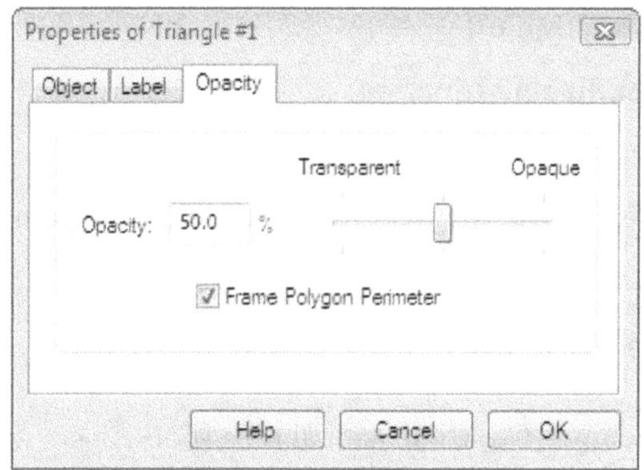

Translucent objects[91] have an Opacity Properties[158] panel.

**Opacity:** Use this panel to change the opacity of a translucent object. You can set the opacity on a scale that goes from 0% (completely transparent) to 100% (completely opaque).

 Note

> In most cases, you cannot make a translucent object completely transparent, because the object would then be invisible. The minimum opacity setting in such cases is 10%. The only objects you can make fully transparent (with an opacity setting of 0%) are angle markers and framed polygons, because even when the interior portion of these objects is transparent, their borders remain visible.

**Frame polygon perimeter:** This checkbox appears only for polygons. Use it to determine whether a polygon displays its frame. If you want to display only the frame, check this box and then set the polygon itself to be completely transparent. (You can set a polygon to be completely transparent only when its frame is showing. When the frame does not appear, the minimum opacity is 10%.)

## 1.21.5  Values

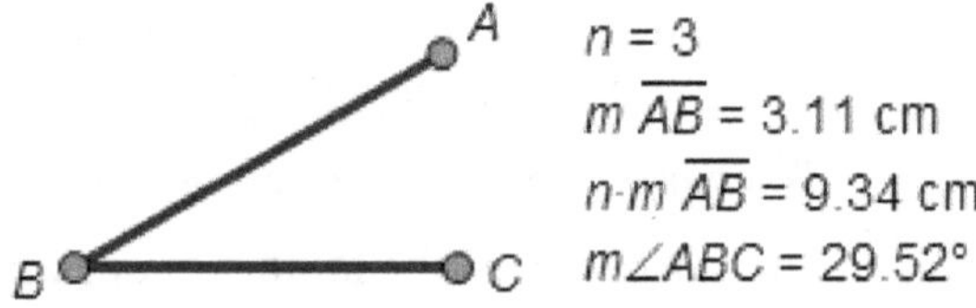


Parameters[36], measurements[36], and calculations[39] display numeric values and have many common characteristics.

The numeric values generated by measurements — and by calculations based on measurements — can be observed in order to discover relationships among objects in your sketch. In addition, the numeric values of all three kinds of objects can be used to define or control the behavior of your sketch in powerful and revealing ways.

Use numeric values to control transformed[192] objects, plotted points[239], calculations[39], functions[45], and iterations[24].

### Units

The units of a value determine how it can be used in a sketch.

- *Distance values* have units of centimeters, inches, or pixels, and can be used for translation.

- *Angle values* have units of degrees or radians, and can be used for rotation or for the angle of a polar translation.

- *Dimensionless values* have no units, and can be used as the scale factor for dilation.

- *Other values* have units that don't match these three categories (for instance, cm$^2$ or pixels/radian).

## Using Values

All calculations, all parameters, and most measurements have a single numeric value. The values of any of these objects can be used in the same ways.

- Change the value's precision using the Value Properties 94 panel. (To set the precision for new measurements, calculations, and parameters, choose **Edit | Preferences | Units** 276.)

- Change the distance units or angle units of this value (and of all values in the sketch) using **Edit | Preferences | Units** 276..

- Change how the value is labeled using the Value Properties 94 and Label Properties 82 panels.

- Include the value in a calculation 39 or function 45 by clicking the object while using the Calculator 257.

- Choose **Transform | Mark Distance** 199 to use a distance value as the marked distance for translation.

- Choose **Number | Tabulate** 228 to create a table which lists how one or more values change over time.

- Choose **Transform | Mark Angle** 195 to use an angle value as the marked angle for translation and rotation.

- Choose **Transform | Mark Scale Factor** 197 to use a value with no units as the marked scale factor for dilation.

- Choose **Graph | Define Unit Distance** 233 to use a distance value as the unit distance for a new coordinate system.

- Select two values and choose **Graph | Plot as (x, y)** 239 to plot a point with coordinates given by the selected values.

- Use **Transform | Iterate to Depth** 208 to define the depth of an iteration by a selected value.

- Use **Display | Color | Parametric Color** 165 to determine the color of an object.

- Use Hot Text 57 to include the value in a caption 56.

## Radian Values

Radian values are displayed as multiples or fractions of $\pi$ when the value is a multiple of $\pi$ or a fraction of $\pi$ with a denominator of 2, 3, 4, 6, or 12. For instance, 1.57 radians is displayed as $\pi/2$, 6.28 radians is displayed as $2\pi$, and 4.1888$\pi$ is displayed as $4\pi/3$. All other radian values are displayed as ordinary decimal numbers.

*Subtopic:*
  *Value Properties* 94

### 1.21.5.1  Value Properties

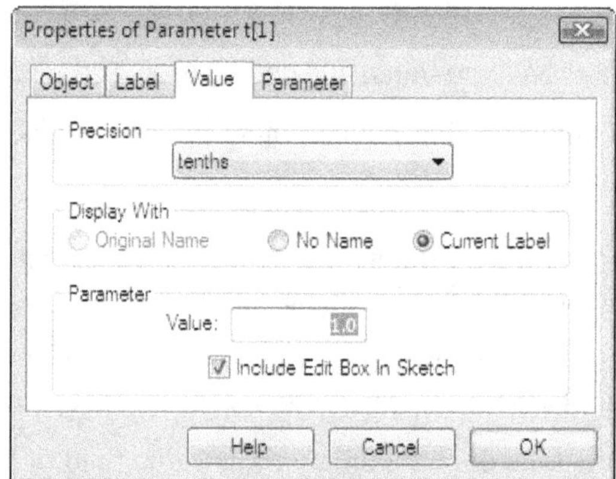

Measurements [36], calculations [39], and parameters [36] have a Value Properties [158] panel. Use this panel to set the precision [277] and the display name of these objects and to set the value of parameter objects.

To open this dialog box, select the object and choose **Edit** [144] | **Properties** [158], or choose **Properties** [158] from the Context [245] menu.

**Precision:** Here you can set the precision with which the measurement is displayed. Choices range from units to hundred-thousandths. This setting determines only how Sketchpad rounds the value when the value is displayed on-screen. (Sketchpad stores the actual value with considerably more accuracy, and this setting doesn't cause any loss of accuracy in the value Sketchpad uses internally.) Choose **Edit | Preferences | Units** [276] to set the default precision used for newly created values.

> Original Name:  m$\overline{AB}$ = 4.00 cm
>
> No Name:  4.00 cm
>
> Current Label:  Length = 4.00 cm

**Display With:** Use this to set the on-screen name that will appear before the value. For instance, if you've measured the length of the segment connecting $A$ and $B$ and changed the measurement's label to *Length*, the three possible ways to display the value are illustrated here.

**Parameter:** Here you can set the value of a parameter [36] and determine whether the parameter can be edited directly in the sketch. (This section appears only for parameters.) You can also choose **Edit | Edit Parameter** [157] or double-click with the **Arrow** tool to change a parameter's value.

*See also:*

## 1.21.6  Plots and Sampled Objects

Certain mathematical objects in Sketchpad are plotted by connecting or combining many individual *samples*, just as you might plot a function on paper by connecting many individual plotted points.

In some cases, the samples are points; the points can be connected to make a continuous plot, or displayed separately to make a discrete plot.

In other cases the samples are other geometric objects or even portions of a picture; these samples are always displayed separately.

- A locus [10] is displayed by considering many locations or values of the driver on its domain; the samples are the corresponding locations of the driven object.

- A function plot [52] is displayed by calculating the dependent variable for many values of the independent variable; the samples are the corresponding plotted points.

- A parametric plot [54] is displayed by evaluating two functions for many values of an understood parameter; the samples are the corresponding plotted points.

- A sampled transformed path [89] is displayed by transforming many locations along the pre-image path; the samples are the transformed locations.

- A sampled transformed picture [23] is displayed by dividing the pre-image picture into many small parts; the samples are the transformed images of those small parts.

For all these objects, the Properties dialog box [273] has a Plot panel that allows you to set the number of samples used to plot the object. The greater the number of samples, the smoother and more accurate the plot is. However, the greater the number of samples, the slower Sketchpad can be in calculating and displaying the plot. Choose **Edit | Preferences | Sampling** [283] to set both the maximum allowed number of samples and the default number of samples for to newly constructed objects.

The properties panel for plots and other sampled objects include these settings:

**Number of Samples:** You can set the number of samples for any sampled object.

**Continuous or Discrete:** When the samples are points, they can be connected so that the plot appears continuous, or they can be displayed individually so that the plot appears as discrete points.

**Show Arrowheads and Endpoints:** When the samples are points and the domain is not a closed curve, the ends of the plot can show arrowheads when they're adjustable and endpoints when they're fixed.

**Domain:** When the driver is numeric, you can determine the domain numerically.

**Allowable Distortion:** For a transformed picture, this setting determines how much distortion is acceptable in a sample. Samples with more than this level of distortion are not shown.



### 1.21.6.1  Plot Properties for a Locus or Sampled Transformed Path

Use this Properties [158] panel to set the number of samples used to display a locus [10] or a sampled transformed path [89] and to determine whether the locus or path is displayed in continuous or discrete form, and to determine whether arrowheads and endpoints appear at the ends of the plot. For a parametric locus, use this panel to determine the domain of the locus.

## ▼ Number of Samples

You can set the number of samples that Sketchpad uses to display the locus. When you increase the number of samples, the locus appears smoother but takes longer for Sketchpad to draw. As a result, dragging and animating may become slow.

You can also change the number of samples by selecting the locus and pressing the + or – key.

 Note

To change the maximum number of samples allowed, or to change the initial number of samples when you construct a new locus, choose **Edit | Advanced Preferences | Sampling** 283.

## ▼ Display as Continuous or Discrete

Use this setting to control whether the individual samples are connected (a continuous locus, function plot, or transformed path) or shown as distinct points (a discrete locus, plot, or transformed path). Below is an elliptical point locus displayed both ways. The choice between **Continuous** and **Discrete** appears only for point loci, function plots, parametric plots, and sampled transformed paths, and affects only the appearance of the sampled object. If you construct a point on such an object, the constructed point is not restricted to the sample positions, but is free to move smoothly between samples.

Continuous

Discrete

## ▼ Show Arrowheads and Endpoints

Use this setting to determine whether a point locus, function plot, or transformed path shows arrowheads at its ends.

If you show the arrowheads, you can drag them to change the domain. Without the arrowheads, this Plot Properties dialog box is the only way to change the domain.

### ▼ Set the Domain (Parametric Locus Only)

For a parametric locus, you can type values for the lower and upper bounds of the domain of the parameter (the driver).

If the locus is set to show arrowheads, you can also change the domain by dragging the arrowheads.

#### 1.21.6.2  Plot Properties for a Function Plot or Parametric Plot

Use this Properties 158 panel to set the number of samples used to display a function plot 52 or parametric plot 54, to determine whether the plot is displayed in continuous or discrete form, to determine whether arrowheads and endpoints appear at the ends of the plot, and to set the domain of the plot.

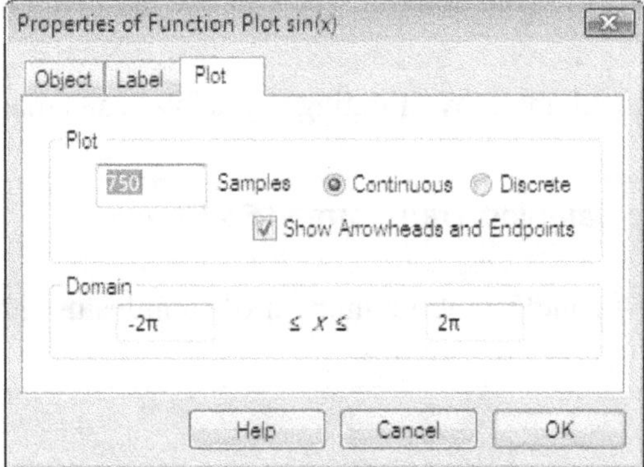


### ▼ Number of Samples

Change this number to determine how many samples that Sketchpad uses to display the function plot. When you increase the number of samples, the function plot appears smoother but takes longer for Sketchpad to draw. As a result, dragging and animating may become slower.

You can also change the number of samples by selecting the function plot and pressing the + or − key.

 Note

> To change the maximum number of samples allowed, or to change the initial number of samples when you construct a new function plot or parametric plot, choose **Edit | Advanced Preferences | Sampling** 283.

### ▼ Display as Continuous or Discrete

Use this setting to control whether the individual samples are connected (a continuous function plot) or shown as distinct points (a discrete function plot). Below is a sine curve displayed both ways.

The choice between **Continuous** and **Discrete** affects only the appearance of the function plot. If you construct a point on a discrete function plot, the constructed point is not restricted to the sample positions, but is free to move smoothly along the plot.

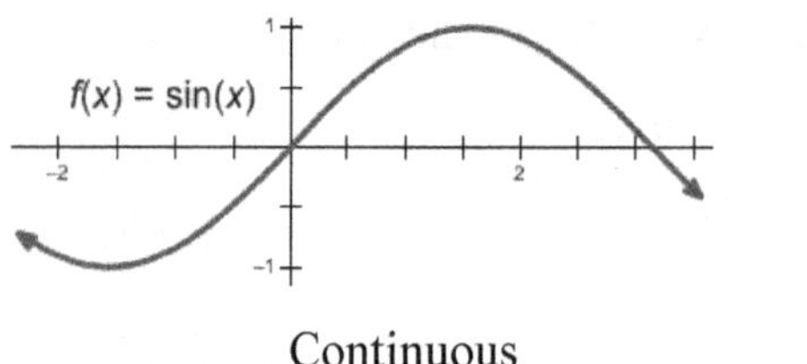

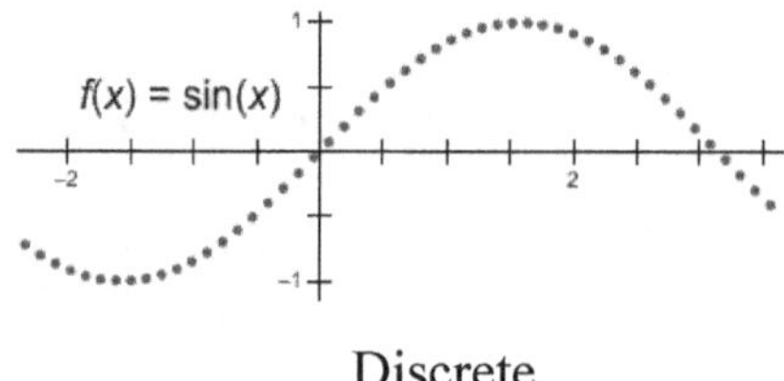

Continuous                                   Discrete

### ▼ Show Arrowheads and Endpoints

Use this setting to determine whether the function plot shows arrowheads at its ends. If you show the arrowheads, you can drag them to change the domain. Without the arrowheads, this Plot Properties dialog box is the only way to change the domain.

### ▼ Set the Domain

When entering limits for the domain, you can use mathematical expressions like 2*3 and $\pi/4$. (Press p for $\pi$.)

You can also set the domain by dragging the arrowheads that appear at the ends of the function plot.

### 1.21.6.3  Plot Properties for a Sampled Transformed Picture

Use this Properties |158| panel to set the number of samples used to display a sampled picture and to set the allowable distortion for the samples.

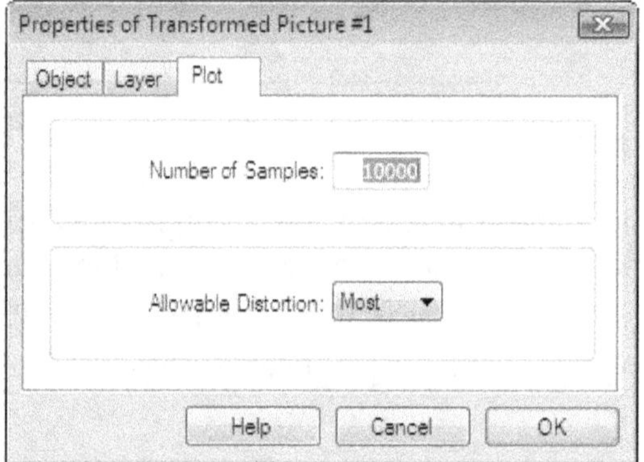

### ▼ Number of Samples

You can set the number of samples that Sketchpad uses to display the sampled picture. When you increase the number of samples, the sampled picture appears smoother but takes longer for Sketchpad to draw. As a result, dragging and animating may become slow. You can also change the number of samples by selecting the sampled picture and pressing the + or − key.

 Note

To change the maximum number of samples allowed, choose **Edit | Advanced Preferences | Sampling** |283|.

### ▼ Allowable Distortion

You can set the allowable distortion for sampled pictures.

If you limit the maximum allowable distortion, Sketchpad will not display individual samples of the sampled picture in which your custom transformation distorts the sample more than a certain amount. This can prevent certain transformations from getting out of hand. For instance, it keeps a single sample from growing as big as the whole sketch and wiping out the more interesting parts of the transformation.

Set **Allowable distortion** to **All** to display the entire picture, no matter how distorted. Set it to **None** to display only those parts of the picture that are not particularly distorted. The middle settings of **Most** or **Some** are satisfactory for most purposes.

## 1.21.7  Text Objects

There are certain objects in Sketchpad that are displayed in text form. These objects are collectively referred to as text objects. Text objects include:

- captions 56
- values 92 (parameters 36, measurements 36, and calculations 39)
- functions 45
- action buttons 66
- coordinates (generated by the **Measure | Coordinates** 223 command)
- equations (generated by the **Measure | Equations** 225 command)

You use similar methods to position the various kinds of text objects, to format them, and to merge a text object to a point.

### ▼ Position Text Objects

When you create a new text object, Sketchpad tries to position it in relation to existing text objects.

- A new parameter, measurement, calculation, or function is created aligned with and slightly below the most recent similar object, provided there's room for the new object without overlapping existing text objects.

- A new action button is created aligned with and slightly below the most recent action button, provided there's room.

You can reposition text objects to display them neatly.

- If you drag a text object near other text objects, it becomes sticky to help you align it to the nearby objects.

- You can press Shift+Enter to arrange selected text objects in an evenly spaced vertical column with their left edges aligned, and you can press Shift+Enter repeatedly to increase the vertical spacing.

### ▼ Format Text Objects

You can format one or more selected text objects.

- To change the font, size, style, or color of the selected text objects, use the Text Palette 270.

 Note

The new font, size, style, or color will become the default for future objects of the same type. To change the format of the selected object(s) without affecting future objects, hold the Shift key while you choose the new format. You can also use Text Preferences panel to determine the default text style for future objects.

- To make all selected text objects larger or smaller, choose **Display | Text**[166] **| Increase/ Decrease Size,** or use the equivalent keyboard shortcuts: hold Alt (Windows) or ⌘ (Mac) and press the > or < key.

- To display selected text objects in bold, italic, or underline style, use the keyboard shortcuts[295] Ctrl+B, Ctrl+I, and Ctrl+U (Windows) or ⌘B, ⌘I, and ⌘U (Mac).

- To format parts of a caption rather than the entire caption, double-click the caption with either the **Text**[120] tool or the **Arrow**[104] tool, and then select the text you want to format.

### ▼ Merge a Text Object to a Point

You can merge any text object except an action button to a point so that it moves with the point. For instance, if you want an angle measurement to appear in the middle of the angle it measures, you can follow these steps:

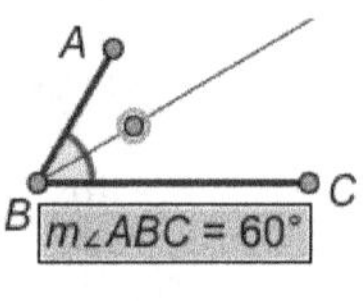


1. Construct the angle bisector by choosing **Construct | Angle Bisector**[181].

2. Construct a point on the angle bisector by choosing **Construct | Point on Bisector**[176].

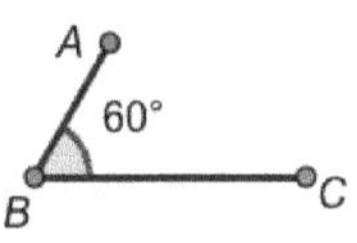


3. Merge the measurement to the point by selecting both objects, holding the Shift key, and choosing **Edit | Merge Text to Point**[153].

4. Hide the point and the angle bisector by choosing **Display | Hide Objects** [167].

The measurement will remain in the middle of the angle even when the angle's position or size changes.

# Chapter 2

# Tools

# 2    Tools

For thousands of years, since the time of Euclid, the fundamental tools of geometry have been the compass and the straightedge. Sketchpad's Toolbox includes these two tools — used to construct circles and straight objects — and several other tools that allow you to select and drag objects, to construct points, to create and manipulate text and labels, and to define and manage custom tools. This section describes how to use each of the tools in Sketchpad's Toolbox.

The Toolbox appears on the left of the screen when you start Sketchpad, and includes nine tools.

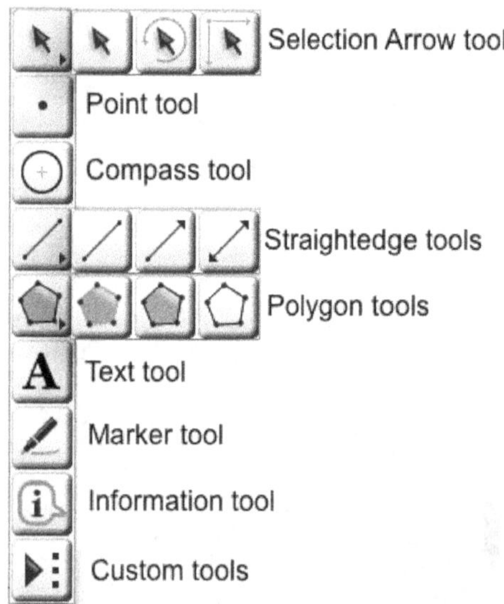


- **Selection Arrow** |104| tools: Use these tools to select and drag objects in your sketch. The three variations of the tool allow you to drag-translate (move), drag-rotate (turn), and drag-dilate (shrink or grow) objects.

- **Point** |112| tool: Use this tool to construct points | 2 |.

- **Compass** |113| tool: Use this tool to construct circles | 6 |.

- **Straightedge** |115| tools: Use these tools to construct straight objects | 5 |. The three variations of the tool allow you to construct segments, rays, and lines.

- **Polygon** |118| tools: Use these tools to construct polygons. The three variations of the tool allow you to construct a polygon, a polygon and the segments that form its edges, or the edge segments only.

- **Text** |120| tool: Use this tool to create and edit text and labels.

- **Marker** |122| tool: Use this tool to create angle markers and tick marks, and to make freehand drawings and notations.

- **Information** |124| tool: Use this tool to explore how a sketch is constructed and how the objects relate to each other.

- **Custom** |125| tools icon: Use this icon to define, use, and manage custom tools.

## ▼ Choosing and Using Tools

To use any of Sketchpad's tools, click the desired tool in the Toolbox. Then move the your pointer over the sketch and click or press and drag to use the tool.

The **Selection Arrow** |104|, **Straightedge** |115|, and **Polygon** |118| tools each come in three variations. To change from one variation to another, press and hold the **Selection Arrow** tool, the **Straightedge** tool, or the **Polygon** tool in the Toolbox until a menu pops out. Choose a different variation of the tool from this menu to make it active.

You can change the active tool by using the keyboard instead of clicking in the Toolbox. Hold the Shift |295| key and press the up or down arrow keys to change the active tool. You can also hold the Shift key and press the right or left arrow keys to switch the **Selection Arrow, Straightedge,** or **Polygon** tool from one variation to another.

Once you've chosen a tool, that tool remains active until you choose a different one, so there's no need to click the same tool repeatedly to use it multiple times.

## ▼ Using Tools to Scroll the Sketch Window

Use any of Sketchpad's tools to scroll the sketch window and see different parts of the sketch. Hold down the Alt key (Windows) or the Option key (Macintosh) and press and drag in the sketch to scroll the entire sketch in the direction in which you drag. This feature works no matter which tool is active in the toolbox.

## ▼ Hiding and Showing the Toolbox

Hide or show the Toolbox by using the **Hide Toolbox** 173 and **Show Toolbox** 173 commands at the bottom of the Display 162 menu.

You may want to hide the Toolbox when making presentations.

## ▼ Moving, Resizing, and Docking the Toolbox

To move the Toolbox, (Windows or Macintosh) grab it by the title bar or (Windows only) by the gray area surrounding the buttons and drag it to a different location on the screen.

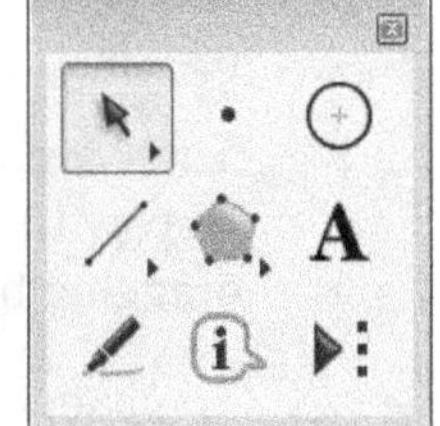

Windows users can dock the Toolbox to the left, top, right, or bottom edge of the application window. When the Toolbox is not docked, Windows users can also resize the Toolbox by dragging a border of the Toolbox window.

Macintosh users can make the Toolbox large or small by choosing **Edit | Preferences | Tools** 280.

## 2.1   Arrow Tools

Use the **Arrow** tools to drag objects, to select and deselect objects [106], and to perform other arrow actions [109].

**Drag Objects:** [105] Dragging objects with the **Arrow** tool is at the heart of Sketchpad's Dynamic Geometry capabilities. You drag objects to investigate mathematical relationships, to explore variations, to test conjectures, and to discover new properties.

**Select and Deselect Objects:** [106] Many of Sketchpad's menu commands act on selected objects. By selecting objects, you focus the software's attention on one or a handful of the many objects that make up a sketch. For example, you'll use selection to identify the objects to be transformed by the Transform menu or the objects to be measured by the Measure menu.

**Perform Other Arrow Actions:** [109] In addition to dragging and selecting, use the **Arrow** tool for several other purposes, such as constructing points of intersection, pressing action buttons, and resizing coordinate systems, function plots, pictures, and loci.

There are three **Arrow** tools: the **Translate Arrow** tool, the **Rotate Arrow** tool, and the **Dilate Arrow** tool. These three tools behave identically when you use them to select objects; it's only their dragging behavior that differs.

To change from one **Arrow** tool to another, press and hold the icon in the Toolbox until a menu pops out. Choose a different variation of the tool from this menu to make it the active **Arrow** tool.

If an **Arrow** tool is already active, you can switch to a different Arrow tool by holding the Shift key and pressing either the right or left arrow key.

### ▼ Translate Arrow Tool

Drag selected objects with the **Translate Arrow** tool to translate them — that is, to slide them by any distance in any direction without turning or changing size or shape. This is the default **Arrow** tool.

Use the **Translate Arrow** tool to select and deselect objects, press action buttons, and construct points of intersection just like the other **Selection Arrow** tools.

If you want to construct a translated image of an object instead of translating the original object itself, use the **Transform | Translate** [200] command.

When you translate an independent point that determines a single straight object, hold the Shift [295] key to constrain the straight object to be horizontal, vertical, or at a multiple of 15°.

### ▼ Rotate Arrow Tool

Drag selected objects with the **Rotate Arrow** tool to rotate them — that is, to turn them around a center point by any desired angle without changing their distance from the center, their size, or their shape.

The center point used for rotation is the point you marked most recently using the **Transform | Mark Center** [194] command. If you haven't marked a center point, Sketchpad automatically marks the point nearest to the center of the screen.

You can also mark a point as the center of rotation by double-clicking it with an **Arrow** [104] tool.

Use the **Rotate Arrow** tool to select and deselect objects, press action buttons, and construct points of intersection just like the other **Selection Arrow** tools.

If you want to construct a rotated image of an object instead of rotating the original object itself, use the **Transform | Rotate** [203] command.

### ▼ Dilate Arrow Tool

Drag selected objects with this tool to dilate them — that is, to stretch or shrink them farther from or closer to a center point by any desired amount without changing their direction from the center or their shape.

The center point used for dilation is the point you marked most recently using the **Transform | Mark Center** [194] command. If you haven't marked a center point, Sketchpad automatically marks the point nearest to the center of the screen.

You can also mark a point as the center of dilation by double-clicking it with an **Arrow** [104] tool.

Use the **Dilate Arrow** tool to select and deselect objects, press action buttons, and construct points of intersection just like the other **Selection Arrow** tools. You can also use the **Dilate Arrow** tool to zoom in and out [111] on a sketch.

If you want to construct a dilated image of an object instead of dilating the original object itself, use the **Transform | Dilate** [206] command.



## 2.1.1 Drag Objects

Drag objects in your sketch for several purposes: to reposition the objects, to resize them, to change the shape of a construction, and to investigate the geometry embedded in the sketch, thereby discovering and revealing mathematical relationships between them.

Mathematically, moving an object in your sketch transforms that object, and Sketchpad's dragging behavior is based on three geometric transformations: translation, rotation, and dilation. To allow you to use each of these transformations, Sketchpad has three **Arrow** tools: the **Translate Arrow** [104] tool, the **Rotate Arrow** [104] tool, and the **Dilate Arrow** [104] tool.

To drag an object, position the tip of the Arrow tool over the object, and then press and drag. (The **Arrow** flips horizontal when it's pointing at an object, and the status line [253] describes the object that will be dragged.)

- If the object was not selected, only that object moves. (Any other selected objects are deselected.)

- If the object was already selected, it and all other selected objects move to follow your mouse.

When you drag an object, other related objects [80] are adjusted to maintain their relationship to the dragged object. For instance, if you drag one endpoint of a segment, the segment adjusts to the new position of the endpoint you're dragging. Similarly, if you drag the segment itself, both endpoints move with it. If you drag one vertex of a triangle, the two sides attached to that vertex adjust, and the third side remains unchanged. If you select and drag all three vertices, the entire triangle moves as a unit.

### ▼ Drag a Straight Object to a Specific Angle

While you're dragging an endpoint or control point of a straight object, you can hold the Shift key to make the straight object horizontal, vertical, or at an angle of 15°, 30°, 45°, 60°, or 75°. Finish your drag before you release the Shift key.

## 2.1.2    Select and Deselect Objects

Click the **Arrow** tool in your sketch to select and deselect objects. You can drag selected objects, apply menu commands to them, or manipulate or modify them with the Motion Controller 263 or the Text Palette 270.

All three **Arrow** tools behave identically when you use them to select and deselect objects.

If you drag an object and other objects move unexpectedly, or an expected menu command is not available, check to make sure you have selected the correct objects for the operation you want to perform. If a menu command appears gray and cannot be chosen, it may mean that your current selection is not appropriate for that command. (Check the status line 253 to see the number and kind of objects that are selected.)

Selected objects appear outlined, or with special marks, as in the following illustrations.

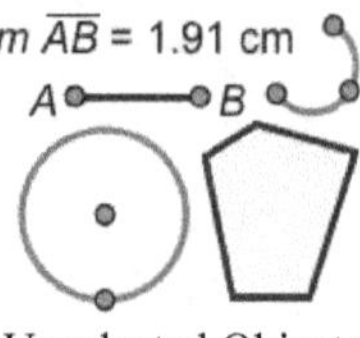


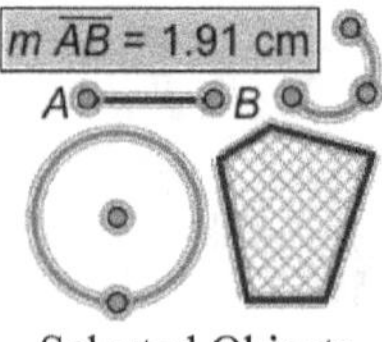


Unselected Objects          Selected Objects

You can also select objects using **Edit | Select All** 151, **Edit | Select Parents** 152, and **Edit | Select Children** 152.

### ▼ Select or Deselect a Single Object

- Select an unselected object by positioning the tip of the **Arrow** tool over the object and clicking it. (The pointer flips horizontal when it's pointing at an object, and the status line 253 describes the object that will be selected by clicking.)

- Deselect a single selected object by positioning the tip of the **Arrow** tool over the object and clicking it. (The pointer flips horizontal and the status line 253 describes the object that will be deselected.)

### ▼ Deselect All Objects

- Deselect all objects by clicking in empty space in the sketch. (If you've turned on **Double-click deselection** in **Edit | Preferences | Tools** 280 you must double-click to deselect all objects.)

- Deselect all objects by pressing Esc 294 repeatedly.

## ▼ Prevent Accidental Deselection

On occasion you may try to select an object by clicking it, but accidentally deselect all objects instead. This is particularly frustrating if you are in the middle of selecting a number of objects.

There are several steps you can take to prevent accidental deselection:

- Make sure the pointer has turned horizontal before you click. (The horizontal pointer indicates that you're pointing at an object.)

- Check the status line 253 before you click. (The status line identifies the object you will select or deselect by clicking.)

- Increase **Selection magnetism** by choosing **Edit | Preferences | Tools** 280.

- Increase the thickness of points and path objects (segments, rays, lines, circles, and other paths) by using the **Display | Point Style** 162 and **Display | Line Style** 163 commands.

- Turn on **Double-click deselection** by choosing **Edit | Preferences | Tools** 280. With this setting turned on, you must double-click in empty space to deselect all objects.

## ▼ Select Multiple Objects by Clicking

Select multiple objects by clicking each object in turn.

## ▼ Select Multiple Objects with a Selection Rectangle

Use a selection rectangle to select multiple objects located near each other in a sketch.

1. Imagine a rectangle that touches or encloses every object you want to select.

2. Position the tip of the **Arrow** in empty space at one corner of this rectangle.

3. Press and drag diagonally toward the opposite corner. A shaded rectangle appears, and every object that the rectangle touches or encloses is selected. If you want previously selected objects to remain selected, hold down the Shift key while starting to drag the rectangle.

4. When all desired objects are selected, release the button. If you started your rectangle in the wrong place to select the desired objects, begin again from step 2.

A properly positioned selection rectangle is useful for selecting multiple objects for certain commands. For instance, select all three sides of a triangle to construct the three midpoints, or select a straight object and a point to construct a perpendicular.

## ▼ Select All Objects

- Select all objects by using **Edit** 144 | **Select All** 151 when the **Arrow** tool is active.

- Select all points, segments, rays, lines, circles, polygons, text objects, or angle markers and tick marks by choosing the corresponding tool and then using **Edit | Select All** 151. The command changes according to the active tool.

## ▼ Select One of Several Overlapping Objects

Select one of several overlapping or coincident objects by clicking repeatedly until the desired object is selected.

Here are some helpful tips:

- When you point at overlapping objects, the status line 253 at the bottom of the Sketchpad window tells you which object you are about to select. If the object described isn't the one you want to select, try moving the **Selection Arrow** around until the status line describes the desired object.

- If possible, point at a portion of your object that doesn't overlap other objects. In the following figure, for example, segment *CD* has been constructed collinear to line *AB*. You can select segment *CD* by clicking on the segment itself between points *C* and *D;* but to select line *AB,* it's easiest to click on the portion of the line outside *C* and *D*.

  When you click overlapping objects, Sketchpad always selects points in preference to any other objects and always selects path objects in preference to interiors.

- If clicking selects the wrong object, click again in the same spot. A second click on similar overlapping objects will deselect the first object and select the next one. Keep clicking until you get the object you want.

- If clicking selects an object which isn't of interest to you and isn't important to the appearance of the sketch, choose **Display | Hide** 167, and then click again in the same spot.

- If all else fails, select a related object, use **Edit | Properties** 158 command to view Object Properties 79, and select the desired object using the Parents or Children pop-up menu.

To select more than one object of several located near each other on the screen:

- Use a selection rectangle to select all the overlapping objects.

- Select the first object, and then hold the Shift key while clicking additional objects.

### ▼ Select the Parents or Children of an Object

Select related objects 80 by using **Edit | Select Parents** 152 or **Edit | Select Children** 152 or by navigating to them using the Object Properties 79 dialog box.

The **Select Parents** 152 and **Select Children** 152 keyboard shortcuts can be very useful when used together. For instance, if you have one side of a triangle selected, you can quickly select all three sides as follows:

1. Select the two adjacent vertices using the keyboard shortcut for **Select Parents** 152: Alt+Up (Windows) or ⌘↑ (Mac).

2. Select all three sides using the keyboard shortcut for **Select Children** 152: Alt+Down (Windows) or ⌘↓ (Mac).

Thus with one side selected you can quickly select all three sides by holding down the Alt or ⌘ key and pressing the up and then down arrow keys. Similarly, with one side selected you can quickly select all three vertices by holding down the Alt or ⌘ key and pressing ↑, ↓, and ↑.

### ▼ Select an Action Button

Select or deselect an action button 66 by clicking its handle, not its body. (Clicking the body of an action button doesn't select it, but instead performs the action associated with the button.)

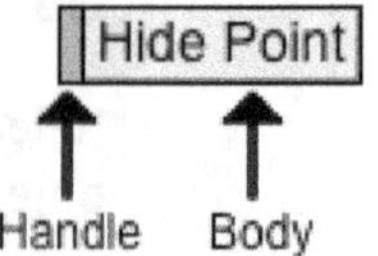

#### ▼ Select a Moving Point

Select a moving point by choosing it from the Motion Controller's [263] Target [265] menu.

Alternatively, click the Motion Controller's Pause [265] button, select the object, and then click the Pause button again to restart the motion.

## 2.1.3 Other Arrow Actions

In addition to dragging and selecting objects, the **Arrow** tool has several special actions when used with specific types of objects.

#### ▼ Mark a Center or Mirror

Double-click a point to mark it as the center for rotations and dilations. This has the same effect as selecting the point and choosing **Transform | Mark Center** [194].

#### ▼ Edit a Calculation or Function

Double-click a calculation or function to edit it. This has the same effect as selecting the object and choosing **Edit | Edit Calculation** [157] or **Edit | Edit Function** [157].

#### ▼ Change the Value of a Parameter

Double-click a parameter to change its value.

#### ▼ Drag Labels and Edit Text

Use the **Arrow** tool to:

- Press and drag a label [81].

- Double-click a label to edit the label with **Label Properties** [82].

- Double-click a caption to edit the caption [82].

#### ▼ Construct a Point of Intersection

In most cases, click the **Arrow** tool on the intersection of two path objects to construct a point at the intersection.

Clicking the **Arrow** tool at a potential point of intersection has the same effect as clicking the **Point** [112] tool or using the **Construct | Intersection** [177] command. So if you want to select an intersection that you haven't yet constructed, use the **Arrow** tool to construct and select it in one action.

You can also click other construction tools (such as the **Straightedge** [115] tool, the **Compass** [113] tool, or the **Polygon** [118] tool) on an intersection to construct the point of intersection.

You cannot construct the intersection of a point locus with another point locus or with a function plot.

### ▼ Press an Action Button

Click the **Arrow** tool on the body of an action button 66 to perform that button's action. (Different buttons have different actions: Some may cause an animation to start, others may hide or show objects, and so forth.)

When you point at the body of an action button, the cursor turns into a pointing finger indicating you are about to press the button. If you want to select or drag the button rather than perform its action, click the handle of the button rather than its body.

If an action button performs an action that takes time to complete, it remains pressed until the action completes. Click the **Arrow** tool on the body of a pressed button to release it, halting the action in process.

### ▼ Change an Axis' Scale

If a coordinate system 41 is constructed based on one or two unit points, you can rescale its axes 41 using the **Arrow** tool. (You cannot drag to rescale a coordinate system constructed using one or two unit distances, as those distances determine the scale.)

There are two ways to rescale an axis of a unit-point-based coordinate system:

- Drag any visible number on the horizontal or or vertical axis. (When you position the Arrow tool over such a number, the cursor turns into a bidirectional arrow, indicating you can drag to grow or shrink the scale of that axis.)

- Drag the unit point of the horizontal or vertical axis. (If the grid is square, there's only one unit point, and both axes are scaled together.)

Rescaling an axis lets you zoom in or zoom out on the coordinate system.

### ▼ Resize a Picture

Use the **Translate, Rotate,** or **Dilate Arrow** tools to scale, rotate, or dilate a picture 23 that's not attached to points or that's attached to a single point. Select the picture and drag one of its resize handles with the appropriate tool. ( To maintain the picture's shape, use the **Dilate Arrow** tool, or hold Shift while using the **Translate Arrow** tool.)

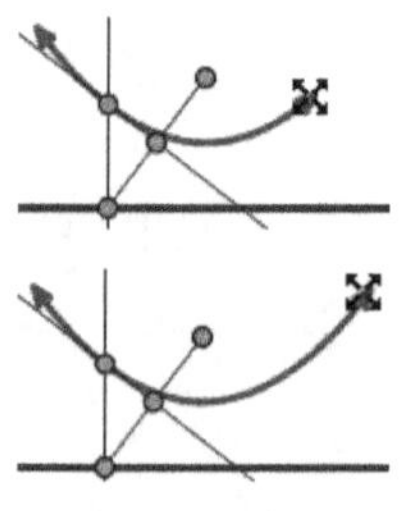

If the picture is attached to two or three points, resize it by moving the points to which it's attached.

### ▼ Resize a Plot or Locus

Function plots 52, parametric plots 54, and certain loci 10 can display an arrowhead at one or both endpoints, indicating that the plot or locus extends farther in the indicated direction than is presently displayed. To resize the plot or locus, drag the arrowhead with the **Arrow** tool. (As you point at the arrowhead, the cursor becomes a multidirectional arrow. Press and drag in the direction in which the arrowhead points to extend the locus or function plot; drag in the opposite direction to contract it.)

To show or hide the arrowheads for a plot or locus, choose **Edit | Properties |**

**Plot** [95] and check or uncheck **Show arrowheads and endpoints.**

You can also change the domain of a plot or a parametric locus numerically in its Plot Properties [97] panel.

### ▼ Resize a Caption

To resize a caption, select it and drag the resize handle that appears at the bottom right corner.

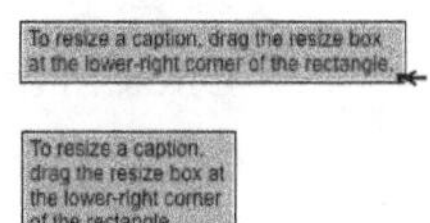


### ▼ Add or Remove Table Data

To add a row to a table [39] of measured values, double-click the table with the **Arrow** tool. To remove the most recently added row to a table with more than one row, double-click the table while holding down the Shift key.

| Area $\odot CD$ | (Radius $\odot CD)^2$ | $\dfrac{\text{Area } \odot CD}{(\text{Radius } \odot CD)^2}$ |
| --- | --- | --- |
| 2.54 cm$^2$ | 0.81 cm$^2$ | 3.1416 |
| 14.79 cm$^2$ | 4.71 cm$^2$ | 3.1416 |
| 36.46 cm$^2$ | 11.61 cm$^2$ | 3.1416 |
| 51.13 cm$^2$ | 16.27 cm$^2$ | 3.1416 |
| 92.80 cm$^2$ | 29.54 cm$^2$ | 3.1416 |

You can also add or remove rows from a table using the **Number | Add Table Data** [228] and **Number | Remove Table Data** [229] commands.

## 2.1.4    How to Zoom In and Out on a Sketch

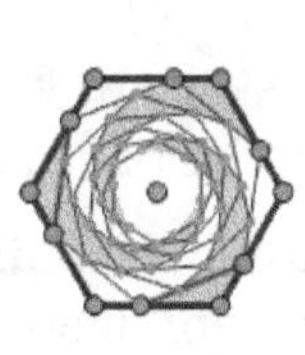
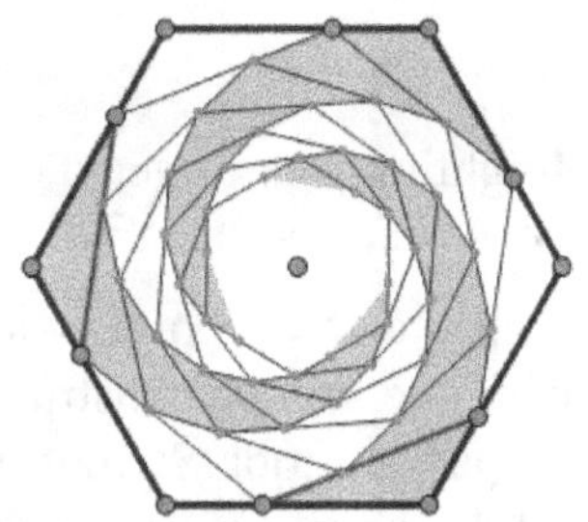

Sometimes a sketch may have so much detail that it becomes hard to select the object you want. Other times, a sketch may get spread out so that you can't see everything in the sketch window. If you want a Zoom command at such a time, you won't find it in Sketchpad's Display [162] menu. However, you don't really need this command because Sketchpad has the **Dilate Arrow** [104] tool, which is the mathematical equivalent of zooming.

To zoom in or out on a sketch by expanding it or shrinking it:

1. Choose **Edit | Select All** [151] to select everything in your sketch.

2. Use the **Dilate Arrow** [104] tool to drag any selected object toward or away from the marked center. As you drag, your sketch zooms in or out.

    If you haven't marked a center point, Sketchpad marks one for you.

3. Increase or decrease the size of labels and text objects by using the **Display | Text** [166] submenu.

## 2.2    Point Tool

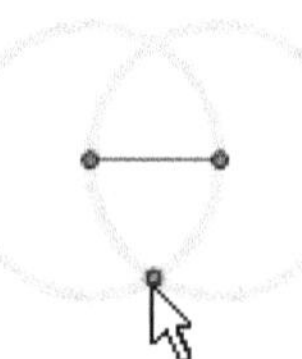Use the **Point** tool to draw or construct independent points [2], points on paths [176], and points at intersections [177].

Click in an empty area of your sketch to create an independent point.

Click a path object [89] — such as a segment, a circle, or the edge of a polygon — to construct a point on the path. When the **Point** tool is in the right place to construct a point on a path, the path is highlighted, appearing thicker and in a special color.

A point constructed on a path can move anywhere along the path, but nowhere else.

Click the intersection of two path objects — such as a segment and a circle, or a line and a function plot — to construct a point of intersection. When the **Point** tool is in the right place to construct an intersection, both paths are highlighted, appearing thicker and in a special color.

In addition to highlighting your targeted objects, Sketchpad displays a message in the status line [253] telling you when you can construct a point on a path or at an intersection.

There are several ways to create points without using the **Point** tool. For example, clicking with the **Arrow** [104] tool on an intersection constructs a point of intersection there in the same way the **Point** tool does. The **Compass** [113] tool, **Straightedge** [115] tools, **Polygon** [118] tools, and most **Custom** [125] tools may construct their own points as part of their operation. The Construct [174] menu commands **Point on Object** [176], **Midpoint** [176], **Intersection** [177], and other menu commands also construct points.

## 2.3    Compass Tool

Use the **Compass** tool to construct circles[6] determined by two points — the center point and another point through which the circle passes. This second point is called the *radius point,* because it determines the radius of the circle.

### ▼ Construct a Circle

1. Choose the **Compass**[113] tool if it's not already active.

2. Click to locate the center of your circle. (You can click in empty space, on an existing point, on a path object[89] such as a segment or another circle, or on an intersection.)

3. Click again to locate the radius point.

Another way to use the Compass[113] tool is to press the button at the center point, drag, and release at the radius point.

### ▼ Attach a Circle to an Existing Object

You can attach either the center point or the radius point to an existing object. To do so, you can click on

- an existing point

- a path object (such as a segment, a line, a circle or an arc)

- an intersection of two path objects

When the **Compass**[113] tool is in the right place to click on an existing object, that object is highlighted (appearing thicker and in a special color) and identified in the status line[253].

Be careful when you want to attach the radius point of a circle to an existing object. It's not enough to position the circle so that the circle appears to go through the desired point. You must actually position the tool itself over the point or object to which you want to attach before clicking or releasing to locate the radius point.

In this illustration, even though the circle appears to pass through the vertices of the triangle, the radius point will not be connected to the triangle. The result is that the radius point will be independent, and dragging either the radius point or any part of the triangle will show that the circle isn't attached to the vertices.

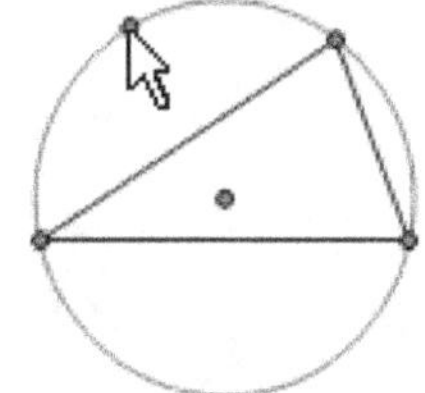

Radius point not
attached to vertex

In this illustration, the tool is positioned at a vertex, and the vertex is highlighted. Thus, that vertex will be the radius point of the circle, and the size of the circle will be linked to the position of the vertex no matter how various parts of the figure are dragged.

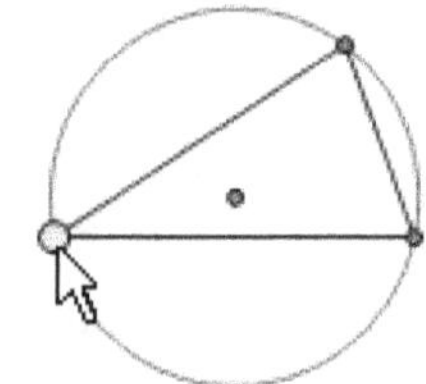

Radius point attached to
vertex

### ▼ Related Commands

There are two Construct|174| menu commands that construct circles without using the **Compass** tool.

- **Construct | Circle by Center+Point** |182| constructs a circle determined by two selected points. The first point is the center point, and the second is the radius point.

- **Construct | Circle by Center+Radius** |183| constructs a circle determined by a selected point and a selected distance. The selected point is the center point, and the selected distance determines the radius. (The selected distance can be either a segment| 5 | or a distance value|92| .)

### ▼ Construct a Circle with its Radius Point at a Specific Angle

While you're constructing a circle, you can hold down the Shift key to make the direction from the center to the radius point horizontal, vertical, or at an angle of 15°, 30°, 45°, 60°, or 75°. Construct the radius point before you release the Shift key.

## 2.4 Straightedge Tools

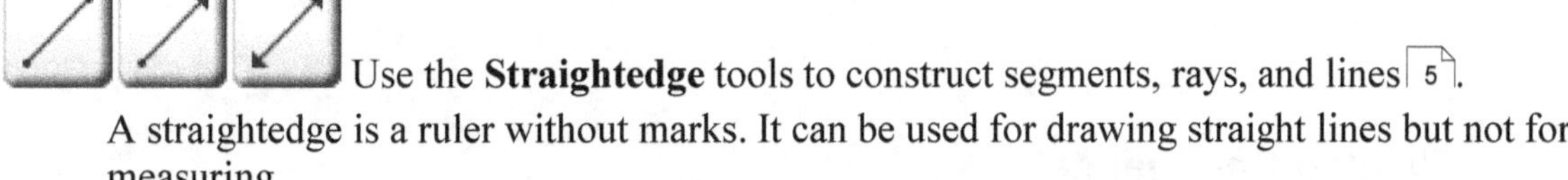 Use the **Straightedge** tools to construct segments, rays, and lines [5].
A straightedge is a ruler without marks. It can be used for drawing straight lines but not for measuring.

There are three Straightedge tools: the **Segment** tool, the **Ray** tool, and the **Line** tool.

When Sketchpad starts, the active **Straightedge** tool is the **Segment** tool. Choose a different **Straightedge** tool by pressing and holding the **Straightedge** tool icon in the Toolbox. When you press and hold, a menu pops out and you can choose any of the three tools.

If a **Straightedge** tool is already chosen, you can switch to a different **Straightedge** tool by holding the Shift key and pressing either the right or left arrow key.

Each straight object constructed by one of these tools is determined by two points. Use the tool in either of two ways:

- Click the location of the first point and then click the location of the second point, or

- Press at the location of the first point, drag to the location of the second point, and release.

### ▼ Segment Tool

Use the **Segment** tool to construct segments.

1. Choose the **Segment** tool in the Toolbox [102] if it's not already active.

2. Click to locate the first endpoint of the segment. (Click in empty space, on an existing point, on a path object [89] such as a segment [5] or circle [6], or on an intersection [177].)

3. Click again to locate the second endpoint.

### ▼ Ray Tool

Use the **Ray** tool to construct rays.

1. Choose the **Ray** tool in the Toolbox [102] if it's not already active.

2. Click to locate the endpoint of the ray. (Click in empty space, on an existing point, on a path object [89] such as a segment [5] or circle [6], or on an intersection [177].)

3. Click again to locate the point through which the ray travels. (This point is called the *through point*.)

To construct a ray that bisects the angle formed by three selected points or by a selected angle marker [60], choose **Construct | Angle Bisector** [181].

### ▼ Line Tool

Use the **Line** tool to construct lines.

1. Choose the **Line** tool in the Toolbox [102] if it's not already active.

2. Click to locate a first point through which the line travels. (Click in empty space, on an existing point, on a path object [89] such as a segment [5] or circle [6], or on an intersection [177].)

3. Click again to locate a second point through which the line travels.

To construct a line that's parallel or perpendicular to an existing straight object, use **Construct | Parallel** [179] or **Construct | Perpendicular** [180].

## ▼ Related Commands

There are several commands in the Construct [174] menu that construct straight objects without using the **Straightedge** tool.

- **Construct | Segment** [178], **Construct | Ray** [178], and **Construct | Line** [178] commands construct straight objects determined by two or more selected points.

- **Construct | Perpendicular Line** [180] and **Construct | Parallel Line** [179] construct a line through a selected point that is perpendicular or parallel to a selected straight object.

- **Construct | Angle Bisector** [181] constructs a ray that bisects the angle formed by three selected points or by an angle marker [60].

## ▼ Construct a Straight Object at a Specific Angle

While you're constructing a straight object, you can hold down the Shift key to make the object horizontal, vertical, or at an angle of 15°, 30°, 45°, 60°, or 75°. Construct the second point before you release the Shift key.

## ▼ Attach a Straight Object to an Existing Object

To attach either the first or second determining point of a straight object to an existing object, use the **Straightedge** tool to click:

- an existing point [2];

- a path object [89] such as a segment [5], a line [5], a circle [6], or an arc [7]; or

- an intersection [177] of two path objects.

When the **Straightedge** tool is in the right place to click an existing object, that object is highlighted (appearing thicker and in a special color) and identified in the status line [253].

Be careful when you want to attach a defining point to an existing object. It's not enough to position the tool so that the ray or line appears to go through the desired location. You must actually position the tool itself over the object to which you want to attach before clicking or releasing the button to locate the defining point.

For example, you may want to draw a diagonal line passing through the intersection of a horizontal line and a vertical line. In the first illustration the diagonal line appears to pass through the intersection, but because the pointer is positioned to click in empty space, the second defining point of the line won't be attached. In the second illustration the pointer is over the intersection and the intersecting lines are highlighted. Because the pointer is positioned correctly, the diagonal line will be attached to the intersection.

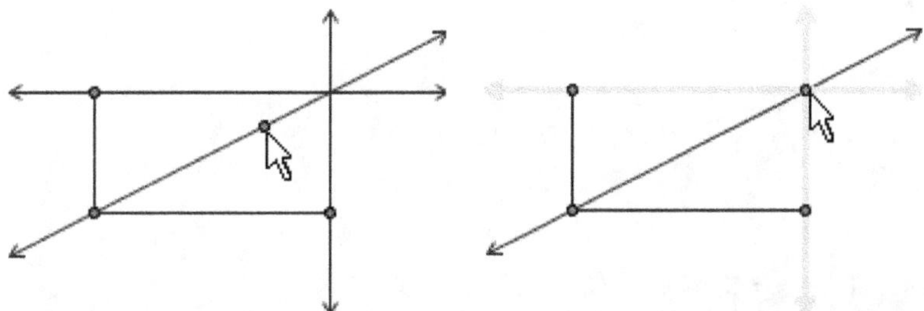

Line not attached to intersection      Line attached to intersection

In another example, you may want to attach a ray's through point to a circle so you can animate the through point around the circle. Positioning the pointer to click as shown in the first illustration won't work because the through point will not be attached to the circle. Instead, click as shown in the second illustration (with the pointer over the circle and the circle highlighted) to attach the point to the circle.

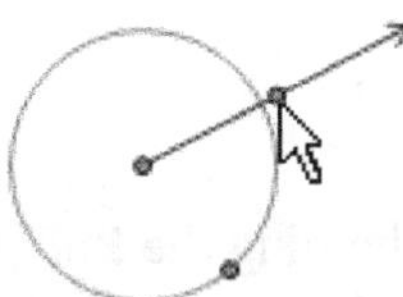 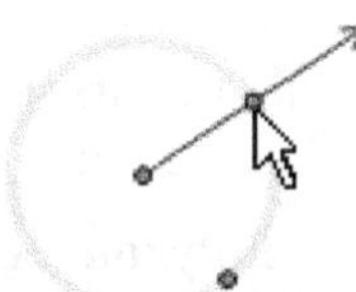

Ray not attached to circle      Ray attached to circle

*See also:*
*How to Construct a Segment of Fixed Length* 182
*How to Construct an Angle of Fixed Measure* 204
*How to Construct Congruent Segments or Angles* 184

## 2.5 Polygon Tools

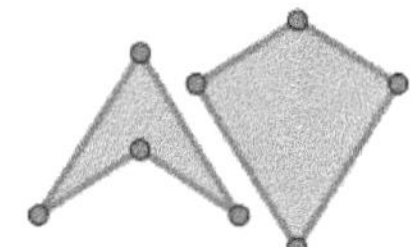

Use the **Polygon** tools to construct a polygon [8], a polygon together with its segment edges, or the segment edges only.

There are three different versions of this tool: the **Polygon** tool, the **Polygon and Edges** tool, and the **Polygon Edges** tool.

To use any of these three tools:

1. Click three or more points. Click an existing point to use it, or click anywhere else to construct a new point.

2. Finish by clicking the first point again. (You can also finish by double-clicking the last point instead of single-clicking it.)

Choose a different **Polygon** tool by pressing and holding the **Polygon** tool icon in the Toolbox. When you press and hold, a menu pops out and you can choose any of the three tools.

If a **Polygon** tool is already active, you can switch to a different **Polygon** tool by holding the Shift key and pressing either the right or left arrow key.

### ▼ Polygon Tool

The **Polygon** tool constructs only the polygon [8].

You can display the interior of the polygon, the frame, or both the frame and the interior by choosing **Edit | Properties | Opacity** [92].

You can measure the area [219] and perimeter [218] of the polygon.

### ▼ Polygon and Edges Tool

The **Polygon and Edges** tool constructs the polygon and the segments [5] that form its edges.

Use the segments to construct midpoints, perpendiculars, or other objects that require a segment.

### ▼ Polygon Edges Tool

The **Polygon Edges** tool constructs only the segment edges, without constructing the polygon.

If you are constructing segment edges for appearance only, you can instead construct the polygon and show just its frame.

### ▼ Related Commands

Use **Edit | Properties | Opacity** [92] to show or hide the frame of the polygon, and to determine the opacity of the interior. (To show only the frame, first show the frame and then set **Opacity** to 0%.)

There are several commands in the Construct 174 menu that construct a polygon or the segment edges of a polygon.

- **Construct | Polygon Interior** 187 constructs the polygon determined by three or more selected points. The command name changes to describe the object that will be constructed. For instance, with five points selected, the command becomes **Construct | Pentagon Interior** 187.

- **Construct | Segments** 178 constructs the polygon segment edges determined by three or more selected points.

## ▼ Construct a Polygon with its Edges at Specific Angles

While you're constructing a polygon, you can hold down the Shift key to make the direction from the previous vertex to the new vertex horizontal, vertical, or at an angle of 15°, 30°, 45°, 60°, or 75°. Construct the new vertex before you release the Shift key.

## 2.6    Text Tool

 Use the **Text** tool to create, show, hide, and edit labels; to create, edit, and resize captions; and to change the text displayed with measurements, calculations, and parameters.

Text plays an important role in Sketchpad. Labels [81] allow you to communicate with others about specific objects in your sketch, and provide a correspondence between measurements and the objects they measure. Captions [56] allow you to identify and describe both your sketch as a whole and individual parts of your sketch. Measurements [36], calculations [39], parameters [36] and functions [45] display important mathematical information in text form. The **Text** tool allows you to show, hide, edit and reposition labels of objects, to edit captions, and to modify the text displayed with measurements, calculations, parameters, and functions.

The **Text** tool has five possible appearances when you move the mouse over the sketch. These appearances depend on what it's pointing at and indicate the effect of using the tool.

---

### ▼ Show or Hide a Label

When you point the **Text** tool at an object that has a label to show or hide, the pointer appears as a filled hand.

- Click to show the object's label.

- Click again to hide the object's label.

You can also show and hide labels using an object's Label Properties [82] or by selecting one or more objects and choosing **Display | Show/Hide Labels** [168]. And you can show and change labels for several objects at once by selecting them and choosing **Display | Label Objects** [168].

Sketchpad has a default labels [81] for each kind of object. For instance, the labels of points start with *A, B,* and *C.*

Note

Objects like points and circles can either show or not show a label. Some other objects — like action buttons or parameters — always show their label. And still other objects — like captions and pictures — never show a label.

By default, objects are labeled automatically when you measure them. (Use **Edit | Preferences | Text** [279] to turn off this feature.) You can also change Text Preferences so that all new points are labeled when they're created.

---

### ▼ Position or Change a Label

When you point the **Text** tool at an object's label or at a measurement, parameter, calculation, or function, the tool appears as a labeled hand.

- Double-click to change the label. (Double-clicking is the same as selecting the object and choosing **Display | Label** [168].)

- Press and drag to reposition the label of an object.

You can also change the font, size, style, and color of selected objects' labels with the Text Palette [270] or by choosing **Display | Text** [166].

---

Choose **Display | Label**[168] to change the labels of several selected objects at once.

 ▼ **Create a Caption**

When you point the **Text** tool at an empty place in the sketch, the tool appears as an open hand.

1. Press and drag to create a new caption.

2. Begin typing your new caption.

3. When you finish typing your caption, press the Esc[294] key or click anywhere outside the caption.

▼ **Edit a Caption**

When you point the **Text** tool at a caption[56], the tool appears as an I-beam.

Click to place an insertion point and begin editing the caption. While you're editing:

- Use various caption-editing keyboard shortcuts[297] to select text or move the insertion point.

- Press and drag to select a block of text to format or replace.

- Double-click to select the word under the cursor.

- Double-click and drag to select text word by word.

- Click an object in the sketch to insert the object's label or value as Hot Text[56].

- Drag the caption's resize handle to resize the caption and reflow its contents.

- Hold the Alt key (Windows) or Option key (Macintosh) while pressing one of the arrow keys to set the text alignment for the caption. Use the left arrow for left alignment, the right arrow for right alignment, and either the up or down arrow for centered alignment.

When you create a caption, by default Sketchpad displays the Text Palette[270]. You can change the font, size, style, or color of selected caption text by using this palette or by using text commands in the Display[162] menu. The Text Palette also contains symbolic notation tools[272] for adding mathematical text to your caption. (You can turn off automatic display of the Text Palette using **Edit | Preferences | Text**[279], and instead use **Display | Show Text Palette**[172] to summon the Text Palette as needed.)

To finish editing, press the Esc[294] key or click anywhere outside the caption.

 ▼ **Resize a Caption**

When you point the **Text** tool at a caption's[56] resize handle, the tool appears as an arrow. A caption's resize handle is displayed at the lower-right corner of the caption whenever the caption is selected or being edited.

Press and drag the handle to resize the caption and reflow the text.

## 2.7    Marker Tool

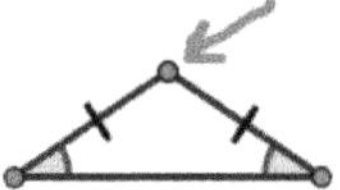

Use the **Marker** tool to place marks in your sketch.

- Call attention to its geometric features by placing angle markers ⌐60⌐ and tick marks ⌐63⌐.
- Create a drawing ⌐18⌐ to embellish your sketch and emphasize important elements.

### ▼ Create an Angle Marker

Use the **Marker** tool to create an angle marker ⌐60⌐ in either of two ways.

- Press the vertex of an angle and drag into the interior of the angle. If several angles overlap, drag farther to select wider angles.
- Press the initial side of the angle and drag to the other side of the angle.

When you use either of these methods, the pointer changes shape when it's over an object on which you can press and drag. (You can also use these two methods, while you're editing a caption, to create an angle marker and insert a Hot Text ⌐57⌐ link to the angle.)

Once you've created an angle marker, you can change its appearance:

- Click it with the **Marker** tool to change the number of strokes it displays.
- Use the the Display ⌐162⌐ menu to change the **Point Style** ⌐162⌐, **Line Style** ⌐163⌐, or **Color** ⌐164⌐.
- Use its Marker ⌐65⌐ properties to change the **Angle definition.**
- Use its Opacity ⌐92⌐ properties to change the transparency of the sector.

### ▼ Create a Tick Mark

Click the **Marker** tool on a path to create a tick mark ⌐63⌐. The path can be a straight object ⌐90⌐, a circle ⌐6⌐, an arc ⌐7⌐, the perimeter of a polygon or other interior ⌐8⌐, a point locus ⌐10⌐, or a function plot ⌐52⌐.

Once you've created a tick mark, you can change its appearance:

- Click it with the **Marker** tool to change the number of strokes it displays.
- Drag it with the **Marker** tool to move it to a different location along its path.
- Use the the Display ⌐162⌐ menu to change its **Line Style** ⌐163⌐ or **Color** ⌐164⌐.
- Use its Marker ⌐65⌐ properties to change its **Tick mark style** to show a crossbar, an open arrow, a hollow arrow, or a solid arrow.

### ▼ Create a Drawing

Press the **Marker** tool in white space and drag to draw in your sketch. You can add handwritten text, emphasize existing objects, sketch approximate functions or curves, or make any kind of drawing you like. A drawing created with the **Marker** tool behaves like any other picture ⌐18⌐ in your sketch, and so you can select, relocate, resize, and even transform it.

- To create a drawing, press and drag in a location that doesn't trigger special marking behavior. Empty space always works as a starting point for a drawing. (Starting on a straight object or on the vertex of an angle will produce a tick mark or an angle marker rather than a drawing.)

- When you draw several ink strokes one after the other, they all become part of the same drawing object in Sketchpad. If you want to create separate pictures, start a new picture by pressing Esc or by clicking the Marker tool again.

- If you use a pressure-sensitive tablet, stylus, or electronic whiteboard, vary the pressure you apply to the **Marker** tool to control the width of its ink.

- Use **Edit | Preferences | Tools** 280 to set the type and width of ink. Use **Display | Color** 164 to set the color of the ink.

- After you draw a shape, you can define it as a function and then evaluate it, use it to define other functions, and so forth. If the drawing doubles back on itself, failing the vertical-line test, the value of the function is based on the highest part of the drawing.

## 2.8    Information Tool

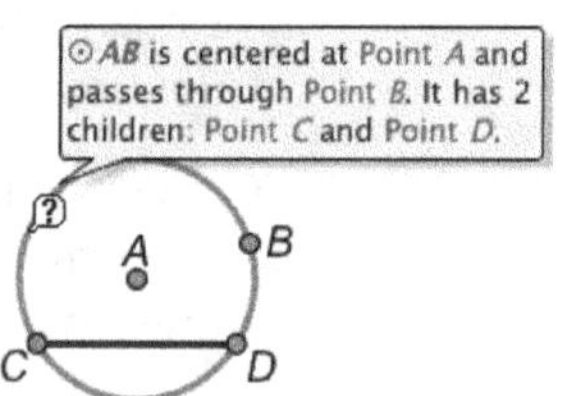


Use the **Information** tool to view descriptions of objects' mathematical definitions and of the relationships between objects in a sketch.

- Click an object to show its information balloon.

- Click the name of an object in its information balloon to show its Properties dialog box.

- Click the names of other objects (parents and children) in the information balloon to show balloons for those objects.

When a hidden object's balloon is shown, the object temporarily becomes visible in the sketch. Clear the **Hidden** checkbox in its balloon to show it permanently.

### ▼ Showing and Hiding Balloons

Old balloons fade away when you switch tools, or as you show new balloons.

- To prevent old balloons from fading when you show new balloons, hold the Shift 295 key. This works whether you show the new balloon by clicking the **Information** tool on an object in the sketch or by clicking the name of another object in a balloon's description.

- To remove all balloons, press the Esc 294 key.

### ▼ Controlling Balloon Behavior with the Context Menu

Information balloons have a Context 245 menu. To use the Context menu, right-click (Windows) or Ctrl-click (Mac) inside the balloon.

**Properties:** This command shows the Properties dialog box for the balloon's object.

**Describe Parents/Children/Both:** These commands change the balloon to show only the object's parents, only the children, or both. (These commands change the display of the current balloon only. To change this setting for future balloons, choose **Edit | Preferences | Tools** 280 and change the settings for the **Information** tool.)

**Larger/Smaller Text:** This command changes the text size used in this balloon and in new balloons.

**Close this Balloon:** This command closes the current balloon.

## 2.9    Custom Tools

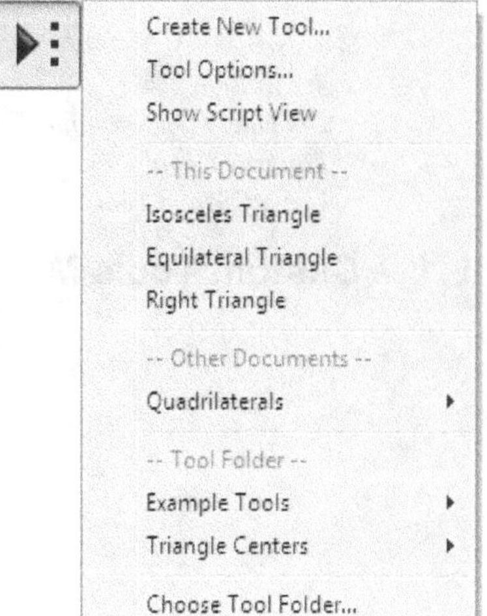

The **Custom** tool icon shows commands that allow you to define and use custom tools. When you press and hold the **Custom** tool icon, the Custom tools menu 126 appears.

Custom tools are tools that you create yourself or that other Sketchpad users create for you. In just the way that Sketchpad's **Compass** 113 tool constructs a circle given its center and radius point, custom tools that you create can construct figures of arbitrary complexity. For example, you can make a custom tool that constructs the perpendicular bisector of a given segment, or one that constructs the circumcircle of a given triangle, or one that constructs a square given two adjacent vertices. A more advanced custom tool might create a fractal, or a tangent to an arbitrary point on a function plot, or a complex tessellation. Any custom tool you define can be used an unlimited number of times in an unlimited number of sketches. By defining new custom tools, you extend the built-in tools available to you in Sketchpad. Since you can create any type or number of custom tools you wish, the possibilities for extending Sketchpad's tools are limitless.

Tools that you create reside inside the document in which you create them. You can always use them in that document (unless you remove them from that document) just as if you were using a built-in tool like the **Compass** 113 or **Straightedge** 115. When a document containing custom tools is open, you can also use those tools in any other open document.

Finally, you can designate or create a special Tool Folder 133 to store frequently used tools. Tools stored in your Tool Folder will be available whenever you start Sketchpad, even when the documents that contain them are not open in Sketchpad.

When you choose a Tool Folder, you can copy a group of example tools into the folder you choose. Here are constructions created by several of these example tools:

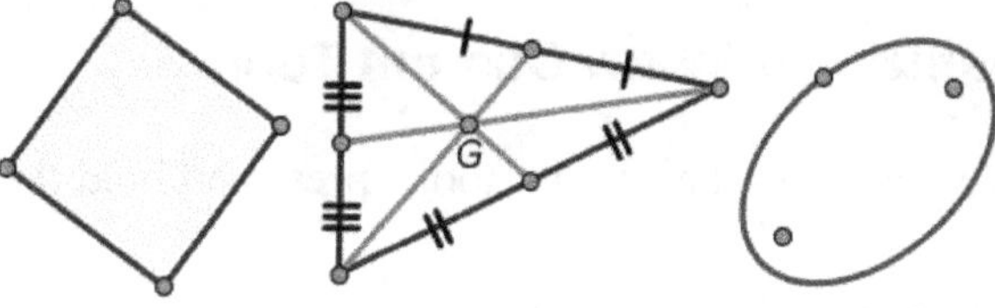

## 2.9.1    Custom Tools Menu

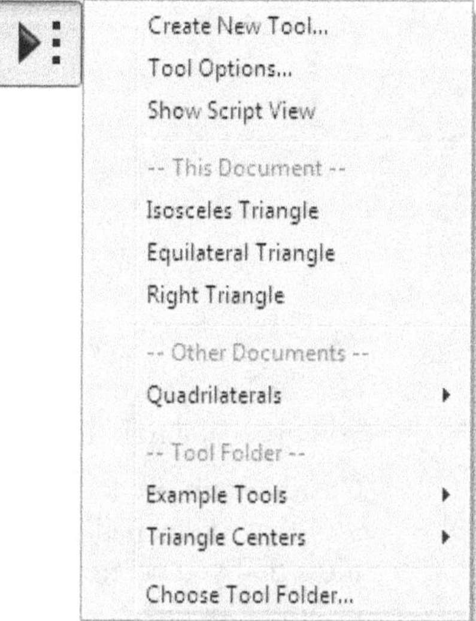


When you press and hold the **Custom** 125 tool icon in the Toolbox 102, Sketchpad displays the Custom Tools menu.

- Use the first part of this menu to create and organize your custom tools.

- Use the middle part of this menu to choose a custom tool to use in your sketch. This part of the menu shows all custom tools in the active document, in other open documents, and in your Tool Folder 133. (If there are no tools in any of these places, this part of the menu does not appear.)

- Use the bottom command on this menu to choose a Tool Folder 133 in which to store frequently-used custom tools.

Hold the Alt key (Windows) or ⌘ key (Mac) when you pull down the Custom Tools menu to use one of the listed custom tools as a command.

Hold the Shift key when you pull down the Custom Tools menu to show the Script View as you choose a listed custom tool.

### ▼ Create, Organize, and View Custom Tools

To define a new tool or organize your tools, press and hold the **Custom** 125 tool icon and use these commands.

**Create New Tool:** This command allows you to make a new custom tool 130 based on your selections in the sketch.

**Tool Options:** This command displays the Tool Options 254 dialog box. Use this dialog box to organize, rename, copy, or remove the custom tools contained in your sketch.

**Show Script View:** This command shows or hides the most recently chosen custom tool's script  — a step-by-step description of what the tool constructs. (The command becomes **Hide Script View** if the script is already showing.)

## ▼ Choose a Custom Tool

To choose a tool, press and hold the **Custom** 125 tool icon and choose a tool from the middle part of the menu.

**This Document:** This section of the menu lists all of the custom tools defined in the current (active) document. If the active document does not yet contain any tools, this part of the menu does not appear. When you define a new custom tool, it appears first in this part of the menu.

**Other Documents:** This section of the menu lists all of the other open documents that contain custom tools. Each entry in this part of the menu lists an open document that contains tools and displays a submenu of each of the tools in that document. If no other open documents contain tools, this part of the menu does not appear.

**Tool Folder:** This section of the menu lists any tools from documents that were stored in your Tool Folder 133 when you started Sketchpad. This part of the menu appears only if you've chosen a Tool Folder within which there are documents containing tools.

Once you've chosen a custom tool, it becomes active until you choose a different tool, just as if you'd chosen the **Compass** 113 or **Straightedge** 115 tool.

Hint

> If you want to use the same custom tool several times in a row, you don't need to choose it again from the Custom Tools menu. Just keep using it.
>
> If you've switched from a custom tool to some other tool, such as the **Arrow** 104 tool, you can switch back to the last custom tool you used by clicking — rather than pressing and holding — the **Custom** 125 tool icon. This activates the most recently chosen, checkmarked tool. If you want to switch to a different custom tool, press and hold the icon to display the Custom Tools menu.
>
> To show the Script View 286 at the same time as you choose a custom tool, hold the Shift key when you pull down the Custom Tools menu.

## ▼ Choose Tool Folder

Use the **Choose Tool Folder** 133 command to choose a Tool Folder 133 in which to store frequently-used custom tools. When you store custom tools in this folder, they become available in the Tool Folder section of the Custom Tools menu.

Hold the Shift key while pressing the Custom Tool menu to change the Choose Tool Folder command to **Forget Tool Folder** 133. Use this command to have Sketchpad forget about the current Tool Folder.

## ▼ Forget Tool Folder

Hold the Shift key while pressing the Custom Tool menu to change the **Choose Tool Folder** 133 command to **Forget Tool Folder** 133. Use this command to have Sketchpad forget about the current Tool Folder.



## 2.9.2    Use a Custom Tool

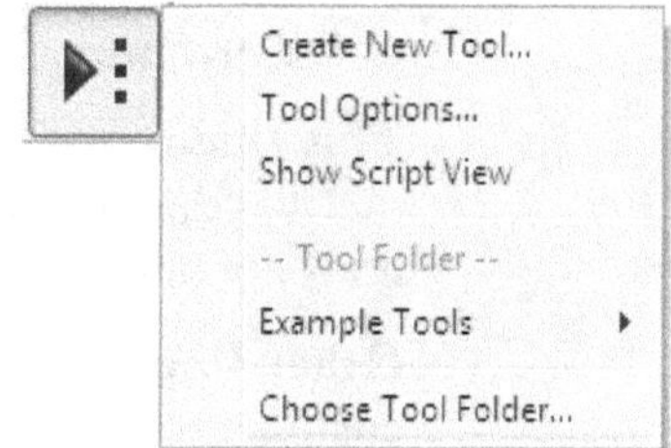


Custom tools are easy to use, even before you learn how to make one yourself.

In this example, you'll first set up a Tool Folder in which to store documents that contain custom tools. (When you set up the Tool Folder, it will contain an example document that comes with Sketchpad, but later on you can add your own documents, with your own tools, to this folder.)

After you've set up the Tool Folder, you'll actually use a tool. In this example, you'll use a tool that constructs a square.

1. Press and hold the **Custom** tools icon in the Toolbox 102 until the Custom Tools menu 126 appears. From this menu, choose the command **Choose Tool Folder** 133.

2. With the dialog box that appears, navigate to a folder in which to store your custom tools. Use an existing folder, or create a new one.

3. Make sure there's a check next to **Create Example Tools document in chosen Tool Folder.** Then click **Choose.** A dialog box appears to indicate that the Example Tools document has been created. (If you see an error message, make sure that you have write access to the folder you've chosen.)

Now that you've chosen a Tool Folder, tools in this folder will be available every time you start Sketchpad (until you change or reset the Tool Folder). You don't need to repeat steps 1–3 each time you want to use a tool.

4. Press and hold the **Custom** tools icon again. This time, from the Custom Tools menu 126 choose **Example Tools | Square Given Two Vertices.** This custom tool is now your active tool.

5. Click the **Square Given Two Vertices** tool in two different places in the sketch. A square appears. (Instead of clicking twice, you can press and drag if you like.)

6. Continue using the custom tool to make lots of squares. You can click in empty space, on existing points, on paths, or on intersections just as you can with the **Point** 112, **Compass** 113, and **Straightedge** 115 tools.

7. When you're finished making squares, click any other tool in the Toolbox 102, or press the Esc 294 key.

To resume making squares, click the **Custom** tool icon again. You don't need to press and hold to choose from the Custom Tools menu the next time — just click the **Custom** tool icon to activate it. Press and hold the **Custom** tool icon in the Toolbox only when you want to switch to a different custom tool from the menu.

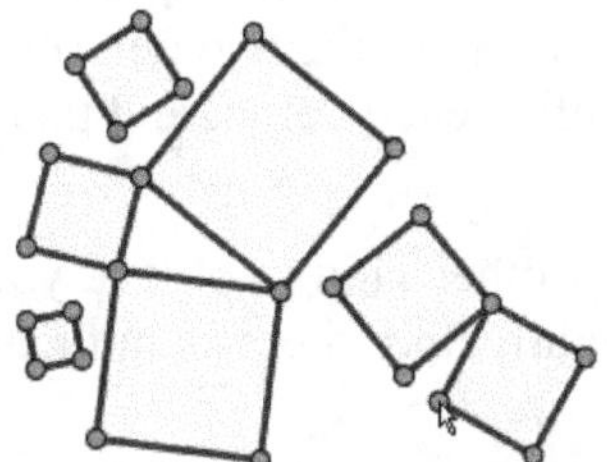

### ▼ Givens and Results of a Tool

Each time you click as you use a tool, you specify a given object — an object that is used to determine the rest of the objects the tool produces. The objects produced by the tool (that depend on the given objects) are the results of the tool.

The **Square** tool has two givens — two adjacent vertices of the square. The remaining vertices, the sides, and the square itself are the tool's results.

Think of the various objects and relationships that make up a tool as a family tree ⬚80. In this family tree, the givens are the ultimate ancestors — objects that have children but no parents. The other objects — those that have parents — are the results of the tool. The Script View ⬚286 shows lists the givens and the results ⬚288 of the currently chosen custom tool.

### ▼ Matching Given Objects

Tools can use as their givens various kinds of objects: points, straight objects, circles, measurements, functions, and so forth. When you use a tool, a message appears on Sketchpad's status line ⬚253 at the bottom of the window describing what kind of given object you need to match next. (In Windows, the status line appears at the bottom-left of the window frame. On Mac, the status line appears at the bottom-right corner of the sketch.)

For example, if a tool uses a point, a segment, and a distance measurement as its three givens, the status line first says "1. Match Point…" to indicate what object you must match first. After you match the point, the status line says "2. Match Segment…" to indicate that you must next specify a segment. And finally, it says "3. Match Distance Measurement…" when it's time for you to click on a distance measurement as the last given object.

You can always match a given by clicking a sketch object of the correct kind. For some givens — points, straight objects, circles, and polygons — you can also match the given by constructing it.

If the Script View ⬚286 is showing, you can see its given objects highlighted as you match them.

If a given object for a tool is a point, you can specify it in either of two ways.

- To specify an existing point, click it in the sketch. The point you click on is used as the given.

- To construct a new point, click somewhere else in the sketch — in empty space, on a path, or at an intersection. The point you construct is used as the given.

If a given object for a tool is a circle, a straight object, or a polygon, you can specify it in either of two ways.

- To specify an existing object of the correct kind, click it in the sketch. The object you click on is used as the given.

- To construct a new object, click twice (or press and drag) in the sketch to construct a circle or straight object, or click at least three times for a polygon. If you're constructing a circle given, the first point is the center and the second point is the radius point. If you're constructing a straight object, the two points are the two determining points of the straight object. If you're constructing a polygon, click each vertex and then click again on the first vertex to signal the end of the polygon.

If the given for a tool is any other kind of object, you must click on a matching object in the sketch. If no object of the correct kind exists in the sketch, you must create such an object before you can use the tool.

If a given tool object in the tool always matches the same sketch object, you can make this match occur automatically 318.

## ▼ Use a Custom Tool as a Command

Use any custom tool as a command 316 by selecting appropriate objects as prerequisites and holding the Alt key (Windows) or ⌘ key (Mac) while you choose the tool from the Custom Tools menu.

*See also:*
*Make a Custom Tool* 130
*The Tool Folder* 133
*Advanced Tool Topics* 316
*Use a Custom Tool as a Custom Command* 316

## 2.9.3   Make a Custom Tool

You define new custom tools by example: you create a construction you want to turn into a tool, select the objects that make up the construction, and define a tool based on that example. The objects you select must be related to each other in such a way that at least one selected object is completely determined by other selected objects. To define the tool, choose **Create New Tool** 126 from the Custom Tools menu.

The selected objects produced by a tool are the *results* of the tool. The selected objects that don't depend on any others, but upon which the results depend, are the *givens* of the tool. Any unselected objects that relate the selected givens to the selected results are *intermediate objects*. (Intermediate objects are not visible when you use the tool; only the objects that were selected when you defined the tool are visible results when you use the tool.)

Follow these steps to make a new custom tool:

1. Make a construction to serve as an example of the construction you want the tool to produce. Use any of Sketchpad's tools or menus to create this exemplar.

2. Select the given objects (usually, independent points) and the desired resulting objects you'd like the tool to produce. The order in which you select the givens determines the order in which you'll match givens when using the tool.

3. If there are other objects that relate the givens to the final results, you can select them or not. If you do select them, they will be shown when you use the tool. If you don't select them, they will still be constructed when you use the tool, but they will be hidden rather than visible.

4. Press and hold the **Custom** Tool icon. Choose **Create New Tool** from the Custom Tools menu 126 that appears.

5. A dialog box appears in which you can type a name for the tool. Type a name and click **OK.** (If the name you type is already used for an existing tool, a dialog box appears allowing you to decide whether to replace the existing tool with the new one.)

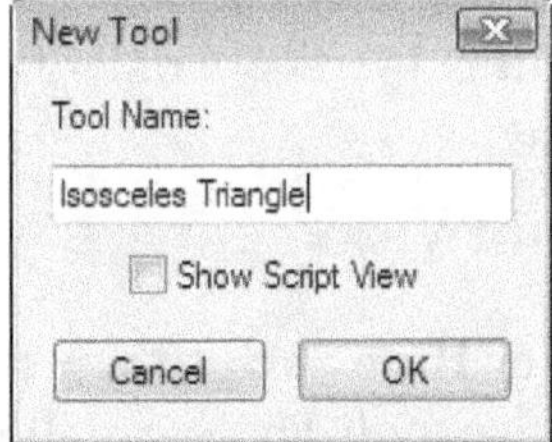



Your tool is added to the Custom Tools menu, and is ready to use.

To see the givens and results of the tool, show the Script View 286.

If a given tool object in the tool should always match the same sketch object, you can make this match occur automatically 318.

You can store the document containing the tool in the Tool Folder 133 to make it permanently available in the Custom Tools menu.

### ▼ Specify the Results of a Tool

When you make a tool, any selected objects that depend on the givens 288 become results 288 of the tool, and will be shown when you use the tool. Any unselected objects that depend on the givens will not be shown when you use it.

For example, if you make a tool that constructs the perpendicular bisector of a segment, the segment is the given object and the perpendicular bisector is a result. If you select the midpoint of the segment when you make the tool, the midpoint is also a result, and is shown when you use the tool. If you don't select the midpoint, it's an intermediate object and is hidden when you use the tool.

| *Making the Tool* | *Using the Tool* |
|---|---|
| Midpoint selected | Midpoint shown |
| Midpoint not selected | Midpoint hidden |

### ▼ Manage Custom Tools

After you've created one or more custom tools in a document, you can rearrange them, copy them to other documents, rename them, or remove them by choosing **Tool Options** 126 from the Custom Tools menu 126. This command opens the Document Options 254 dialog box to a view of your document's tools.

You can store your document in the Tool Folder 133 to make the document's tools available every time you use Sketchpad.

### ▼ Show or Hide the Script View

Choose **Show Script View** or **Hide Script View** from the Custom Tools menu 126 to show or hide the script view 286 of the active tool. This view allows you to see the given objects and the steps the tool takes to construct its results, change the properties of the steps, and observe and control the tool as it functions.

*See also:*
*Custom Tools Menu* 126
*Use a Custom Tool* 128
*Advanced Tool Topics* 316
*Document Options and Tool Options* 254
*How to Make a Perpendicular Bisector Tool* 132


## 2.9.4    How to Make a Perpendicular Bisector Tool

If you're doing an investigation in which you need to construct several perpendicular bisectors, you can make a perpendicular bisector custom tool to simplify your work. Suppose you want to construct the perpendicular bisectors to all three sides of a triangle. Rather than do the perpendicular bisector construction three times, you can do it once and make it into a tool. Then use the tool for the other two perpendicular bisectors. This example shows you how.

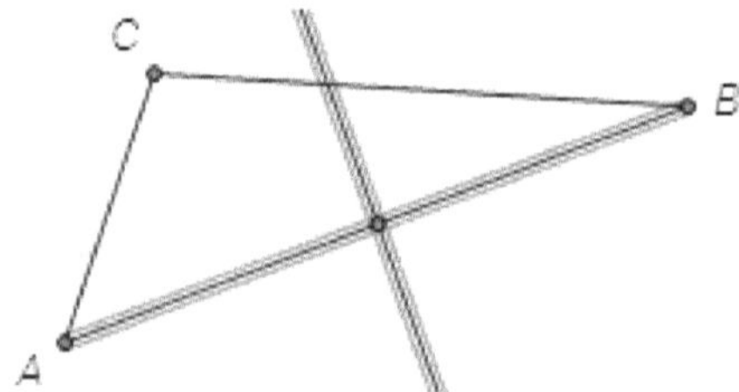


First you'll create a **Perpendicular Bisector** tool.

1. In a new document, use the **Segment** 115 tool to construct the edges of $\triangle ABC$.

2. On segment $AB$, construct the midpoint 176 and the perpendicular 180 to segment $AB$ through the midpoint.

3. Select segment $AB$, the midpoint, and the perpendicular line. Segment $AB$ will be the given 288 object, and the midpoint and perpendicular will be results 288. (If you had also selected points $A$ and $B$, they would have been the givens, and the segment would have been an intermediate result.)

4. Press and hold the **Custom** 125 tool icon and choose **Create New Tool** 126 from the Custom Tools menu that appears.

5. Type "Perpendicular Bisector" to name your tool and click **OK**.

Next you'll use your new **Perpendicular Bisector** tool.

6. Click the **Custom** tool icon to choose your new tool. (You don't need to pull down the Custom Tools menu, because your new tool is already the active custom tool.)

7. Click each of the other two sides of the triangle. The tool constructs the perpendicular bisectors on those two sides.

8. Press the Esc [295] key or choose a different tool from the Toolbox [102] to stop using your custom tool.

Because the given of this tool is a segment, match it either by clicking a segment, as you did above, or by clicking twice (or pressing and dragging) to construct a new segment to match the given. This means you can use the tool to construct a triangle from scratch with the perpendicular bisectors constructed automatically.

9. Click the **Custom** tool icon to choose your **Perpendicular Bisector** tool.

10. Press in empty space in the sketch and drag to construct one side of a triangle.

11. Press and drag two more times to construct the remaining two sides of the triangle.

You've just constructed a triangle, as easily as you'd have done it with the **Segment** [115] tool, but with perpendicular bisectors on all three sides.

Use your new tool to construct a quadrilateral with the perpendicular bisectors of all four sides. Then construct one diagonal, along with its perpendicular bisector. What do you notice about the perpendicular bisector of the diagonal as you drag the quadrilateral into different configurations?

*See also:*
  *Script View* [286]
  *Advanced Tool Topics* [316]

## 2.9.5    The Tool Folder

The Tool Folder is a special folder in which you can store frequently used tools. When Sketchpad starts, it checks the Tool Folder and puts every tool in this folder into the Custom Tools menu [126] that appears when you press and hold the **Custom** [125] tools icon. (When you first install Sketchpad, there is no Tool Folder until you designate one by using the **Choose Tool Folder** command from the Custom Tools menu [126] menu.)

### ▼ Choose Tool Folder

The **Choose Tool Folder** command in the Custom Tools [126] menu allows you to choose a location in which to store documents containing custom tools that you use frequently. The tools contained in these documents appear automatically in the Custom Tools menu when you start Sketchpad, so they are always available.

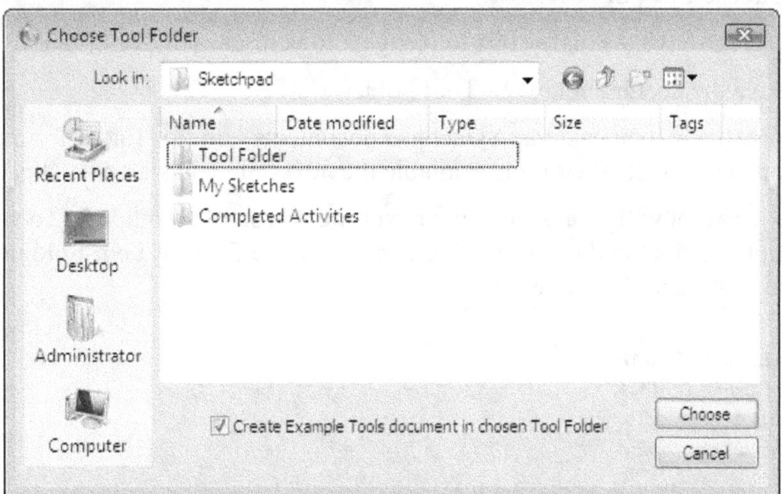

When you choose this command, a dialog box appears. Choose a folder from the dialog box, or create a new folder. You can give the folder any name you like.

If the **Create Example Tools document in chosen Tool Folder** box is checked, Sketchpad will copy into the folder a Sketchpad document called **Example Tools.gsp.** This document contains a number of tools that show the variety and usefulness of custom tools.

Once you've chosen a Tool Folder, use it to save your own Sketchpad documents containing tools. Every tool in this folder will be available whenever you start Sketchpad, without any need to open the documents containing the tools.

Sketchpad scans the Tool Folder when it starts, so if you save a new document containing a tool in the Tool Folder, it won't appear right away. To make the new tool appear, you can either restart Sketchpad, or use the **Choose Tool Folder** command again, choosing the same folder as before. (When you choose the command again, Sketchpad scans the chosen folder to find all the tools located there.)

On Windows you can also use a command-line flag 326 to set a specific tool folder that takes precedence over the chosen Tool Folder.

## ▼ Forget the Tool Folder

Hold the Shift key while pulling down the Custom Tools menu 126 to change the **Choose Tool Folder** command to **Forget Tool Folder.** Choose this command to forget the current Tool Folder, so that there's no active Tool Folder. When Sketchpad forgets the Tool Folder, it also forgets the tools contained in that Tool Folder. (The folder and its tools are not deleted; they remain where they were, even though Sketchpad no longer remembers them.)

If you want to choose a different Tool Folder, there's no need to forget the current one first; just use the **Choose Tool Folder** command to identify a different folder.

## ▼ Store a Tool in the Tool Folder

To store a tool in the Tool Folder:

1. Create or copy 254 the tool(s) you want into a new document. (You can copy tools between open documents with Tool Options 254.)

2. Choose **File | Save As** 140.

3. Use the Save As dialog box to navigate to your chosen Tool Folder.

4. Save the document in your Tool Folder.

Alternatively, you can drag Sketchpad documents into this folder from other locations on your computer to make their tools available the next time you start Sketchpad.

The new tool will be available the next time you start Sketchpad. (To scan the folder for tools without quitting and restarting Sketchpad, use the **Choose Tool Folder** command again, and choose the same folder as before.)

*See also:*

# Menus

# 3    Menus

This section describes the commands in Sketchpad's menus.

- The File 137 menu allows you to create, save, and print entire documents.

- The Edit 144 and Display 162 menus contain commands that alter the appearance, format, or definition of existing objects in your active sketch.

- The Construct 174, Transform 192, Measure 216, Number 226, and Graph 233 menus all allow you to define new mathematical content in the active sketch, often by creating new objects related to selected objects.

- The Window 243 menu (Microsoft Windows and Macintosh OS X only) lets you manage open document windows on your desktop.

- The Help 244 menu contains links to the Sketchpad Learning Center, to this Reference Center, to the Sketchpad Resource Center, and to the Picture Gallery.

- The Context 245 menu appears when you right-click (Windows) or Ctrl+click (Macintosh) in the sketch, and presents options relevant to the clicked object, or to the sketch document if the click was in empty space.

- The Custom Tools 126 menu appears when you press and hold the **Custom** tool icon at the bottom of the Toolbox.

The Macintosh edition of Sketchpad contains a Sketchpad application menu in addition to the menus listed here. This application menu contains standard commands used by all Mac applications, and isn't discussed here.

# 3.1　File Menu

The File menu provides commands for managing your Sketchpad documents. Many of these are standard commands that appear in most software applications. This chapter briefly describes the familiar aspects of these commands and provides more detail on ways in which Sketchpad treats them differently. If you are unfamiliar with any of the basic commands, you'll probably want to start by looking in the manual that came with your computer.

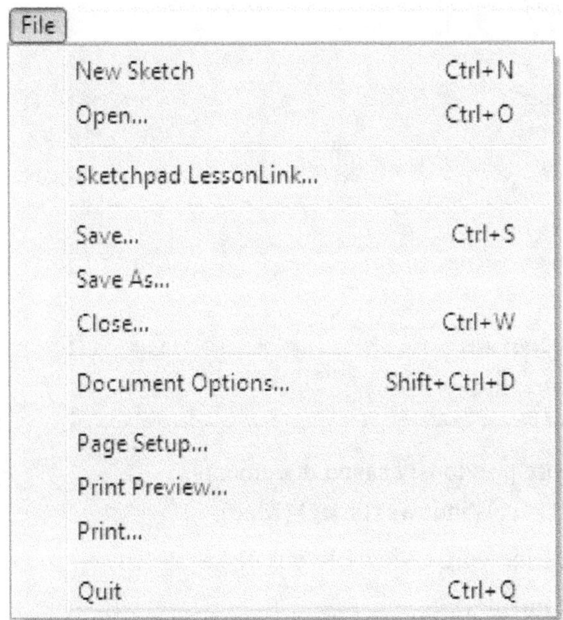


Detailed information is available for each command:
New Sketch 137
Open 138
Sketchpad LessonLink 138
Save 139
Save As 140
Close 142
Document Options 142
Page Setup 142
Print Preview 142
Print 143
Quit 143

## 3.1.1　New Sketch

This File 137 menu command opens a new, blank document. A new document window appears on top of all other windows and becomes the active window. The new document is untitled until you name it by saving it.

The keyboard shortcut for **New Sketch** is Ctrl+N (Windows) or ⌘N (Mac).

## 3.1.2   Open

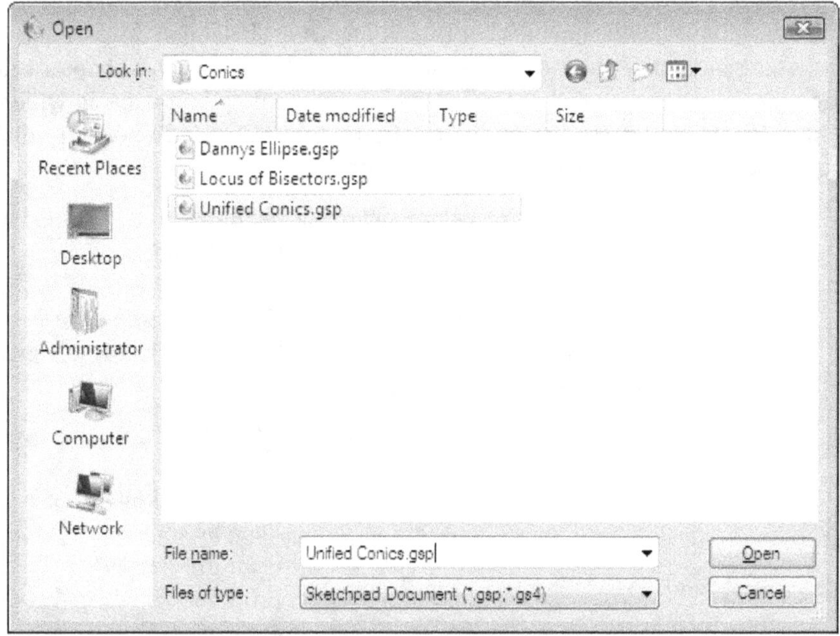

This File 137 menu command opens one or more previously saved documents.

The keyboard shortcut for **Open** is Ctrl+O (Windows) or ⌘O (Mac).

When you choose **Open,** a dialog box appears showing a view of your disk.

1. Navigate to the folder containing your document(s).

2. Click a name to highlight the document you want to open.

3. Hold down the Shift key (Macintosh) or Ctrl key (Windows) while clicking additional document names to open more than one document at a time.

4. Click **Open** or double-click the name of the document.

The chosen documents open on your desktop.

In Windows, you can use a command-line flag 326 to determine what folder appears in the dialog box the first time you choose the **Open** command.

### 3.1.3   Sketchpad LessonLink

This File 137 menu command provides students with easy access to activity sketches from Sketchpad LessonLink™.

Sketchpad LessonLink is a service that enables teachers to easily assign Sketchpad activities to a class. The teacher chooses one or more activities for the class (from a library of more than 500 activities). Sketchpad LessonLink then creates a web page from which students can easily download the worksheets and sketches for those activities. The web page is identified by a *ClassPass* that students use to access it.

As an example, we've created a student web page that has several activities for a variety of age levels. (The ClassPass for this web page is *ReferenceCtr.*) To view this student web page, use this link: http://www.keymath.com/classpass/ReferenceCtr.

Students can download the activity sketches from the student web page, but it's often more convenient for them to open them directly from within Sketchpad, with no need to use a browser to download them first. To open the sketch for any of the sample activities, follow these steps:

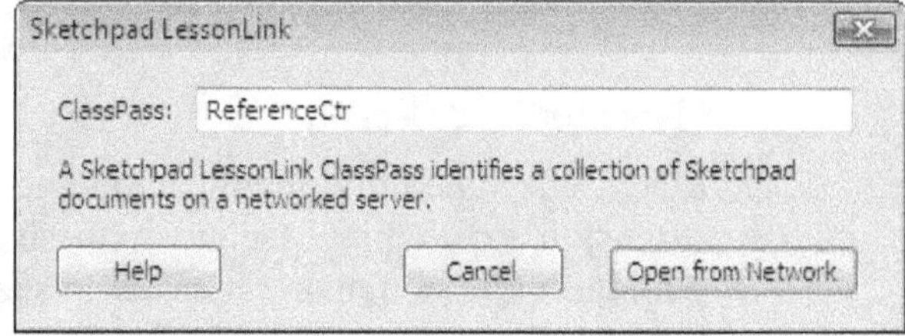


1. Start Sketchpad.

2. Choose **File | Sketchpad LessonLink.**

3. In the dialog box that appears, enter this ClassPass: *ReferenceCtr*

4. Choose from the list of sketch documents that appears.

The activity sketch appears in a new window.

### ▼ Teachers: Collect and Organize Activities with Sketchpad LessonLink

Sketchpad LessonLink is an online, searchable subscription-based library of more than 500 Sketchpad activities, aligned to leading math textbooks and state standards for grades 3-12. Powerful search options make it easy for teachers to find the perfect demonstration or student activity to develop or reinforce concepts and skills. Pre-built sketches, teaching notes, student worksheets, and tips on using Sketchpad provide teachers with everything they need in one place.

With Sketchpad LessonLink teachers can organize the activities they want to use into folders and publish those folders for students. When a teacher publishes a folder, a ClassPass is assigned so students can use the **Sketchpad LessonLink** command to open the sketches. Alternatively, students can access a Student Web Page containing the activities, the worksheets, and relevant Sketchpad Tips.

### ▼ Students: Use a ClassPass to Download and Use Sketchpad Activities

When a teacher publishes a Sketchpad LessonLink folder, it's assigned a ClassPass that students can use to access the activities. There are two ways students can use the ClassPass:

- Within Sketchpad, choose **File | Sketchpad LessonLink,** enter the ClassPass, and choose from a list of the sketch documents for all the activities on the student web page. The chosen sketch document opens immediately in Sketchpad.

- From a web browser, go to the student web page. (For example, if the ClassPass is ReferenceCtr, the web page is http://www.keymath.com/classpass/ReferenceCtr.) From this web page students can see a short description of each activity, read any teacher comments entered for the activity, download the sketch, download the student worksheet, and view Sketchpad Tips that are relevant for the activity.

For more detailed information on Sketchpad LessonLink, go to http://www.keypress.com/sll.

## 3.1.4  Save

This File 137 menu command saves whatever changes have been made to the current document since the last time it was saved. If the document is being saved for the first time, the **Save** command prompts you for the location in which to save (see **Save As** 140). This command is enabled only if you've made changes in the document since the last time you saved it.

The keyboard shortcut for **Save** is Ctrl+S (Windows) or ⌘S (Mac).

Note for Mac Users

Mac users: When first saving a document, Sketchpad suggests a filename ending with the extension **.gsp.** While Macintosh doesn't require file extensions, naming Sketchpad documents with this extension makes it easier to share them with Microsoft Windows users or users on the Internet.

## 3.1.5    Save As

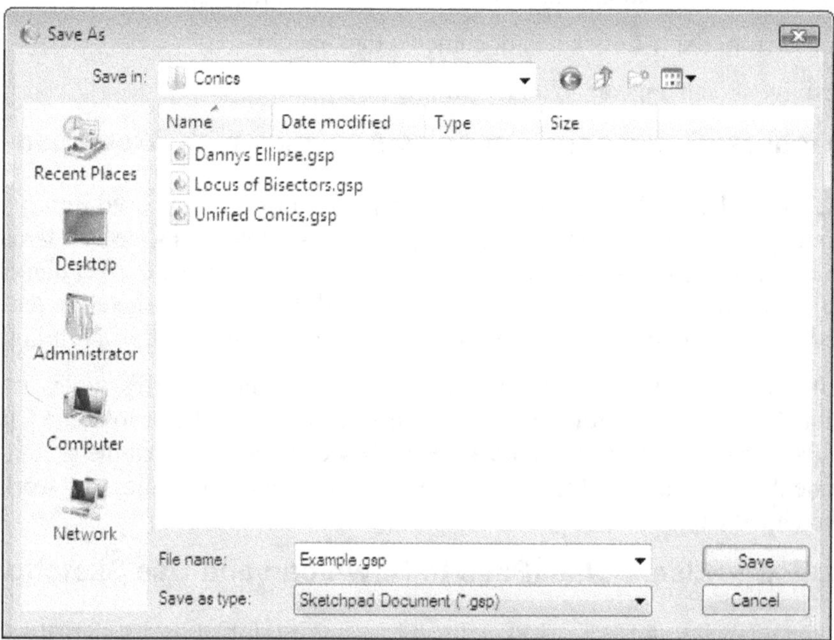

This File 137 menu command names and saves the active document in a location that you specify.

When you choose **Save As,** a dialog box like the one above appears.

1. Navigate your folders to locate the folder in which you want to save your document, or create a new folder, if needed.

2. Type a name for the document.

3. Click **Save.**

### ▼ Save Different Copies of a Document

Use **Save As** to save an additional copy of a previously saved document, with or without changes. For example, suppose you have a document called **Pythag.gsp** to which you want to add action buttons,and you also want to keep a copy of the original version. Make sure you've saved the sketch in the original form, and then make the changes and choose **Save As.** Enter a different name for the modified version, such as **Pythag2.gsp.** Now you'll have both versions saved as separate documents.

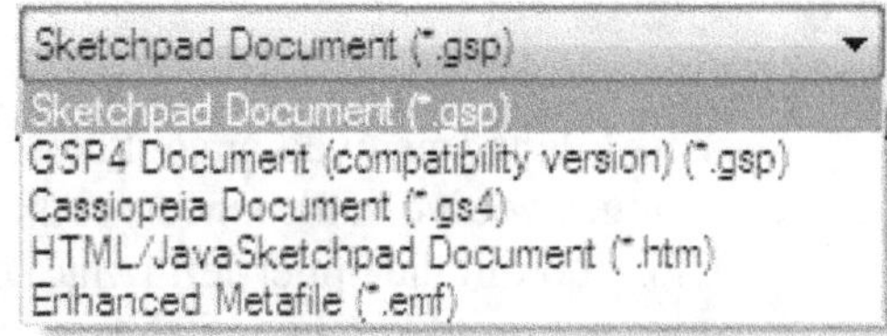

Your document is normally saved as an ordinary Sketchpad document. This format saves your entire document, and the document can be reopened only by Sketchpad Version 5 and later.

In addition, there are several other formats in which you can save documents.

### ▼ Save as GSP4 Sketchpad Document

Use **Save As** to save a copy of your document in a form that can be opened by Sketchpad Version 4. Choose **GSP4 document** from the **Format** list (Macintosh) or **Save as type** list (Windows) before you click the **Save** button.

> Some objects cannot be saved in GSP4 format. These objects, and any other objects that depend on them, will be automatically excluded from the saved GSP4 sketch.

### ▼ Save as Cassiopeia™ Sketchpad Document

Use **Save As** to save a copy of your document in a form that can be used with the version of Sketchpad available for Casio's Cassiopeia Computer Extender handheld computer. Choose **Cassiopeia document** from the **Format** list (Macintosh) or **Save as type** list (Windows) before you click the **Save** button.

> Some objects cannot be saved in Cassiopeia format. These objects, and any other objects that depend on them, will be automatically excluded from the saved Cassiopeia sketch.

### ▼ Save as HTML/JavaSketchpad Document

Use **Save As** to save a copy of your document in a form that can be used on the World Wide Web with JavaSketchpad® [307]. Choose **HTML/Java Sketchpad document** from the **Format** list (Macintosh) or **Save as type** list (Windows) before you click the **Save** button.

Hold the Shift key while pulling down the File menu to change the **Save As** command to **Save As HTML.**

 Note

> When you save as HTML, the folder containing your saved document must also have **jsp5. jar** (the JavaSketchpad applet). Sketchpad automatically tries to save this applet into the same folder; if you later move the HTML document, be sure to move **jsp5.jar** with it. See Essential JavaSketchpad Folder Structure [308] for more details.

### ▼ Save as Enhanced Metafile (Windows only)

Use **Save As** to save a graphics file that can be used by many other Windows programs. The graphics file includes all objects visible in the sketch window. Choose **Enhanced Metafile** from the **Save as type** list before you click the **Save** button.

*See also:*
*Advanced Graphics Export* [320]

### 3.1.6    Close

This File|137| menu command closes the current document window. You'll be prompted to save any changes you've made since you last saved. The **Close** command does the same thing as clicking the Close box in the upper-left (Mac) or upper-right (Windows) corner of the window's title bar. **Close** does not quit Sketchpad itself.

The keyboard shortcut for **Close** is Ctrl+W (Windows) or ⌘W (Mac). You can also use Ctrl+F4 in Windows.

*See also:*
*Quit command*|143|

### 3.1.7    Document Options

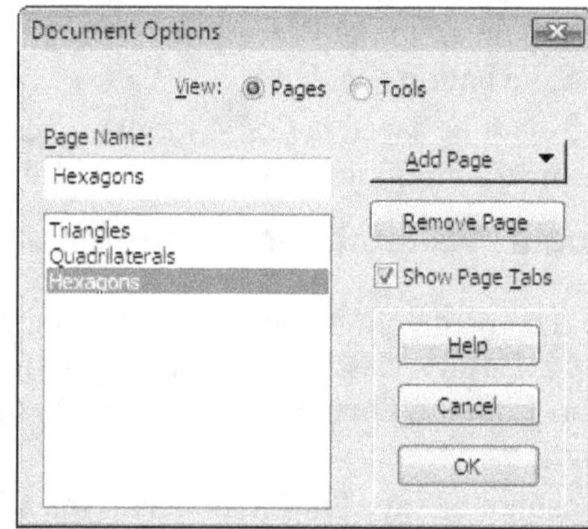

This File|137| menu command opens the Document Options|254| dialog box. Use this dialog box to manage the pages and custom tools|125| contained in the currently active document|250|.

The keyboard shortcut for **Document Options** is Shift+Ctrl+D (Windows) or Shift-⌘D (Mac).

### 3.1.8    Page Setup

This File|137| menu command sets up the page size, orientation, and other printing options for your document. This dialog box differs depending your operating system and the printer you've chosen as your default printer.

*See also:*
*Print Preview command*|142|
*Print command*|143|

### 3.1.9    Print Preview

This File|137| menu command displays a preview of your document as it will appear when printed. This dialog box allows you to change the scale of your printout. If the printout will be more than a single sheet of paper, you can either click **Fit to Page** to shrink the document to a single sheet, or you can view the different sheets to decide which pages to print.

*See also:*
*Page Setup command*|142|
*Print command*|143|

## 3.1.10 Print

This File 137 menu command prints the current page of the active document on the default printer. The Print dialog box allows you to specify which sheets to print and allows you to print multiple copies. Depending on your printer and operating system, it may also allow you to change printers, decide between color and black-and-white, save output as PDF or Postscript, and make other adjustments.

*See also:*
   *Print Preview command* 142
   *Page Setup command* 142

## 3.1.11 Quit

This command closes all open documents and exits Sketchpad. You'll be prompted to save any unsaved work in open documents before Sketchpad quits.

The keyboard shortcut for **Quit** is Ctrl+Q (Windows) or ⌘Q (Mac).

This command appears in the File 137 menu in Windows, and on the Sketchpad menu on Macintosh.

## 3.2    Edit Menu

The Edit menu contains commands for undoing and redoing recent operations, for managing the clipboard, for creating action buttons, for selecting objects in your sketch, and for modifying various elements and properties of your sketch and of Sketchpad itself.

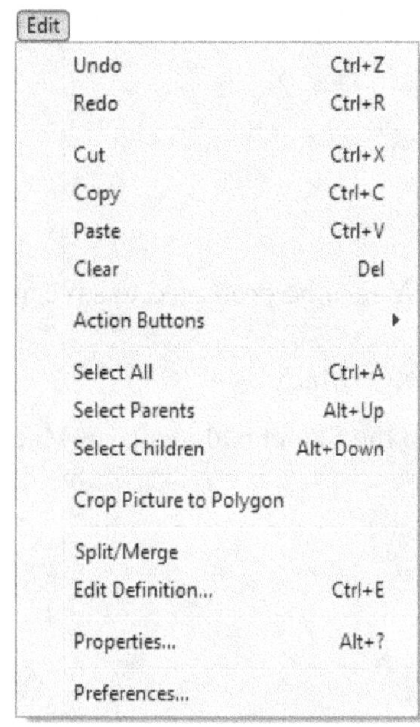

Detailed information is available for each command:

Undo 144
Redo 145
Cut 145
Copy 146
Paste 147
Clear 147
Action Buttons 148
Select All 151
Select Parents 152
Select Children 152
Crop Picture to Polygon 152
Split/Merge 153
Edit Definition 157
Properties 158
Preferences 159
Advanced Preferences 160

### 3.2.1    Undo

This Edit 144 menu command undoes the most recently performed action.

The keyboard shortcut for **Undo** is Ctrl+Z (Windows) or ⌘Z (Mac).

When you hold down the Shift key, the command becomes **Undo All.**

Use this command in combination with **Redo** 145 to move backward and forward through your recent Sketchpad actions. Sketchpad's capability to undo/redo is unlimited: you can use it to undo your actions, one at a time, all the way back to the point at which you created or opened the sketch. Similarly, after undoing, you can redo those actions to restore your sketch to the state it was in before you started undoing.

Unlimited undo/redo is helpful for correcting mistakes — undoing something you didn't mean to do — or for going back and trying a different approach to a construction — testing a different hypothesis. Unlimited undo is also a way to review your work or someone else's work step-by-step: Undo back to the beginning, and then redo one step at a time to review each action.

Undo many steps quickly by repeatedly pressing the keyboard shortcut for **Undo.** Or undo back to the beginning with a single command, by holding down the Shift key and then choosing **Undo.**

## 3.2.2  Redo

This Edit[144] menu command redoes an action you have undone. If you've undone several steps, you can redo each of those steps.

The keyboard shortcut for **Redo** is Ctrl+R (Windows) or ⌘R (Mac).

When you hold down the Shift key, the command becomes **Redo All** and redoes all previously undone actions.

**Redo** is available only immediately after using **Undo**[144]. If you take any other action after undoing operations, you can no longer redo the original operations.

Use **Redo** in combination with **Undo** to move backward and forward through your recent Sketchpad actions.

## 3.2.3  Cut

This Edit[144] menu command removes from the sketch any object that is selected[106], along with any objects that depend on it[80]. All removed objects are placed on the clipboard and can be pasted into the same sketch, a different sketch, or as a picture into another application.

The keyboard shortcut for **Cut** is Ctrl+X (Windows) or ⌘X (Mac).

This command completely removes the selected objects, and any objects that depend on them, from your sketch. Use **Display | Hide**[167] instead if you don't want to remove dependent objects from your sketch, or if you may want to make the selected objects visible again at a later time.

If you cut objects and paste them into the same sketch or a different sketch, Sketchpad pastes the actual selected objects.

If you cut objects and paste them into a different application, Sketchpad pastes a picture of the selected objects.

Note: Pictures on the Clipboard

On Macintosh, when you cut objects to paste into another application, Sketchpad places a PDF picture of the selected objects on the Clipboard. PDF is a higher-resolution, better-quality picture than the older PICT format; however, older applications may not recognize it.

In Windows, when you cut objects to paste into another application, Sketchpad places an EMF+ (Enhanced Metafile) picture of the selected objects on the Clipboard. EMF+ produces a better-quality picture than the older WMF format. For older applications that don't recognize EMF+, Sketchpad also places a WMF picture and a DIB (device-independent bitmap) picture. Neither of these formats produces as high-quality an image as EMF+.

If you cut while you're editing a caption[56] or other text[120], **Cut** puts the selected text on the clipboard. It can then be pasted back into the same sketch, a different sketch, or into another application.

Note: Text on the Clipboard

When you cut text or a text object (caption, parameter, measurement, calculation, button, or function) to the clipboard and paste it into a different application, the resulting text on the clipboard uses Unicode to incorporate appropriate mathematical symbols. Such objects are available on the clipboard in several formats in order to make available as much of the mathematical information and formatting as possible. The program into which the clipboard text is pasted determines which format is used.

- The original text, with full mathematical formatting, is used when you paste into Sketchpad (version 5) or when you paste into Fathom Dynamic Data (version 2).

- Unicode text is used when you paste into modern text editors (Word, Wordpad, Notepad, BBEdit, TextEdit, etc.). This includes any Unicode mathematical symbols, but not styling (such as bold and italics), layout (such as fractions, overbars) or custom symbols not in Unicode.

- Non-Unicode text is used when you paste into older programs. Various Unicode symbols, such as the parallel or perpendicular symbols, may be converted into plain text.

*See also:*
  *Copy command* 146
  *Paste command* 147
  *Clear command* 147
  *Advanced Graphics Export* 320

## 3.2.4 Copy

This Edit 144 menu command places a copy of each selected 106 object on the clipboard. The contents of the clipboard can then be pasted 147 into the same sketch, a different sketch, or into another application.

The keyboard shortcut for **Copy** is Ctrl+C (Windows) or ⌘C (Mac).

If you copy objects and paste them into the same sketch or a different sketch, Sketchpad pastes the actual selected objects.

If you copy objects and paste them into a different application, Sketchpad pastes a picture of the selected objects.

Note: Pictures on the Clipboard

On Macintosh, when you copy objects to paste into another application, Sketchpad places a PDF picture of the selected objects on the Clipboard. PDF is a higher-resolution, better-quality picture than the older PICT format; however, older applications may not recognize it.

In Windows, when you copy objects to paste into another application, Sketchpad places an EMF+ (Enhanced Metafile) picture of the selected objects on the Clipboard. EMF+ produces a better-quality picture than the older WMF format. For older applications that don't recognize EMF+, Sketchpad also places a WMF picture and a DIB (device-independent bitmap) picture. Neither of these formats produces as high-quality an image as EMF+.

If you copy a single table 39, you can paste the table's data into another application such as Fathom Dynamic Data or Microsoft Excel.

If you copy while you're editing a caption 56 or other text 120, **Copy** puts the selected text on the clipboard. It can then be pasted back into the same sketch, a different sketch, or into another application.

Note: Text on the Clipboard

When you copy text or a text object (caption, parameter, measurement, calculation, button, or function) to the clipboard and paste it into a different application, the resulting text on the clipboard uses Unicode to incorporate appropriate mathematical symbols. Such objects are available on the clipboard in several formats in order to make available as much of the mathematical information and formatting as possible. The program into which the clipboard is pasted determines which format is used.

- The original text, with full mathematical formatting, is used when you paste into Sketchpad (version 5) or when you paste into Fathom Dynamic Data (version 2).

- Unicode text is used when you paste into modern text editors (Word, Wordpad, Notepad, BBEdit, TextEdit, etc.). This includes any Unicode mathematical symbols, but not styling (such as bold and italics), layout (such as fractions, overbars) or custom symbols not in Unicode.

- Non-Unicode text is used when you paste into older programs. Various Unicode symbols, such as the parallel or perpendicular symbols, may be converted into plain text.

*See also:*
  *Cut command* 145
  *Paste command* 147
  *Clear command* 147
  *Advanced Graphics Export* 320

## 3.2.5 Paste

This Edit 144 menu command pastes the contents of the clipboard into the active sketch.

> The keyboard shortcut for **Paste** is Ctrl+V (Windows) or ⌘V (Mac).

If the clipboard contains sketch objects, these objects are inserted into the sketch.

If the clipboard contains a picture 18, the picture is inserted into the sketch. When you paste a picture, if you have one, two, or three points selected, the picture will be attached to those points 19.

If you have a picture selected, or a transformed image of a picture, the command becomes **Paste Replacement Picture** and the pasted picture from the clipboard replaces the original picture.

> You can also replace a picture by dragging a picture from another application and dropping it onto the existing picture.

Note: Transparency when Pasting Pictures

> When you paste a picture, pure white portions of the picture are fully transparent, so that the background of the sketch (including other objects behind the picture) shows through. Such transparent areas remain fully transparent no matter how you set the opacity of the picture.

> To preserve white portions of the picture as white rather than transparent, hold the Shift key while choosing **Edit | Paste Picture** or dropping the picture into your document. (Once you've pasted the picture, use its Opacity Properties 92 to adjust the opacity of the entire picture.)

If the clipboard contains text and you're editing a caption 56 or other text, the text from the clipboard is inserted into the text you're editing.

*See also:*
  *Cut command* 145
  *Copy command* 146
  *Clear command* 147

## 3.2.6 Clear

This Edit 144 menu command removes from the sketch any selected object 106, as well as any objects which depend on it. The removed objects are not placed on the clipboard.

> Pressing the Delete or Backspace key performs the same action as choosing **Edit** 144 | **Clear.**

This command completely removes the selected objects, and any objects that depend on them, from your sketch. Use **Display | Hide** 167 instead if you don't want to remove dependent objects from your sketch, or if you may want to make the selected objects visible again at a later time.

*See also:*
*Cut command* 145
*Copy command* 146
*Paste command* 147

## 3.2.7    Action Buttons

The **Edit** 144 | **Action Buttons** submenu contains commands for creating various kinds of action buttons 66. Action buttons are sketch objects that you can press to perform a previously defined action.

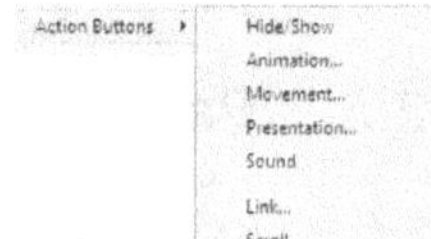


Sometimes you'll create an action button for your own convenience in working with a sketch, and sometimes you'll create a button to help present a mathematical feature of your sketch to someone else.

For example, instead of hiding a group of objects by selecting them and choosing **Display | Hide Objects** 167, create a Hide/Show 149 button that hides the objects with a single click. Instead of using the **Arrow** 104 tool to drag point $A$ toward point $B$, create a Movement 150 button to do it for you.

Most of these commands display a Properties 158 dialog box panel that allows you to specify certain details about the action the button will perform.

| Command | Prerequisites | Action | Button Description and Properties |
|---|---|---|---|
| **Hide/Show** 149 | One or more objects | Hide or show the selected objects | Hide/Show Buttons and Properties 67 |
| **Animation** 149 | One or more geometric objects or parameters | Animate the selected objects | Animation Buttons and Properties 69 |
| **Movement** 150 | One or more pairs of points or values. The first object of each pair must be free to move or vary. | For a pair of points, move the first point toward the second; for a pair of values, vary the first value toward the second | Movement Buttons and Properties 71 |

| Presentation [150] | One or more action buttons | Present the action buttons simultaneously or sequentially | Presentation Buttons and Properties [72] |
|---|---|---|---|
| Sound [150] | One or two functions or function plots | Make a sound defined by the function(s) | Sound Buttons [74] (No Properties) |
| Link [151] | None | Link to another page, another sketch, or a location on the Internet | Link Buttons and Properties [75] |
| Scroll [151] | One point | Scroll the window based on the location of the point | Scroll Buttons and Properties [76] |

### 3.2.7.1 Hide/Show

| |
|---|
| Hide Point A |
| Show Objects |

***Selection prerequisite:*** *One or more objects*

This Edit | Action Buttons [148] submenu command creates an action button [66] that hides or shows the selected object(s).

Hold the Shift key while choosing this command to create two different buttons: one that always hides and one that always shows. (With the Shift key down, the command changes from **Hide/Show to Hide & Show.**)

After you create the button choose **Edit | Properties | Hide/Show** [67] to view and change the button's properties.

### 3.2.7.2 Animation

| |
|---|
| Animate Points |

***Selection prerequisite:*** *One or more objects that can be animated (Only geometric objects* [2] *and parameters* [36] *can be animated* [267].*)*

This Edit | Action Buttons [148] submenu command creates an action button [66] that animates the selected object(s). When you choose this command, the Animate [69] panel of the Properties [158] dialog box appears, allowing you to set the speed and direction for each animated point and for each animated parameter.

After you've created the button, choose **Edit | Properties | Animate** [69] to display the panel again to make further adjustments.

*See also:*

*Principles of Animation* 267

### 3.2.7.3   Movement

| Move A → B |

***Selection prerequisite:*** *One or more pairs of points or values. In a pair of values, the first value must be a parameter* 36.

This **Edit | Action Buttons** 148 submenu command creates a Movement button 71 that moves the first object of each selected pair (the moving point or parameter) toward the second object (the destination point or value). When you choose this command, the Move 71 panel of the Properties 158 dialog box appears, allowing you to set the speed and behavior of the movement. After you've created the button, choose **Edit | Properties | Move** 71 to display the panel again to make further adjustments.

> If the command is not available, make sure you've selected an even number of objects, that each pair is a pair of either points or values, and that the first object of any pair of values is a parameter.

### 3.2.7.4   Presentation

| Present 2 Actions |
| Sequence 2 Actions |

***Selection prerequisite:*** *One or more action buttons*

This **Edit | Action Buttons** 148 submenu command creates a Presentation button 72 that activates the selected buttons' actions, either simultaneously or sequentially, to form a presentation.

When you choose this command, the Presentation 72 panel of the Properties 158 dialog box appears, allowing you to specify whether to present the selected actions simultaneously or sequentially, as well as to set other properties of the Presentation.

After you've created the button, choose **Edit | Properties | Presentation** 72 to display the panel again to make further adjustments.

### 3.2.7.5   Sound

frequency = | 440 |

$f(x) = \sin(2 \cdot \pi \cdot frequency \cdot x)$

| Hear Function f |

***Selection prerequisite:*** *One or two functions or function plots*

This **Edit | Action Buttons** 148 submenu command creates a Sound button 74 that plays the sound defined by the selected function(s).

The volume is determined by the amplitude of the function, and the pitch is determined by the period.

> If there are two selected functions, the first plays on the left channel and the second on the right channel.

Note

The amplitude of the function should be between −1 and 1. Larger amplitudes may result in clipping and distortion of the sound.

### 3.2.7.6 Link

| Page 2 |
| http://www.keypress.com/ |

*Selection prerequisite: None*

This Edit | Action Buttons 148 submenu command creates a Link button 75 that links to a different page in the current document, or to a web site or other location defined by a URL.

When you choose this command, the Link panel 75 of the Properties 158 dialog box appears, allowing you to determine whether the button links to a different document page or to a URL.

If the button links to a document page, you can also specify an action button 148 on that page that will be activated when the link occurs.

After you've created the button, choose **Edit | Properties | Link** 75 to display the panel again to make further adjustments.

### 3.2.7.7 Scroll

| Scroll |

*Selection prerequisite: One point*

This Edit | Action Buttons 148 submenu command creates a Scroll button 76 that scrolls the window based on the position of the selected point.

When you choose this command, the Scroll panel 76 of the Properties 158 dialog box appears to allow you to determine whether the button scrolls the window to put the selected point at the top-left corner of the window, or to put the selected point in the center of the window.

After you've created the button, choose **Edit | Properties | Scroll** 76 to display the panel again to change how the button scrolls.

Use Scroll buttons when you want to be able to "jump" to a point anywhere in the scrollable plane.

## 3.2.8 Select All

This Edit 144 menu command selects all objects that match the active tool in the Toolbox 102.

The keyboard shortcut for **Select All** is Ctrl+A (Windows) or ⌘A (Mac).

If a **Selection Arrow** 104, **Information** 124, or **Custom** 125 tool is active, this command selects all objects 2 in the sketch.

If another tool from the Toolbox 102 is active (**Point** 112, **Compass** 113, **Segment, Ray, Line** 115, **Polygon** 118, **Text** 120, or **Marker** 122), all matching objects are selected. For example, if the **Ray** tool is active, then the command becomes **Select All Rays**.

*See also:*

### 3.2.9 Select Parents

This Edit 144 menu command selects the parents 80 of each selected object.

The keyboard shortcut for **Select Parents** is Alt+Up (Windows) or ⌘↑ (Mac). (Think "Up the family tree.")

The parents of an object are those objects upon which the object directly depends. For example, a segment's parents are the endpoints used to define it and a midpoint's parent is the segment on which it's constructed.

If the selected object has no parents (in other words, if it's an independent object), it remains selected. If the parents of the selected object are hidden, then the selected object is deselected (leaving nothing selected).

*See also:*


### 3.2.10 Select Children

This Edit 144 menu command selects the children 80 of each selected object.

The keyboard shortcut for **Select Children** is Alt+Down (Windows) or ⌘↓ (Mac). (Think "Down the family tree.")

The children of an object are those objects that directly depend on the object. For example, a circle constructed by the **Circle by Center+Radius** 183 command is a child of both the point and the segment used to define it.

If the selected object has no children, it remains selected. If the children of the selected object are hidden, then the selected object is deselected (leaving nothing selected).

*See also:*


### 3.2.11 Crop Picture to Polygon

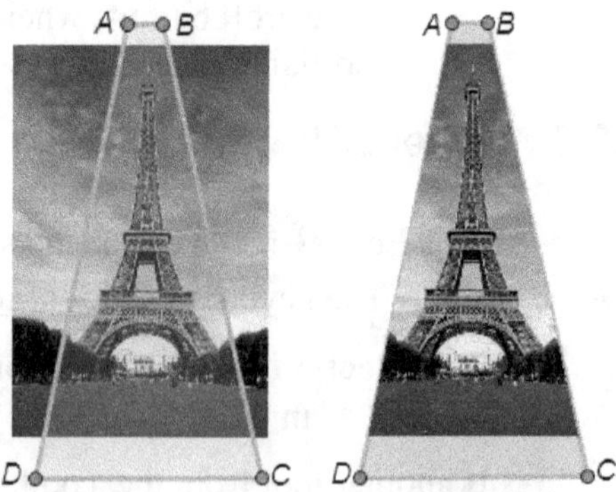


**Selection prerequisite:** *One picture* 18 *and one polygon* 8

This Edit 144 menu command constructs a picture that is the image of a given picture cropped to its intersection with a given polygon. Use this command to isolate interesting portions of larger pictures. (For example, you could isolate some pattern or shape you'd like to tessellate using transformations.) Dynamically change the cropped picture by changing how the cropping polygon overlaps the original picture (for example, by dragging the polygon's vertices). A cropped picture does not exist when its cropping polygon doesn't intersect the picture.

When you create a cropped picture, Sketchpad hides the original picture and places the cropped picture on a layer above the polygon, so you can better see the cropped result. You can reveal the original picture using **Display | Show All Hidden** 167 or the **Information** 124 tool.

The **Crop Picture to Polygon** command is object-sensitive, so it may read **Crop Picture to Triangle** or **Crop Picture to Pentagon,** depending on the selected polygon.

## 3.2.12 Split/Merge

These Edit 144 menu commands allow you to alter the relationships 80 of existing objects by splitting points from their parents, by merging points either with other points or onto paths, and by merging pictures or text to points. These commands allow you to fix construction mistakes, to make significant changes in a sketch without starting over, to explore constructions 156, and to modify geometric and mathematical investigations in flexible and powerful ways.

### ▼ Split a Point from Its Parents

You can split a midpoint 176, point on a path 176, point of intersection 177, or plotted point 239 from its parents. Depending on the relationships, the command changes. (For instance, it might **Split Midpoint from Segment, Split Intersection from Path Objects,** or **Split Plotted Point from Coordinate System.**)

To split a transformed image point from its pre-image point, you must hold the Shift key while choosing the command.

If the selected point is a point on a path 89, it's removed from its path. If the selected point is an intersection point, it's split from the intersecting objects. If the selected point is a plotted point, it's split from the coordinate system. In each of these cases, it becomes an independent point and can be dragged anywhere.

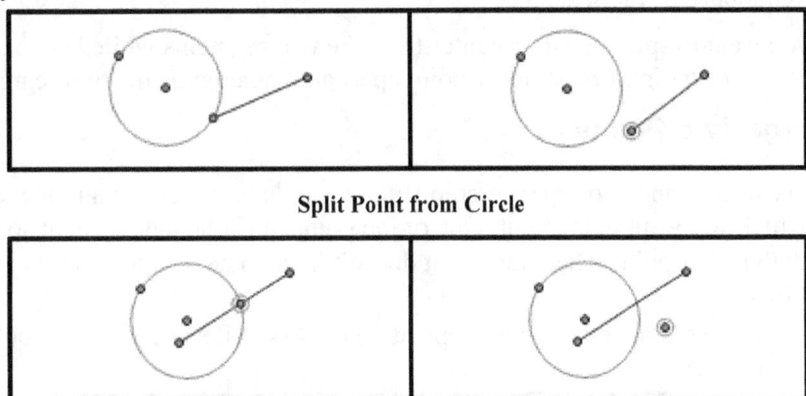

**Split Point from Circle**

**Split Intersection from Path Objects**

### ▼ Split a Point from Its Definition

You can split a transformed image point from its definition. Hold the Shift key to enable this command.

The point is split from its pre-image point, and becomes an independent point that can be dragged anywhere. In the first illustration below, the point on the right is a reflected image. In the second illustration, it has been split from its definition as a reflected image, and can be dragged freely.

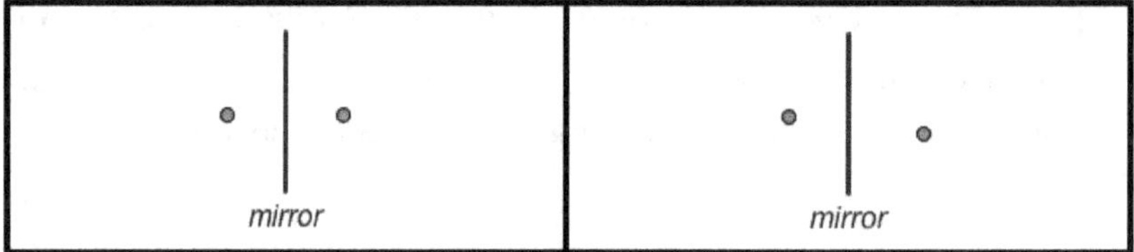


**Split Point from Definition**

### ▼ Split a Point Apart

You can split an independent point with more than one child into multiple points, one for each child. For example, if you have a point that is the center of a circle and is also the endpoint of two segments, you can split it so that the circle center and segment endpoints are now three separate, unrelated points. Select the point you wish to split and choose **Edit** 144 | **Split Point.** The point splits into two or more separate points a small distance from each other.

To split a point apart, you must select an independent point with two or more children.

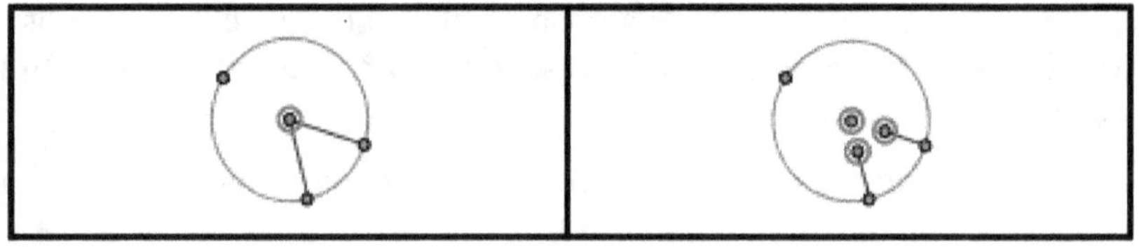

**Split Point**

In this example, the center point has three children: the circle and each of the two segments. The **Split Point** command results in three points: one is the center of the circle and the other two are the endpoints of the segments.

If you want to split the circle center from the two segments while leaving the segments with a common endpoint, first split the point apart and then merge the two segment endpoints.

### ▼ Merge Two Points

You can combine two separate points into a single point. Select an independent point and the point to which you want to merge it. One of the points must be independent so that it's free to merge with the other point. The other point doesn't have to be independent, but it must not depend 80 on the first.

> If the second point could depend on the first, after merging, it would be defined in terms of itself!

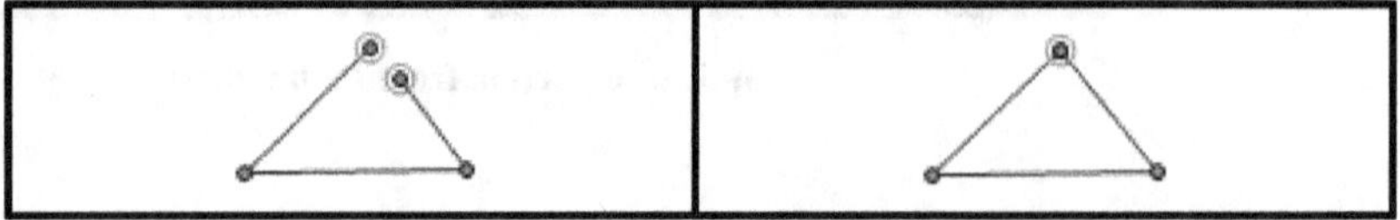

**Merge Points**

Merging two points is a handy way to fix any mistakes you make while using drawing tools.

### ▼ Merge a Point to a Path

You can merge a point [2] to a path [89] (straight object [5], circle [6], arc [7], interior [8], point locus [10], or function plot [52]) using the **Merge** command. Select an independent point and a path that doesn't depend on that point. The point must be independent, and the path must not depend [80] on the point. In this way, for example, you can merge an endpoint of a segment to another segment. After it's merged, the point is attached to the segment and can move along the segment but cannot leave it (unless you later split it from that segment).

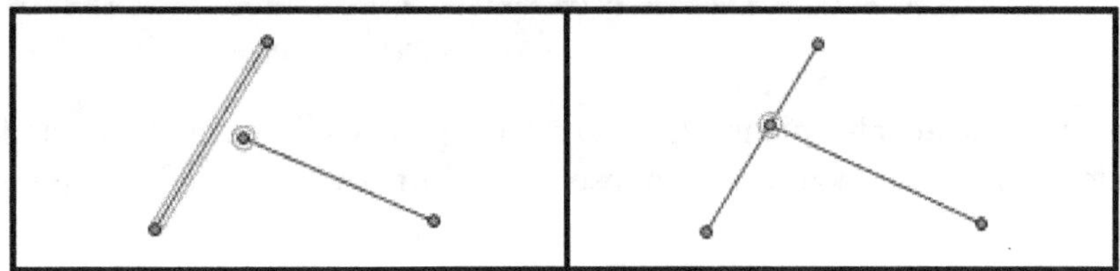

**Merge Point to Segment**

### ▼ Merge Text to a Point

You can merge a text object — a caption [56], measurement [36], calculation [39] or function [45] — to a point. To enable this command, hold the Shift key while choosing the **Merge** command.

1. Select one text object and one point.

2. While holding down the Shift key, choose **Edit | Merge Text to Point.**

A merged copy of your selected text appears, centered on the point you selected. As you move the point (by dragging it [104], animating it [171], or dragging or animating its parents [80]), the merged text moves with it; and as you drag the merged text, the point moves with it. (You may want to hide [167] the point and/or the original measurement to make your sketch more attractive.)

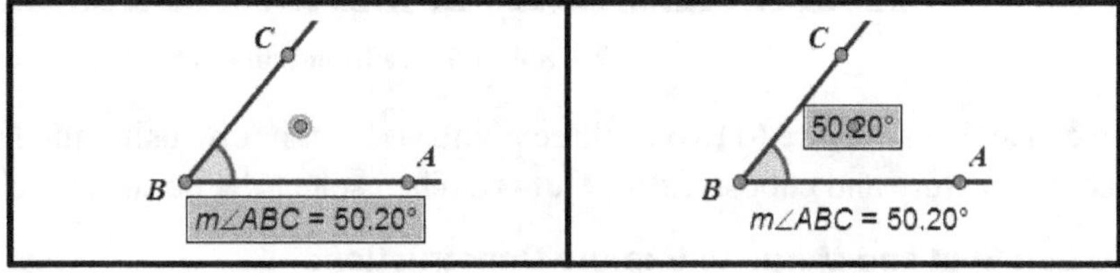

**Merge Text to Point**

A caption merged to a point can be used to provide a richer description of the point (or the object on which the point is constructed) than can be entered in a label, because you can use the style buttons and Symbolic Notation tools from the Text Palette [270]. A measurement merged to a point can provide a useful visualization of measured properties. For instance, you can merge the measure of an angle to that angle's vertex or the length of a segment to a (hidden) point on that segment. Note that the copy of your original text merged to the point is not the measurement itself, however. It's only a display copy of your original measurement. If you want to use the original measurement elsewhere in your sketch (in a calculation, for instance), you should not hide it after merging its text to a point.

Because the merged text is a copy of the original text object, there's no need to split the text from the point. Instead, make sure that both the original text object and the point are visible, and then delete the merged text object.

### ▼ Merge a Picture to a Point

You can merge a picture[18] to a point. Select both the picture and the point, and choose **Edit | Merge Picture to Point.**

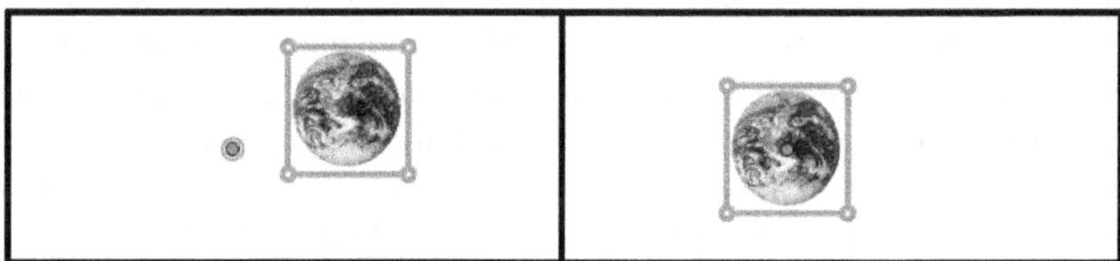

**Merge Picture to Point**

You can also attach a picture to two or three points[19], but not by using the **Merge** command. Instead, select the picture and choose **Edit | Cut**[145]. Then select the points and choose **Edit | Paste**[147].

### ▼ Split a Picture from One, Two or Three Points

You can split a picture[18] from a point. Select a picture attached to one point and choose **Edit | Split Picture from Point.**

The picture is separated from the point and no longer moves with it.

You can split a picture[18] from two or three points. Select a picture attached to two or three points, hold the Shift key, and choose **Edit | Split Picture from Points.**

The picture is separated from the points and no longer moves or changes shape with them.

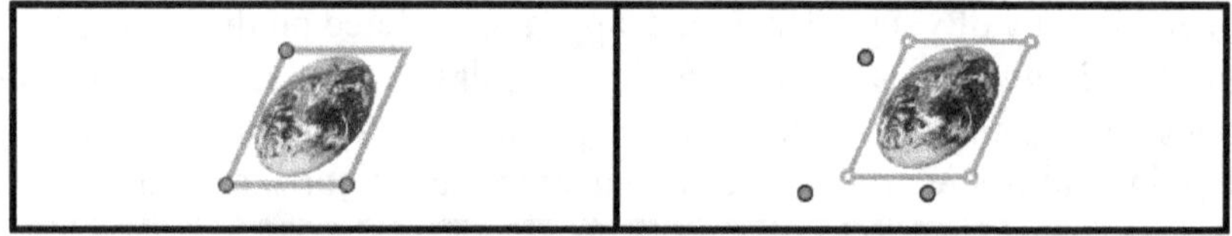

**Split Picture from Point(s)**

You can attach a picture to two or three points[19], but not by using the **Merge** command. Instead, select the picture and choose **Edit | Cut**[145]. Then select the points and choose **Edit | Paste**[147].

#### 3.2.12.1  How to Use Split and Merge to Explore Constructions

The **Edit | Split/Merge**[153] commands are often useful when you're investigating the behavior of a particular construction and you want to see what happens in the case of a slightly different construction. In this extended example, you'll construct the diagonals of a general quadrilateral, and then investigate what happens when you turn the quadrilateral into a parallelogram, a rectangle, and a rhombus.

1. To begin, construct quadrilateral *ABCD,* its two diagonals, and the point of intersection *E* of the diagonals. Then measure[216] the distances[218] and angle [218] shown in the figure at right. Drag the vertices around to see if there's any relationship between the various measurements.

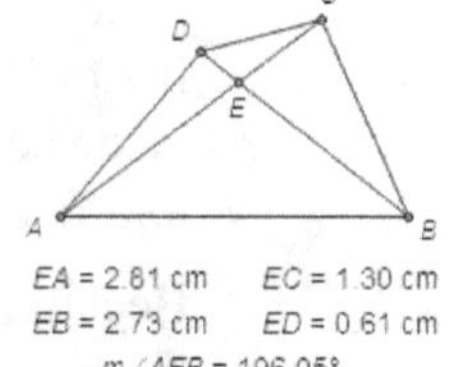

2. Now turn the quadrilateral into a parallelogram. Construct two parallel lines as shown at right: one through *D* parallel[179] to the segment connecting *A* and *B*, and another through *B* parallel to the segment connecting *A* and *D*. Also construct the intersection[177] of the parallels.

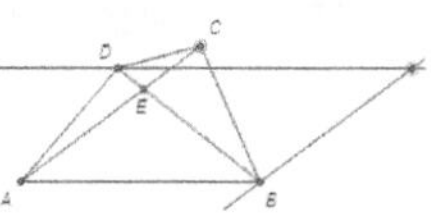

3. Select both point *C* and the intersection, and choose **Edit | Merge Points**[153] to merge the vertex with the intersection point.

4. Hide[167] the parallel lines. Then drag vertices *A, B,* and *D*. How are the measurements now related?

5. Next turn the quadrilateral into a rectangle by making $\angle BAD$ into a right angle. Construct a perpendicular[180] as shown at right and merge[153] point *D* to the perpendicular. Again, drag the vertices and watch the measurements. Do you notice anything different?

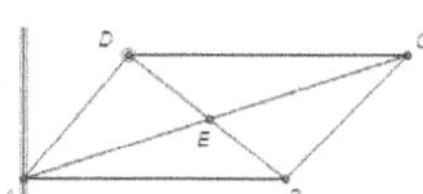

6. Finally, change your quadrilateral into a rhombus. First split[153] *D* from the perpendicular. Then construct a circle centered at [183]*A*[183] and passing through [183]*B*[183], and merge[153] *D* to the circle. Drag the vertices once more to see what relationships the measurements now reveal.

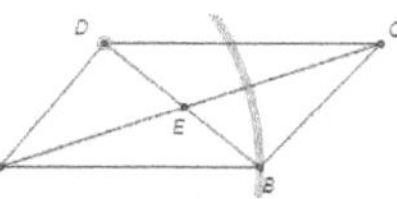

## 3.2.13  Edit Definition

This Edit[144] menu command allows you to edit the definition of a selected calculation[39], function[47], or numeric parameter[36]. When you choose this command, the Calculator[257] appears, allowing you to modify the value or expression.

Editing definitions allows you not only to correct any mistakes you might make, but also to explore a mathematical model over an unlimited range of cases.

This command's name changes to match the selected object.

| *Select:* | *For this command:* |
| --- | --- |
| One calculation[39] | **Edit Calculation** |
| One function[45] | **Edit Function** |
| One parameter[36] | **Edit Parameter** |

You can double-click a calculation or function with the **Arrow**[104] tool as a shortcut for **Edit Definition.**

When you edit the definition of a parameter, the Calculator appears. You can either enter a new value, or you can enter an expression and change the parameter into a calculation.

## 3.2.14  Properties

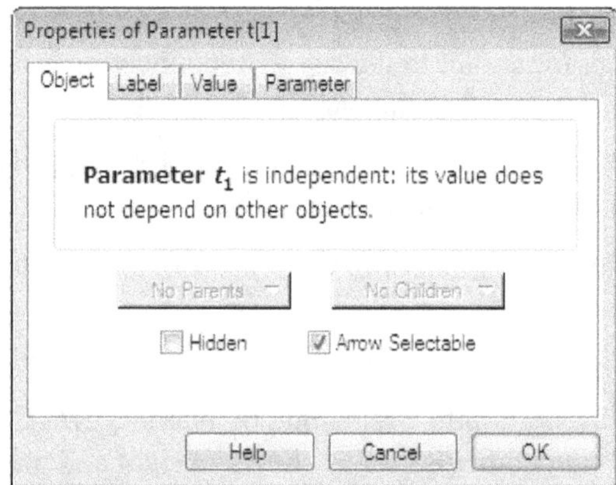

This Edit|144| menu command displays the Properties dialog box|273| and allows you to change a variety of properties of a single selected object.

The keyboard shortcut for **Properties** is Alt+? (Windows) or ⌘? (Mac).

There are two other ways to view an object's properties.

- Use the Context|245| menu. In Windows, right-click the object. On a Mac, hold down the Ctrl key while you click the object. The **Properties** command appears in the resulting Context menu (directly under your mouse if possible).

- Use the **Information**|124| tool. Click the tool on the object, and then click the object's name in the balloon that appears.

The Properties dialog box can include these panels:

| Panel | Objects |
|---|---|
| Object |79| | All objects |
| Label |82| | All objects that can show labels |
| Value |94| | Measurements |36|, calculations |39|, and parameters |36| |
| Parameter |38| | Parameters |36| |
| Table |41| | Tables |39| |
| Iteration |35| | Iterations and iterated objects |24| |
| Opacity |92| | Polygons |8|, circle interiors |8|, arc interiors |8|, pictures |18|, iterations |24| and loci |10| of these objects, and angle markers |60| |
| Axis |44| | Axes |41| |

| Function [51] | Data-defined functions [51] |
|---|---|
| Plot [95] | Loci (including families of functions, families of curves, and families of loci) [10] and sampled transformed paths [89] |
| Plot [97] | Function plots [52] and parametric plots [54] |
| Plot [98] | Sampled transformed pictures [23] |
| Marker [65] | Angle markers [60] and tick marks [63] |
| Hide/Show [67] | Hide/Show buttons [67] |
| Animate [69] | Animation buttons [69] |
| Move [71] | Movement buttons [71] |
| Presentation [72] | Presentation buttons [72] |
| Link [75] | Link buttons [75] |
| Scroll [76] | Scroll buttons [76] |

## 3.2.15  Preferences

This Edit [144] menu command allows you to change a variety of settings that determine how Sketchpad works.

Use the Context [245] menu as a shortcut to **Preferences.** Right-click (Windows) or Ctrl+click (Macintosh) in a blank area to invoke the Context menu.

(Hold the Shift key while pulling down the Edit [144] menu to change **Preferences** to **Advanced Preferences** [160]. The **Advanced Preferences** command allows you to control additional aspects of Sketchpad's operation that rarely need to be changed.)

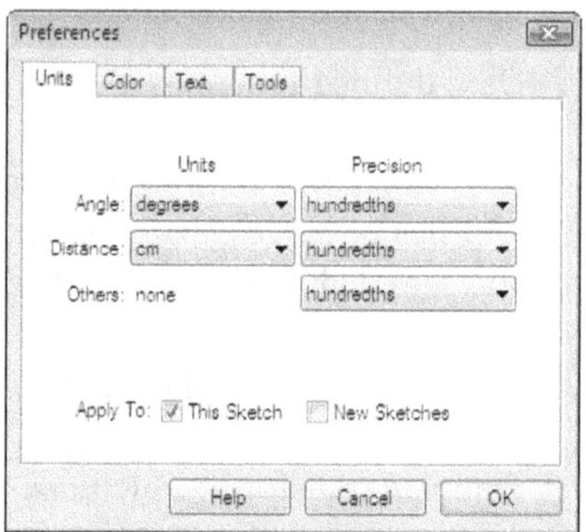

Here are the Preferences panels and the types of settings each controls:

| *Panel* | *Settings* |
|---|---|
| Units 276 | Units and precision of values 92 |
| Color 277 | Color 86 of objects and background, and fading of traces |
| Text 279 | Display of labels 82, behavior of text palette 270, and mathematical italicization |
| Tools 280 | Various settings for the **Arrow** 104, **Polygon** 118, **Marker** 122, and **Information** 124 tools |

When you make changes in Sketchpad's Preferences, use the checkboxes at the bottom of the dialog box to decide whether your changes will apply to the current sketch only, to new sketches only, or to both the current sketch and new sketches. Choose **Apply to: This Sketch** to have your changes affect newly constructed objects in the current sketch. Choose **Apply to: New Sketches** to have your changes affect all new sketches (including new blank pages you add to the current document).

If you want some changes to apply only to the current sketch and some to apply only to new sketches, you'll need to open the Preferences dialog box twice.

*Subtopic:*
  *Advanced Preferences* 160

### 3.2.15.1  Advanced Preferences

This Edit 144 menu command allows you to change a variety of advanced settings 281 that determine how Sketchpad works. Change these settings only after careful consideration; most users never need to modify them or need to modify them only once.

To use this command, hold down the Shift 295 key before activating the Edit 144 menu, causing the **Preferences** 159 command to become **Advanced Preferences.**

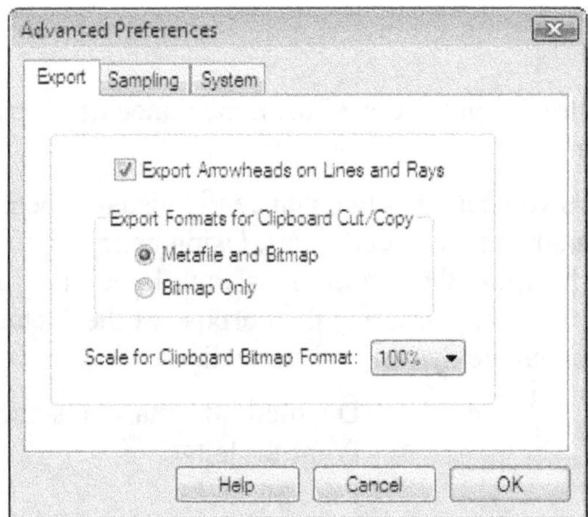

Here are the Advanced Preferences panels and the types of settings each controls:

| *Panel* | *Settings* |
| --- | --- |
| Export 282 | Format of exported images 320 |
| Sampling 283 | Default and maximum number of samples 95 for loci 10, function plots 52, and iterations 24 |
| System 284 | Animation speed 267, screen resolution, language for imported GSP3/4 documents, and color menu |

## 3.3     Display Menu

The Display menu allows you to control the appearance of objects in your sketch and of the tools you use to work with them.

With these commands you can greatly enhance the visual appeal of a sketch and its effectiveness in communicating the mathematics it embodies. Using appropriate line widths and colors, and hiding some objects while showing others help focus attention on the important parts of a sketch. Properly styled labels and captions help describe the purpose of the sketch and the mathematics behind it. Tracing and animation create dynamic visualizations of your sketch's underlying principles.

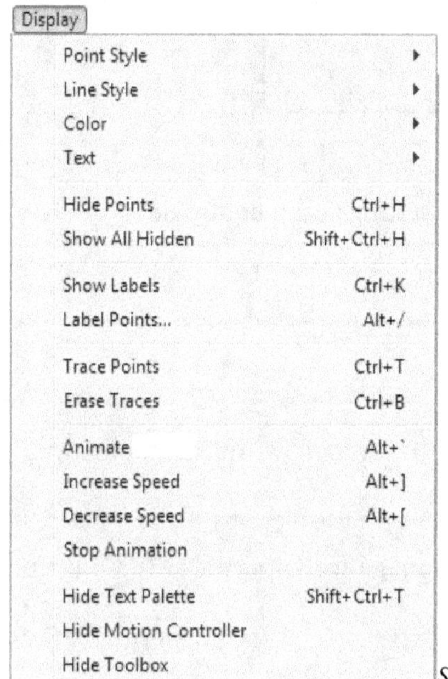


Detailed information is available for each command:
Point Style [162]
Line Style [163]
Color [164]
Text [166]
Hide Objects [167]
Show All Hidden [167]
Show/Hide Labels [168]
Label [168]
Trace [170]
Erase Traces [171]
Animate [171]
Increase Speed/Decrease Speed [172]
Stop Animation [172]
Show/Hide Text Palette [172]
Show/Hide Motion Controller [173]
Show/Hide Toolbox [173]

*See Also:*
*How to Show Just One Hidden Object [167]*
*How to Zoom In and Out on a Sketch [111]*

### 3.3.1     Point Style

***Selection prerequisite:*** *One or more objects, at least one of which must be a point [2], an iterated image [24] of a point, an angle marker [60], a dotted grid, or a discrete function plot [52] or point locus [10]*

This Display [162] menu command sets the point style of each selected object.

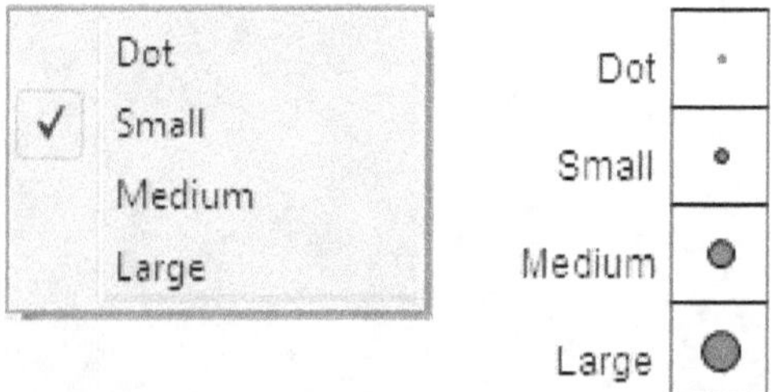


When all selected objects share a common point style, a checkmark appears in the submenu next to that style.

When you change the selected object's point style, Sketchpad remembers your chosen style for future similar objects in the current sketch.

To change an object's point style without changing the setting for future objects, hold down the Shift key while choosing the command. (You can also uncheck **Update automatically when restyling existing objects** in Text Preferences [279] to preserve the default point style.)

The Point Style setting has special results for angle markers [60], for dotted grids, and for discrete loci [10] and function plots [52].

| Point Style | Appearance of Discrete Loci and Function Plots | Appearance of Angle Markers | Appearance of Dotted Grid |
|---|---|---|---|
| Dot | 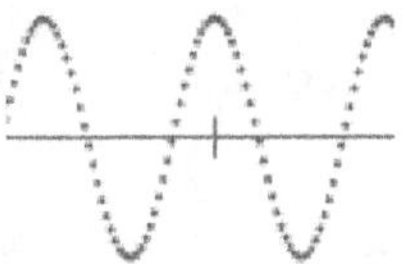 | 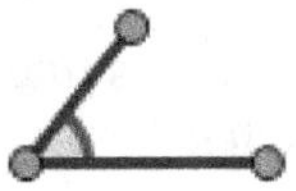 | |
| Small | 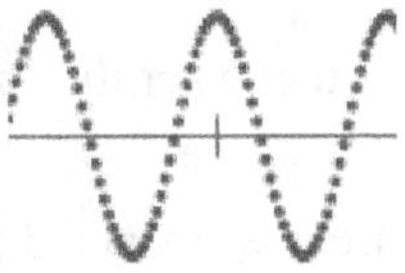 |  | |
| Medium | 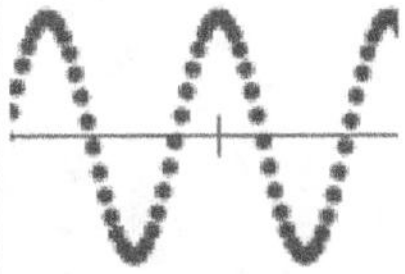 | 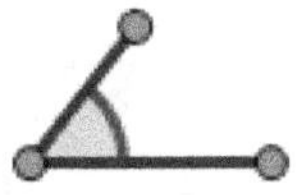 | |
| Thick | 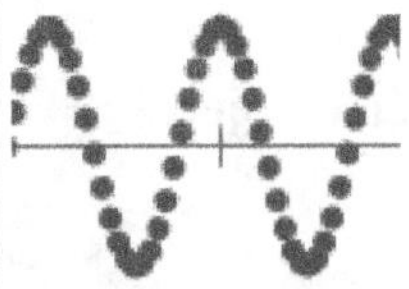 | 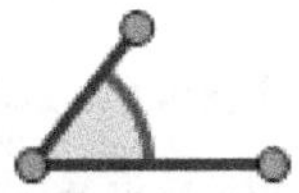 | |

*See also:*
  *Object Display Attributes* [86]

## 3.3.2   Line Style

***Selection prerequisite:*** *One or more objects, at least one of which must be a straight object* [90]*, circle* [6]*, arc* [7]*, framed polygon* [8]*, continuous function plot* [52] *or point locus* [10]*, an iterated image* [24] *of such an object, a tickmark* [60]*, or an angle marker* [60]*.*

This Display [162] menu command sets the line style of each selected object.

Line style includes both width (Hairline, Thin, Medium, Thick) and pattern (Solid, Dashed, Dotted).

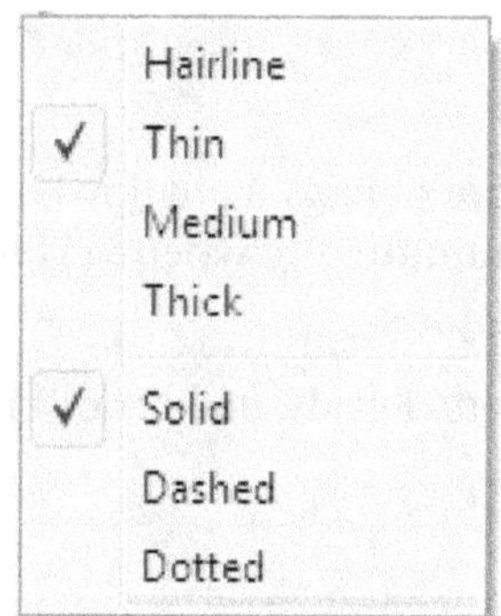

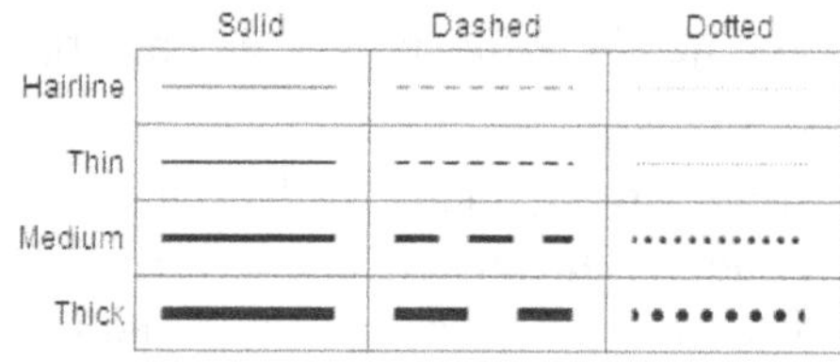

When all selected objects share a common line style, a checkmark appears in the submenu next to that style.

When you change the selected object's line style, Sketchpad remembers your chosen style for future similar objects in the current sketch.

To change an object's line style without changing the setting for future objects, hold down the Shift key while choosing the command. (You can also uncheck **Update automatically when restyling existing objects** in Text Preferences 279 to preserve the default line style.)

For tick marks, angle markers 60 , and framed polygons 8 you can set the line width, but you cannot set a pattern.

You cannot change the width of coordinate grid lines; they are always Hairline width. (But you can choose **Graph | Dotted Grid** 236 to change the coordinate grid to appear as a dotted grid instead of as hairlines.)

*See also:*
  *Object Display Attributes* 86
  *Dotted Grid command* 236

### 3.3.3   Color

This Display 162 menu command sets the color of each selected object.

1.  Select every object whose color you wish to change.

2.  Choose the desired color from **Display | Color.**

When you set the color of an object, Sketchpad remembers your chosen color for new objects of the same kind.

To change an object's color without changing the setting for future objects, hold down the Shift key while choosing the command. (You can also uncheck **Update automatically when restyling existing objects** in Text Preferences 279 to preserve the default color.)

If you set the color of a caption 56 that has different colors for different characters, choosing this command sets all of the characters to the new color you choose.

If you set the color of a labeled 168 geometric object, the color applies only to the object, not to its label.

> To change the color of an object's label, select the object and then choose the new color from the Text Palette 270.

Change the color choices available in the Color submenu by using the System panel 284 of the Advance Preferences 281 dialog box.

*Subtopics:*


*See also:*


## 3.3.3.1   Parametric Color

This **Display** 162 | **Color** 164 submenu command sets the color of selected objects based on the numeric value of either one or three selected values.

> Parametric Color lets you "color by numbers." The number is provided by a Sketchpad value (such as a measurement). When the value changes, the color changes, too.

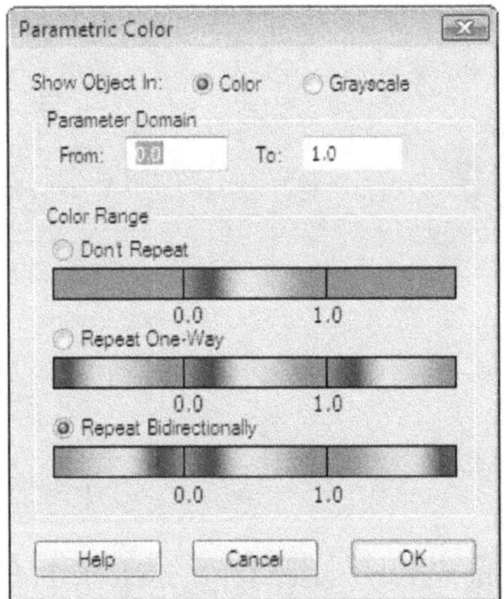
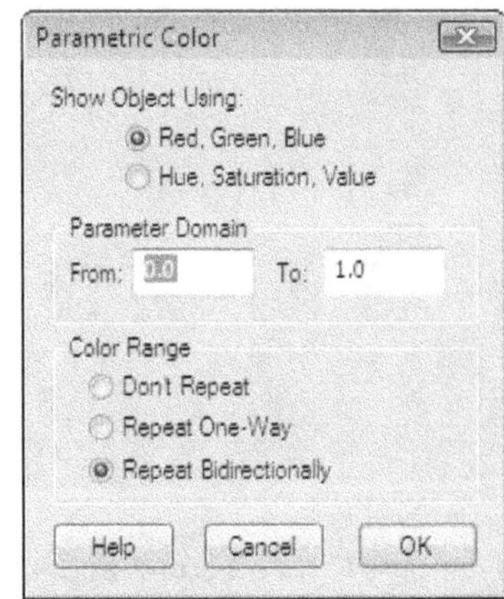

This command is available only when you have selected either one or three numeric values 92 ( measurements 36, calculations 39, or parameters 36), as well as one or more objects 2 that can be colored parametrically (points, circles, arcs, straight objects, interiors, angle markers, or tick marks).

> While you cannot color a locus 10 or an iterated image 24 parametrically, the locus or iterated image of a parametrically colored object will display the range of that object's colors.

If you have selected one measurement, that measurement is used to set a color from the spectrum (ranging from violet to deep red) or a grayscale color. You can set the **Parameter Domain:** the numeric interval that corresponds to one complete cycle of the available colors or shades. You can also set the **Color Range** so that the cycle doesn't repeat, repeats one-way, or repeats bidirectionally.

If you have selected three measurements, all three measurements are used to determine the object's color. As with a single color, you can set the **Parameter Domain** and **Color Range.** You can also decide whether the three measurements are interpreted as **Red, Green, Blue** settings or as **Hue, Saturation, Value** settings.

### 3.3.3.2    Other Color

This **Display** 162 | **Color** 164 submenu command sets the color of the selected objects to a color other than those that appear in the Color submenu. Your system's Color Picker 290 dialog box appears, allowing you to specify any color your computer can display.

> If your computer is set to display 256 or fewer colors, colors may not appear accurately, and you may not be able to select the exact color you want.

*See also:*
> *Edit the Color Menu in Windows* 285
> *Edit the Color Menu on Macintosh* 286
> *System Preferences* 284
> *Color Preferences* 277
> *Display Attributes* 86

## 3.3.4    Text

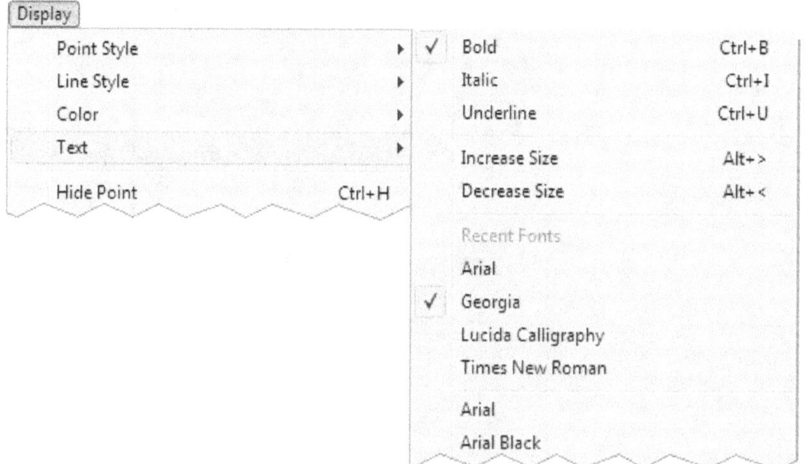

This Display 162 menu command sets the font and style used for the selected objects 106, and increases or decreases the font size.

Choose **Bold, Italic,** or **Underline** to change the text style for the selected objects, or use keyboard shortcuts for these commands: Ctrl+B, Ctrl+I, or Ctrl+U in Windows, or ⌘B, ⌘I, or ⌘U on Mac.

Choose **Increase Size** to increase the font size for each selected object to the next-larger standard font size. Choose **Decrease Size** to reduce the font size for each selected object to the next-smaller standard font size.

> The keyboard shortcut for **Increase Size** is Alt+> (Windows) or ⌘> (Mac). The shortcut for **Decrease Size** is Alt+< (Windows) or ⌘< (Mac).

Choose a font from the lower part of the menu. Recently-used fonts appear at the top of the font list.

To change an object's text style, size, or font without changing the setting for future objects, hold down the Shift key while choosing the command. (You can also uncheck **Update automatically when restyling existing objects** in Text Preferences 279 to preserve the default style, size, or font.)

Alternatively, use the Text Palette 270 to change the font, style, and size of text.

## 3.3.5    Hide Objects

This Display [162] menu command hides each selected object from view without changing its geometric role in the sketch. Use this command to hide objects that shouldn't be seen, but that are still required for geometric purposes in the sketch. When an object is hidden, it remains present in the sketch and can continue to define the position and behavior of other objects that remain visible.

The keyboard shortcut for **Hide** is Ctrl+H (Windows) or ⌘H (Mac).

If you have objects in your sketch that you don't want to see and that are not required for geometric reasons, it's better to clear those objects using the Backspace key or the **Edit | Clear** [147] command than to hide them.

*See also:*
  *Show All Hidden* [167]
  *Object Attributes* [86]
  *Object Properties* [79]
  *Hide/Show Button* [149]
  *How to Show Just One Hidden Object* [167]

## 3.3.6    Show All Hidden

This Display [162] menu command displays and selects all objects previously hidden in a sketch.

You can also show hidden objects by using the **Information** [124] tool or the Properties [158] dialog box, starting from a parent or child [80] of the hidden object. Use the Parents or Children list in the information balloon or the Object panel [79] of the Properties dialog box to navigate to the desired hidden object. Then uncheck the **Hidden** checkbox.

*See also:*
  *Object Properties* [79]
  *How to Show Just One Hidden Object* [167]
  *Hide/Show Button* [149]
  *Hide/Show Properties* [67]
  *Object Attributes* [86]

### 3.3.6.1    How to Show Just One Hidden Object

In a sketch containing many hidden objects, you may want to show just one or two of those hidden objects. The **Display | Show All Hidden** [167] command shows all the hidden objects, and once you've shown them, it can be a lot of trouble to reselect all of them in order to hide them again.

Here's a convenient way to show one particular object among many that have been hidden:

**1.** Choose **Display | Show All Hidden** [167].

All previously hidden objects appear and are selected.

**2.** Using a **Selection Arrow** [104] tool, click on the object you wish to remain visible, deselecting it.

The object you click is deselected; the other just-shown objects remain selected.

**3.** Choose **Display | Hide** [167].

The remaining selected objects are hidden, and the deselected object remains visible.

Another way to hide or show objects is by using the **Hidden** checkbox on the Object panel [79] of the Properties [158] dialog box. To display properties for a hidden object, you must first display properties for a parent or child [80] of the hidden object, then choose the desired object from the Parents or

Children pop-up menu.

Similarly, you can use the **Information** 124 tool starting from a parent or child 80 of the hidden object. Use the list of parents or children in the information balloon to navigate to the desired hidden object. Then uncheck the **Hidden** checkbox. (If the Information balloon doesn't show parents or children, use its Context 245 menu, by right-clicking or Ctrl-clicking, to show them.)

### 3.3.7  Show/Hide Labels

This Display 162 menu command shows or hides the label of each selected object. Most Sketchpad objects can show their labels, except for grids, measurements and calculations, captions, pictures, and loci of objects other than points.

To show or hide labels:

1.  Select every object whose label you want to show or hide.

2.  Choose **Display | Show Labels** or **Display | Hide Labels.**

Using this command is equivalent to clicking on an object with the **Text** 120 tool.

The keyboard shortcut for **Show/Hide Labels** is Ctrl+K (Windows) or ⌘K (Mac).

When you measure a quantity that depends on a particular object, the object's label may automatically be shown if it's needed to display the measurement. For instance, if you measure the angle defined by three points, the labels of those points may be shown. Use the Text panel 279 of Preferences 159 to choose whether labels are automatically shown.

An object is labeled automatically when the label is first needed for some purpose — because you've chosen the **Show Label** command, because you've clicked the **Text** tool on the object, or because you've measured some quantity that depends on the object. Thus the first point you label will become point *A,* even if it's the tenth point you've actually created in your sketch.

Although you can't show or hide the labels of measurements and calculations 92, you can control how they are named when they're displayed, using the Value panel 94 of the Properties 158 dialog box.

*See also:*
  *Text Tool* 120
  *Label Properties* 82
  *Text Preferences* 279
  *Value Properties* 94

### 3.3.8  Label Object

This Display 162 menu command allows you to create or edit the label of one or more selected objects.

The keyboard shortcut for **Label Object** is Alt+/ (Windows) or ⌘ / (Mac).

If you have selected a single object, the Label Objects command opens the Properties 273 dialog box for that object and displays the Label panel 82, where you can type a new label.

If you have selected more than one object, this command allows you to label all the selected objects, in a sequence.

If you hold the Shift key with no objects selected, the command becomes **Reset Next Labels** and allows you to reset the next labels that will used for all objects.

## ▼ Label Multiple Objects

If more than one object is selected, this command displays the Label Multiple Objects dialog box.

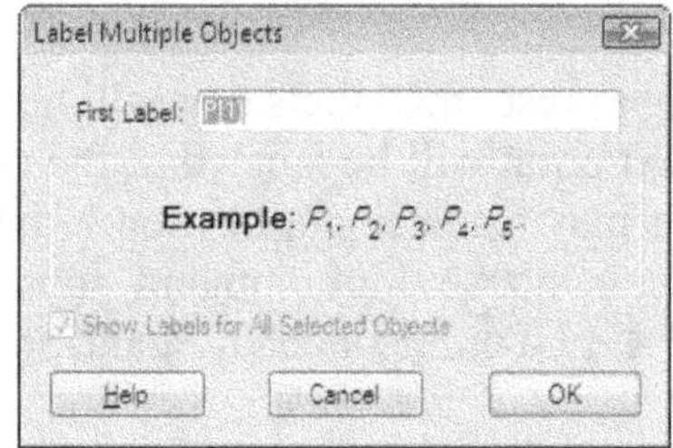

💡 Shortcut

If an object in your sketch already has the desired first label, you can select that object first, and then select the other objects to be relabeled. When you choose the **Label** command, you won't have to type anything into the dialog box.

**First Label:** Type the label you want to assign to the first of the selected objects. Sketchpad will create a sequence of labels to use for the remaining selected objects. The sequence is produced by varying the last element of the first label. The last element can be a letter, a number, or a subscript (indicated by a letter or number enclosed in square brackets).

**Show labels for all selected objects:** If some of the selected objects aren't already showing their labels, use this checkbox to determine whether or not the labels will be shown when you dismiss the dialog box.

Here are some examples of the labels Sketchpad will create, depending on what you type as the first label:

| *Type This:* | *To Produce This Sequence:* |
|---|---|
| *P* | *P, Q, R, S, T, ...* |
| *d* | *d, e, f, g, h, ...* |
| *P*1 | *P*1, *P*2, *P*3, *P*4, *P*5, ... |
| *P*[1] | $P_1, P_2, P_3, P_4, P_5, ...$ |
| *A*98 | *A*98, *A*99, *A*100, *A*101, *A*102, ... |
| *W* | *W, X, Y, Z*1, *Z*2, ... |
| 1-*a* | 1-*a*, 1-*b*, 1-*c*, 1-*d*, 1-*e*, ... |

If you want to produce a sequence of labels in which the element that varies is not the last element of the label, use a custom label sequence  starting with an equal sign and using "{...}" immediately following the element that should be varied. For instance, type =A{...}1 to generate $A_1, B_1, C_1, D_1$, and so forth.

### ▼ Reset Next Labels

The **Reset Next Labels** command appears when you hold the Shift key, and have no objects selected.

Choose this command to reset the labels that will be used by newly-labeled objects. If you choose this command, the next point will be labeled *A,* the next straight object will be labeled *j,* and so forth. Resetting labels may result in duplication of labels. For instance, if your sketch already has labeled points when you choose this command, subsequently labeling additional points may result in two points labeled *A,* two points labeled *B,* and so forth.

## 3.3.9   Trace

This Display 162 menu command turns tracing on or off for each selected object. If all the selected objects are currently being traced, a check mark appears next to the command. Choosing the command turns tracing off for each selected object. If no check mark appears, choosing the command turns tracing on for each selected object.

The keyboard shortcut for **Trace** is Ctrl+T (Windows) or ⌘T (Mac).

When an object is traced, it leaves behind a trail as it moves, no matter how the motion is caused — whether by dragging 104 the traced object, by dragging some object on which the traced object depends 80, or by animating 171.

Tracing can be useful for exploring how an object's movement is constrained or for creating interesting mathematical art.

The faster a traced object moves, the more spread out its trail; the slower the object moves, the denser its trail.

Use Color Preferences 277 to make traces gradually fade over time and eventually disappear. You can also control how fast traces fade — or turn off the fading of traces altogether. If fading is disabled, traces will remain on the screen indefinitely. To remove traces (whether fading or not) from the screen, choose **Erase Traces** 171.

Think of a trace as a temporary display of an object's locus 10. To construct a permanent locus, use the **Construct** 174 | **Locus** 190 command.

There's an important difference between erasing traces and deactivating tracing for an object or objects. Erasing traces (using the **Display | Erase Traces** 171 command) removes all existing traces from the screen but has no effect on whether objects will leave traces in the future. Turning tracing off for an object (using the **Trace** command) prevents that object from leaving traces as it moves in the future, but has no effect on any traces the object has already left on the screen. If you want to remove all existing traces an object has left behind, and also to prevent that object from leaving traces in the future, you must use both commands:

1. Select the object 106 and choose **Display** 162 | **Trace** to turn off future tracing for the object.

2. Choose **Display | Erase Traces** 171 to erase any existing traces from the screen.

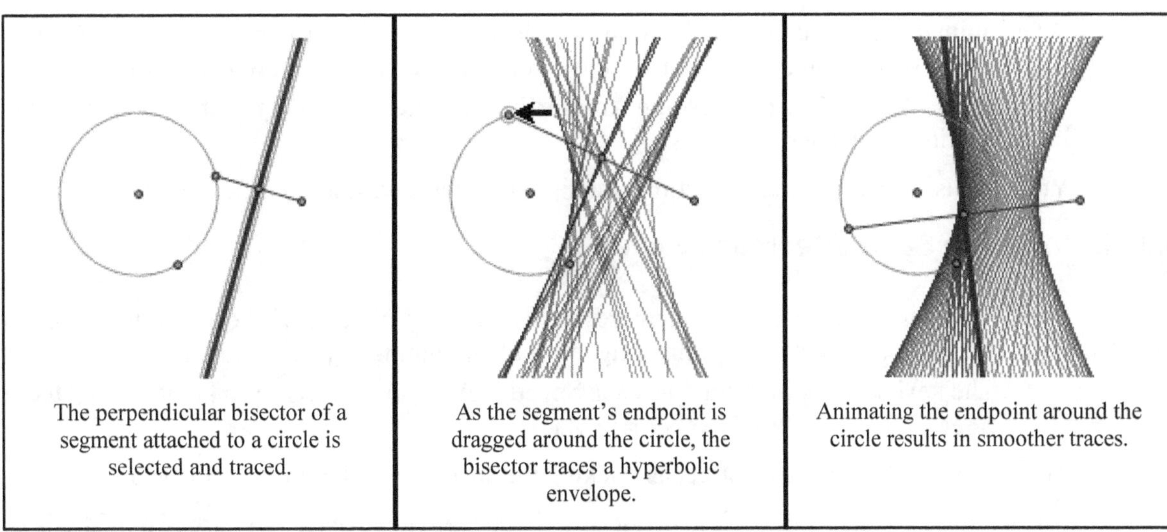

| The perpendicular bisector of a segment attached to a circle is selected and traced. | As the segment's endpoint is dragged around the circle, the bisector traces a hyperbolic envelope. | Animating the endpoint around the circle results in smoother traces. |

## 3.3.10 Erase Traces

This Display 162 menu command removes all visible traces from the screen.

The keyboard shortcut for **Erase Traces** is Shift+Ctrl+E (Windows) or Shift-⌘ E (Mac).

This command erases all traces immediately. If you want to control whether and how quickly traces fade, use the Color 277 panel of the Preferences 159 dialog box.

This command does not prevent traced objects from leaving new traces on the screen as they continue to move. To prevent an object from leaving new traces, you must select that object and turn tracing off for the object by choosing **Display | Trace** 170.

You can also use the Esc 294 key to erase traces.

## 3.3.11 Animate

This Display 162 menu command puts each selected geometric object into animated motion 267.

- Independent points move freely in the plane, in random directions.

- Points on paths 89 move along their paths. Points on straight objects and arcs move bidirectionally. Points on circles and interiors move around their paths.

- Other objects move by moving the objects upon which they depend (that is, their parental objects 80 ).

The keyboard shortcut for **Animate** is Ctrl+` (Windows) or ⌘ ` (Mac). (The ` symbol is called the *alternate quote symbol*.)

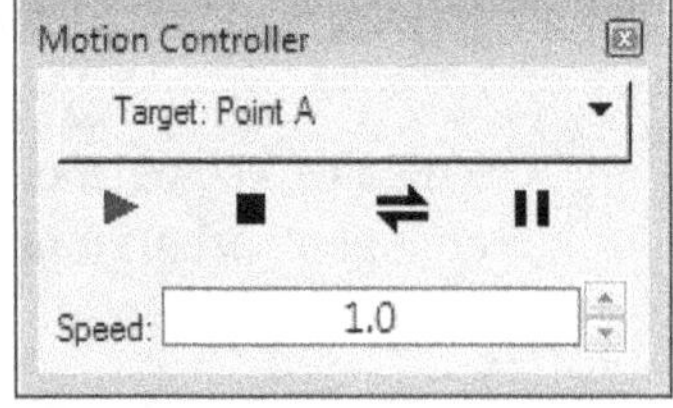

This command has the same effect as pressing the Animate 265 button on the Motion Controller 263. (The Motion Controller appears automatically when you choose the **Animate** command.)

When animations are running and nothing is selected, the **Animate** command becomes **Pause Animation.** When animation is paused, the command becomes **Resume Animation.**

> **Pause Animation** and **Resume Animation** have the same effect as the Pause button on the Motion Controller.

You can also create Animation action buttons[149] to start and stop specific animations.

### 3.3.12  Increase Speed/Decrease Speed

This Display[162] menu command increases or decreases by about 25% the animation speed of each object that is both selected and animating, or of all animating objects if nothing is selected.

> The keyboard shortcut for **Increase Speed** is Alt+] (Windows) or ⌘] (Mac); for **Decrease Speed** it's Alt+[ (Windows) or ⌘[ (Mac).

This command has the same effect as clicking the speed control buttons of the Motion Controller[263].

If it's difficult to select the object whose speed you want to change, there are two methods you can use:

- Select any moving point by choosing it from the Motion Controller's Target menu[265].

- Pause animation[265], select the desired object, and resume animation.

If nothing is selected, these commands change to **Increase All Speeds** and **Decrease All Speeds.**

The easiest way to control animation speed separately for multiple animating objects is to create an Animation action button[149].

*See also:*
> *Animate command*[171]
> *Stop Animation command*[172]


### 3.3.13  Stop Animation

This Display[162] menu command stops the motion of each selected moving object. If nothing is selected but at least one object in the sketch is moving, this command changes to **Stop All Motions** and stops the motion of every moving object in the sketch.

This command has the same effect as pressing the Stop[265] button on the Motion Controller[263].

You can also use the Esc[294] key to stop all motion.

*See also:*
> *Animate command*[171]
> *Animation Button*[149]
> *Movement Button*[150]

### 3.3.14  Show/Hide Text Palette

This Display[162] menu command shows or hides the Text Palette[270]. The Text Palette allows you to change the font, font size, and style for the label[81] or text of each selected object.

If the palette is not currently showing, the command is **Show Text Palette;** if the palette is currently showing, the command is **Hide Text Palette.**

The keyboard shortcut for **Show/Hide Text Palette** is Shift+Ctrl+T (Windows) or Shift-⌘-T (Mac).

Use the Text panel 279 of the Preferences 159 dialog box to determine whether or not the Text Palette appears automatically when you create or edit a caption 56.

You can use **Display | Text** 166 to change the size and font of labels and text objects, but the Text Palette is usually more convenient.

*See also:*
    *Text Tool* 120

## 3.3.15  Show/Hide Motion Controller

This Display 162 menu command shows or hides the Motion Controller 263.

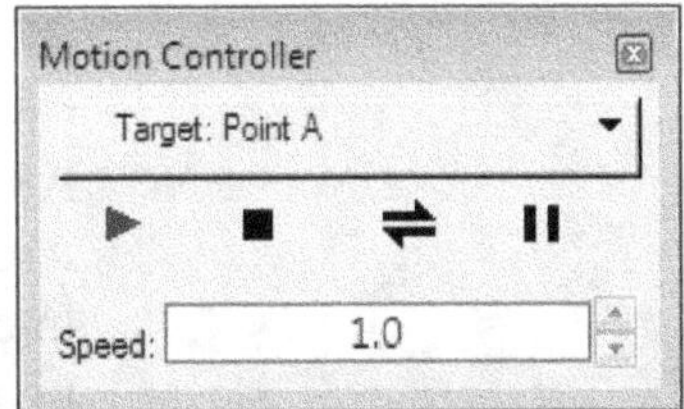


Use the Motion Controller to control the motion of the selected objects in your sketch. The Motion Controller controls the motion of both animated objects and objects set in motion by pressing a Movement button 150.

If the Motion Controller is not currently showing, the command is **Show Motion Controller;** if it is currently showing, the command is **Hide Motion Controller.**

The Motion Controller also appears automatically when you choose **Display | Animate** 171.

*See also:*
    *Animation Button* 149

## 3.3.16  Show/Hide Toolbox

This Display 162 menu command shows or hides Sketchpad's Toolbox 102.

If the Toolbox is not currently showing, the command is **Show Toolbox;** if it is currently showing, the command is **Hide Toolbox.**

Hide the Toolbox to give you more space when you're working on a large sketch, or to remove the distraction of visible tools when they're not needed.

## 3.4    Construct Menu

The Construct menu provides commands for accomplishing many important geometric constructions. Most of these constructions could also be done with Sketchpad's **Compass**|113| and **Straightedge**|115| tools, but the Construct menu often provides simpler and quicker ways of completing a construction.

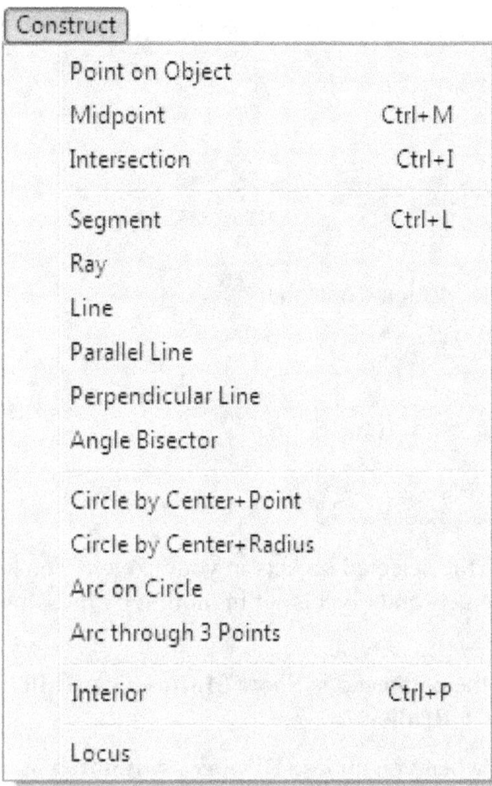


Detailed information is available for each command:

Point on Object|176|
Midpoint|176|
Intersection|177|
Segment, Ray, and Line|178|
Parallel Line|179|
Perpendicular Line|180|
Angle Bisector|181|
Circle By Center+Point|182|
Circle By Center+Radius|183|
Arc On Circle|185|
Arc Through 3 Points|186|
Interior|186|
Locus|190|

To use any Construct menu command, you must select the appropriate objects (the *prerequisites*) in the sketch.

If a command you want to use is unavailable, you might have too few or too many objects selected. Check the list of prerequisites below, and check the status line|253| to make sure you don't have extra unnoticed objects selected. Then select additional required objects or deselect unwanted objects. Alternatively, deselect all objects|106| and then carefully select the appropriate prerequisites.

| Command: | Prerequisites: |
| --- | --- |
| **Point on Object**|176| | • One or more path|89| objects |
| **Midpoint**|176| | • One or more segments|5| |
| **Intersection**|177| | • Two segments, rays, lines|5|, circles|6|, or arcs|7| <br> • One locus|10| or function plot|52| and one segment, ray, line|5|, circle|6|, or arc|7| <br> • Two function plots|52| |

| | |
|---|---|
| | • One polygon [8] and one ray [5] (command becomes **First Intersection**) |
| **Segment, Ray, and Line** [178] | • Two or more points [2] |
| **Parallel Line** [179] | • One point [2] and one segment, ray, or line [5]<br>• Multiple points [2] and one segment, ray, or line [5]<br>• One point [2] and multiple segments, rays, or lines [5] |
| **Perpendicular Line** [180] | • One point [2] and one segment, ray, or line [5]<br>• Multiple points [2] and one segment, ray, or line [5]<br>• One point [2] and multiple segments, rays, or lines [5] |
| **Angle Bisector** [181] | • Three points [2], with the vertex point selected second<br>• Two rays or segments [5] with a common endpoint |
| **Circle by Center+Point** [182] | • Two points [2] |
| **Circle by Center+Radius** [183] | • One point [2] and one segment [5] or distance value [92]<br>• Multiple points [2] and one segment [5] or distance value [92]<br>• One point [2] and multiple segments [5] and/or distance values [92] |
| **Arc on Circle** [185] | • One circle [6] and two points [2] on the circle<br>• One center point [2] and two other points equally distant from the center point |
| **Arc through 3 Points** [186] | • Three noncollinear(noncollinear points do not lie in a straight line) points [2] |
| **Polygon Interior** [186] | • Three or more points [2] |
| **Circle Interior** [186] | • One or more circles [6] |
| **Arc Sector Interior** [186] | • One or more arcs [7] |
| **Arc Segment Interior** [186] | • One or more arcs [7] |
| **Locus** [190] | • One point [2] on a path [89] and one object that depends on the point on path<br>• One parameter [36] and one object that depends on the parameter<br>• One independent point [2], one path [89] that does not depend on the |

| | point, and one object that depends on the independent point |
| --- | --- |

## 3.4.1 Point on Object

*Selection prerequisites: One or more path* [89] *objects*

This Construct [174] menu command constructs a point [2] on each selected path object. The point is randomly placed on the object. It can later be animated or dragged anywhere on the object, but it will not leave the path.

Path objects on which you can construct points include straight objects (segments, rays, and lines) [5], circles [6], arcs [7], interior objects (polygons, and circle and arc interiors) [8], point loci [10], function plots [52], and parametric plots [54]. (When you use an interior as a path, the actual path is the perimeter of that interior. A point constructed on a polygon is free to move around the entire perimeter of the polygon.)

You can also construct a point on an object by clicking the object with the **Point** [112] tool, or with any other tool that constructs points as part of its operation (such as the **Compass** [113] tool, **Straightedge** [115] tools, **Polygon** [118] tools, or most custom [125] tools).

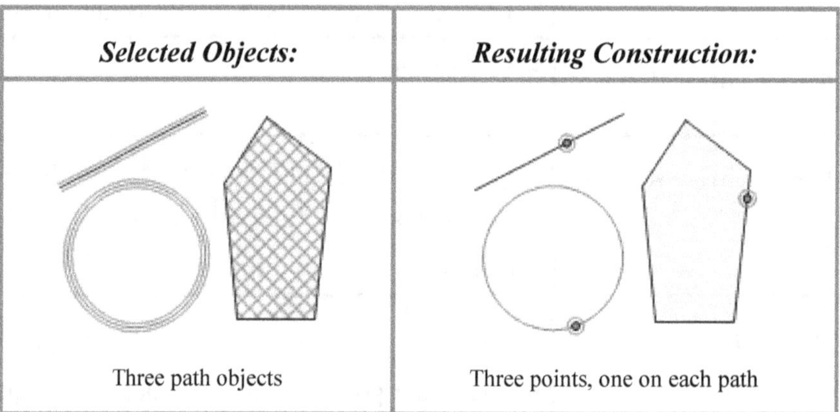

| *Selected Objects:* | *Resulting Construction:* |
| --- | --- |
| Three path objects | Three points, one on each path |

## 3.4.2 Midpoint

*Selection prerequisites: One or more segments* [5]

This Construct [174] menu command constructs a point [2] at the midpoint of each selected segment [5]. As a segment gets longer or shorter (through dragging an endpoint, for example), the midpoint moves accordingly.

The keyboard shortcut for **Midpoint** is Ctrl+M (Windows) or ⌘M (Mac).

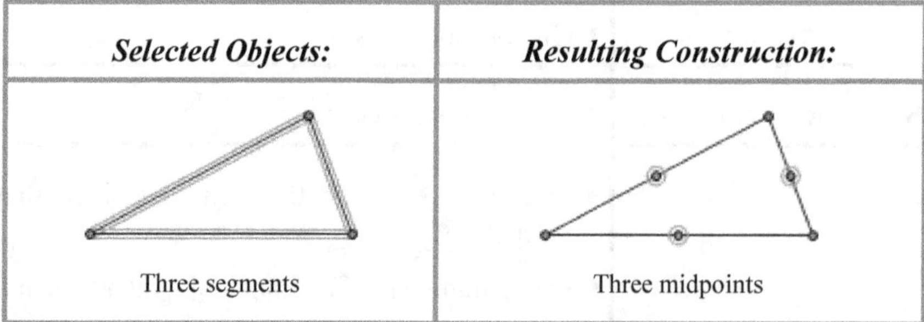

| *Selected Objects:* | *Resulting Construction:* |
| --- | --- |
| Three segments | Three midpoints |

### 3.4.3 Intersection

***Selection prerequisites:*** *Two intersecting objects, which can be any of the following:*

- *Two straight objects* [90]*, circles* [6]*, or arcs* [7]
- *One locus* [10]*, function plot* [52]*, or parametric plot* [54] *and one straight object* [90]*, circle* [6]*, or arc* [7]
- *Two function plots* [52] *on the same coordinate system* [41]
- *One polygon* [8] *and one ray* [5] *(command becomes **First Intersection**)*

This Construct [174] menu command constructs a point [2] at each intersection of the two selected objects. If the objects actually intersect in multiple places, multiple points are constructed; otherwise one point is constructed.

The keyboard shortcut for **Intersection** is Shift+Ctrl+I (Windows) or Shift-⌘I (Mac).

If you later drag the intersecting objects apart, the constructed intersection point disappears, and reappears when you drag the objects so that they're again touching.

You can also construct a point of intersection by clicking the intersection with the **Selection Arrow** [104] tool or with any tool that constructs points as part of its operation (such as the **Point** [112], **Compass** [113], or **Straightedge** [115] tools).

| *Selected Objects:* | *Resulting Construction:* |
| --- | --- |
| One circle, one straight object | Two intersections |
| One circle, one straight object | One intersection |
| One locus, one arc | Two intersections |
| Two function plots | Three intersections |
| One ray, one polygon | First intersection |

### 3.4.4    Segment, Ray, and Line

*Selection prerequisites: Two or more points* 2

This Construct 174 menu command constructs a segment, ray, or line 5 through the selected points. ( The **Ray** command constructs a ray from the first point through the second.) If more than two points are selected, this command constructs the same number of segments, rays, or lines as the number of selected points. (For instance, choosing **Line** with the four points *J, K, L,* and *M* selected will construct the four lines shown in the bottom example.)

The keyboard shortcut for **Segment** is Ctrl+L (Windows) or ⌘L (Mac).

The results of these commands are the same as the results of using the **Straightedge** 115 tools on the selected points. The commands are particularly useful for constructing multiple straight objects and for making sure that your straight objects go through the proper points.

To quickly construct the sides of a pentagon, hold down the Shift key while you click the **Point** 112 tool five times. Then choose the **Segment** command to construct the five sides.

If the pentagon interior has already been constructed, select the interior and choose **Edit | Select Parents** 152. Then choose the **Segment** command.

| Command: | Selected Objects: | Resulting Construction: |
|---|---|---|
| **Segment** | Five points | Five segments |
| **Ray** | Three points | Three rays |
| **Line** | Four points | Four lines |

## 3.4.5  Parallel Line

*Selection prerequisites: One of the following combinations of point(s) and straight object(s):*

- *One point* 2 *and one straight object* 90
- *Multiple points* 2 *and one straight object* 90
- *One point* 2 *and multiple straight objects* 90

This Construct 174 menu command constructs a line 5 through each selected point parallel to each selected straight object.

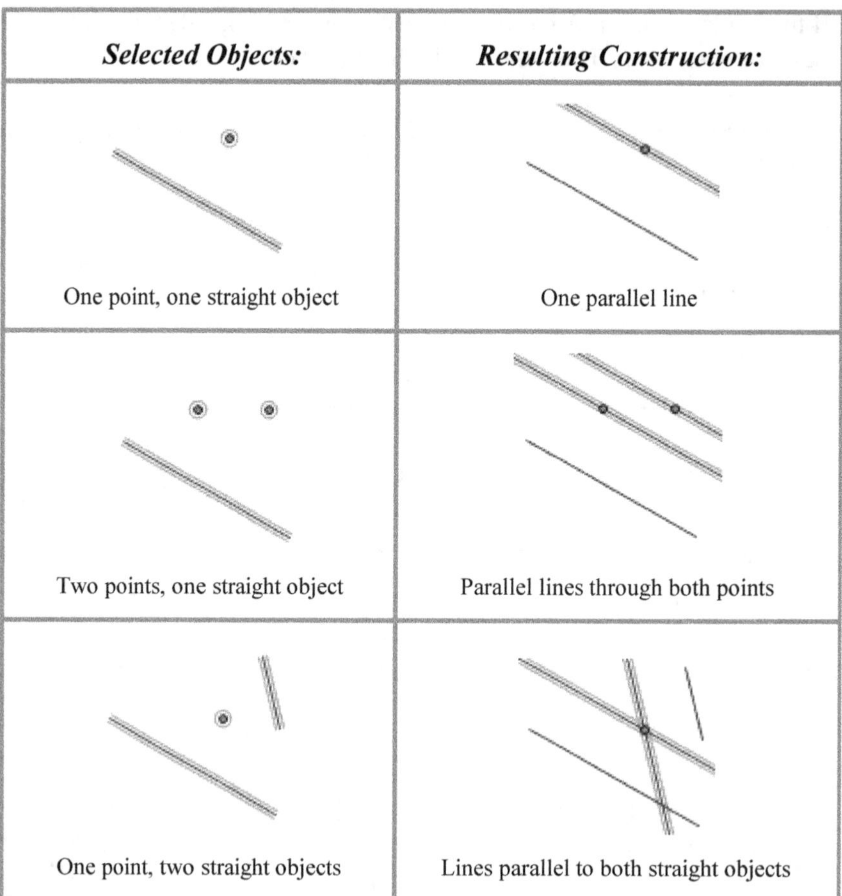

| Selected Objects: | Resulting Construction: |
|---|---|
| One point, one straight object | One parallel line |
| Two points, one straight object | Parallel lines through both points |
| One point, two straight objects | Lines parallel to both straight objects |

## 3.4.6    Perpendicular Line

***Selection prerequisites:*** *One of the following combinations of point(s) and straight object(s):*

- *One point* ⌐2⌐ *and one straight object* ⌐90⌐
- *Multiple points* ⌐2⌐ *and one straight object* ⌐90⌐
- *One point* ⌐2⌐ *and multiple straight objects* ⌐90⌐

This Construct ⌐174⌐ menu command constructs a line ⌐5⌐ through each selected point perpendicular to each selected straight object.

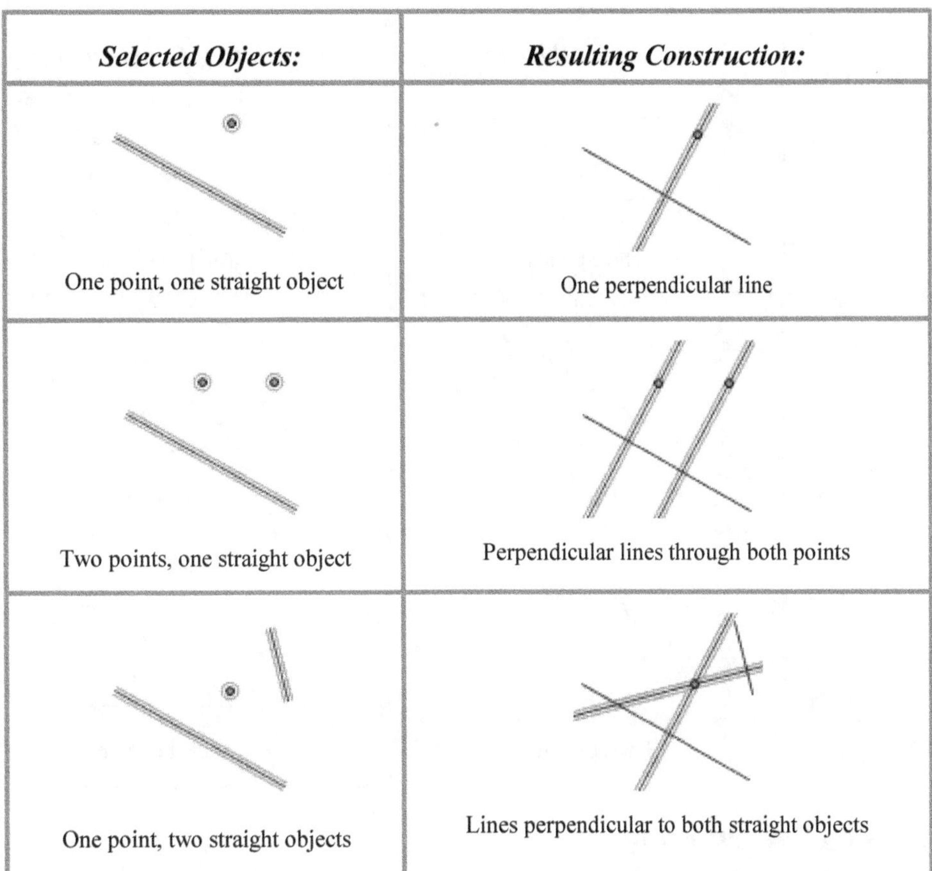

|  | *Selected Objects:* |  | *Resulting Construction:* |
|---|---|---|---|
|  | One point, one straight object |  | One perpendicular line |
|  | Two points, one straight object |  | Perpendicular lines through both points |
|  | One point, two straight objects |  | Lines perpendicular to both straight objects |

*See also:*
  *How to Make a Perpendicular Bisector Tool* 132

## 3.4.7    Angle Bisector

***Selection prerequisites:*** *Either of the following:*

- *Three points* 2 *, with the vertex point selected second*

    The second selected point designates the vertex of the angle. For instance, to bisect $\angle ABC$, select points in the order: first *A*, then *B*, and finally *C*.

- *Two rays* 5 *or segments* 5 *with a common endpoint*

    You can optionally select the common endpoint as well. This makes it easy to use a selection rectangle to select the vertex and both sides of an angle.

This Construct 174 menu command constructs a ray that bisects the minor angle formed by the three selected points or two selected straight objects.

| *Selected Objects:* | *Resulting Construction:* |
|---|---|
| Three points | Angle bisector |
| Two rays | Angle bisector |
| Two segments | Angle bisector |

## 3.4.8    Circle by Center+Point

*Selection prerequisites:* Two points [2]

This Construct [174] menu command constructs a circle [6] with its center at the first selected point and with its circumference passing through the second selected point. The result of this command is the same as the result of using the **Compass** [113] tool on the two selected points.

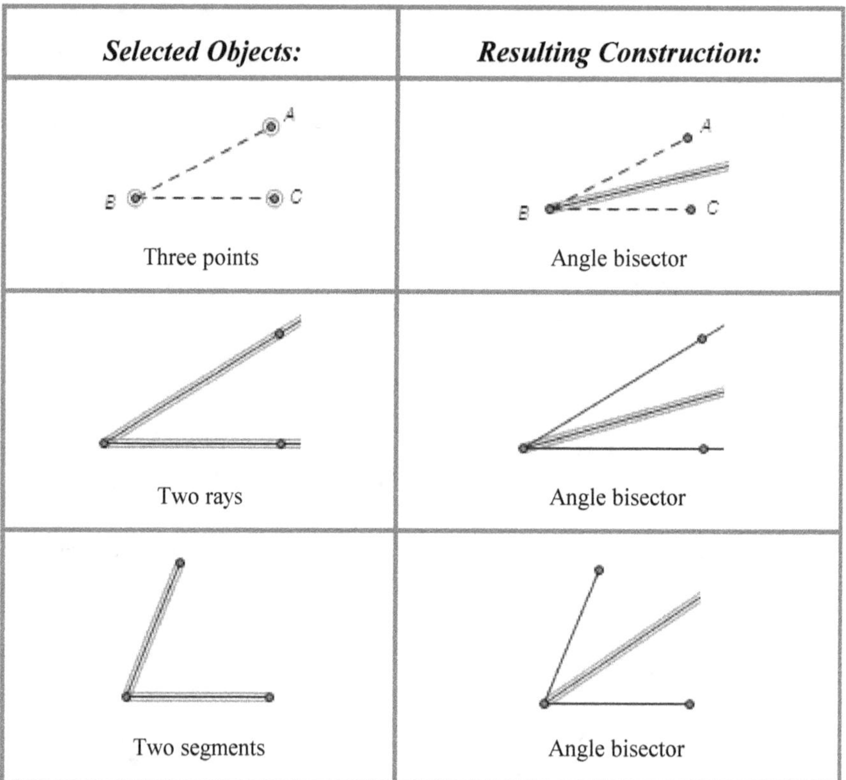

| *Selected Objects:* | *Resulting Construction:* |
|---|---|
| Two points | Circle |

*See also:*
   *Circle by Center+Radius command* [183]

### 3.4.8.1    How to Construct a Segment of Fixed Length

Occasionally you may want to create a segment [5] of fixed length — for instance, a segment that is exactly 0.8 cm in length. While you could create a segment with the **Segment** [115] tool, measure its length, and drag one endpoint until its length was 0.8 cm, this segment would not be constructed to be fixed at 0.8 cm long. That is, dragging an endpoint[104] again would change it from its current length to some other length.

1. Create a new parameter[36] by choosing **Number | New Parameter**[226]. Call the parameter *d*, and set its value to 0.8 cm.

   If the distance units are not cm, use **Edit | Preferences | Units**[276] to set them to cm before doing this step.

2. Use the **Point**[112] tool to construct a point; this will be the first endpoint of the segment.

3. Select the point and the parameter. Choose **Construct | Circle by Center+Point**[182] to construct a circle whose radius is determined by the parameter.

4. Use the **Segment**[115] tool to construct a radius of the circle. This is the segment of fixed length.

5. Hide[167] the parameter and the circle.

6. Drag either endpoint of the segment to see how it behaves.

If you later want to change the length of the segment to some other fixed length, you can show the hidden objects[167] and edit the parameter.[157]

*See also:*
   *How to Construct an Angle of Fixed Measure* [204]
   *How to Construct Congruent Segments* [184]

## 3.4.9 Circle by Center+Radius

*Selection prerequisites: One of the following combinations of point(s) and distance(s):*

- *One point* [2] *and one segment* [5] *or distance value* [92]
- *Multiple points* [2] *and one segment* [5] *or distance value* [92]
- *One point* [2] *and multiple segments* [5] *and/or distance values* [92]

This Construct[174] menu command constructs one or more circles[6] centered at each selected point and with the radius determined by each selected segment or distance measurement.

   This command duplicates the function of a noncollapsible compass, which allows you to set a length and then draw circles centered anywhere with that length as the radius.

| *Selected Objects:* | *Resulting Construction:* |
|---|---|
| One point, one segment | One circle |
| m $\overline{AB}$ = 0.81 cm<br>Two points, one distance measurement | m $\overline{AB}$ = 0.81 cm<br>Two circles centered at the points with radius determined by the measurement |
| One point, two segments | Two circles centered at the point with radii determined by the segments |

*See also:*
   *Circle by Center+Point command* 182

### 3.4.9.1   How to Construct Congruent Segments

Occasionally you may want to create a segment 5 that's congruent to an existing segment. For instance, you might want to make a triangle or other polygon that's congruent to an existing triangle or polygon, and that can be positioned and rotated as you please while always remaining congruent to the original.

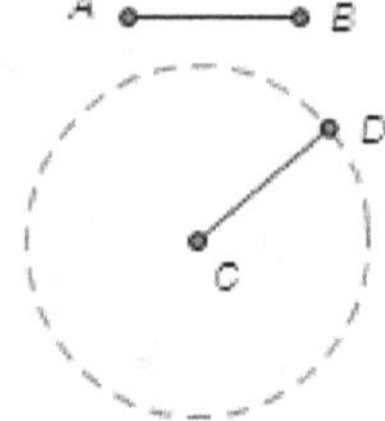


For this construction you'll use the **Construct | Circle by Center+Radius** 183 command to make a new segment *CD* exactly the same length as the original segment *AB*.

1. Use the **Point** 112 tool to construct a starting point $C$. This will be the first endpoint of the new segment.

2. Select point $C$ and the original segment $AB$. Choose **Construct | Circle by Center+Radius** 183 to construct a circle whose radius is determined by the existing segment.

3. Use the **Segment** 115 tool to construct a radius of the circle. This is the segment that will remain congruent to the original. Label the new segment endpoint $D$.

4. Hide the circle.

5. Test your construction by changing the length of the original segment, and by dragging endpoints $C$ and $D$ of the new segment.

*See also:*
   *How to Construct a Segment of Fixed Length* 182
   *How to Construct an Angle of Fixed Measure* 204
   *How to Construct Congruent Angles* 204

## 3.4.10  Arc on Circle

***Selection prerequisites:*** *Either of the following:*

- *One circle* 6 *and two points* 2 *on the circle*

- *One center point* 2 *and two other points equally distant from the center point*

This Construct 174 menu command constructs an arc 7 of the given circle 6 bounded by the two points, or an arc 7 centered on the first selected point and bounded by the other two points.

The arc is constructed counter-clockwise from the first selected bounding point to the second. The order in which you select the bounding points determines whether the constructed arc is a minor or major arc.

| **Selected Objects:** | **Resulting Construction:** |
|---|---|
| One circle, two points on circle | Arc |
| Three points, with $D$ and $E$ equally distant from $C$ | Arc |

*See also:*
   *Arc through 3 Points command* 186

## 3.4.11  Arc through 3 Points

***Selection prerequisites:*** *Three noncollinear*(noncollinear points do not lie in a straight line) *points* [2]

This Construct [174] menu command constructs an arc [7] through the three selected points [2]. The arc starts at the first selected point, passes through the second point, and ends at the third point.

 Note

> If you construct an arc through three points and then drag the three points so they are collinear, the second point determines the appearance of the arc. If the second point falls between the first and third points, the arc has zero angle measure [219] but nonzero arc length [220]; it is displayed as a segment [5] that starts at the first selected point, passes through the second point, and ends at the third point. If the second point is collinear with, but not between, the first and third points, the arc is not well defined: it disappears and any measurements that depend on it are undefined.

| *Selected Objects:* | *Resulting Construction:* |
| --- | --- |
| Three points | Arc |

*See also:*
Arc on Circle command [185]

## 3.4.12  Interior

This Construct [174] menu command constructs the interior [8] defined by the selected object or objects. Depending on your selection, this command may appear as **Polygon Interior** [187], **Circle Interior** [188], or **Arc Interior.** If the command appears as **Arc Interior,** it shows a submenu that allows you to choose between **Arc Sector** [189] and **Arc Segment** [189].

The keyboard shortcut for **Interior** is Ctrl+P (Windows) or ⌘P (Mac).

| Prerequisites: | Selected Objects: | Resulting Construction: |
| --- | --- | --- |
| Three or more points | Five points | **Pentagon Interior** |
| One or more circles | Circle | **Circle Interior** |
| One or more arcs | Arc | **Arc Interior \| Arc Sector** |
| One or more arcs | Arc | **Arc Interior \| Segment** |

*Subtopics:*
   *Polygon Interior command* 187
   *Circle Interior command* 188
   *Arc Sector command* 189
   *Arc Segment command* 189

### 3.4.12.1 Polygon Interior

**Selection prerequisites:** *Three or more points* 2

This Construct 174 menu command constructs a polygon 8 using the selected points as vertices. The order in which you select the points determines the order of the vertices of the polygon.

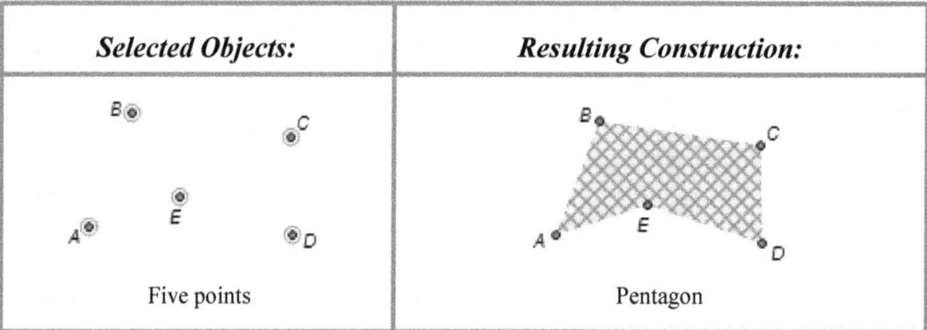

| Selected Objects: | Resulting Construction: |
|---|---|
| Five points | Pentagon |

*See also:*
  *Circle Interior command* 188
  *Arc Sector command* 189
  *Arc Segment command* 189

### 3.4.12.2  Circle Interior

*Selection prerequisites:* *One or more circles* 6

This Construct 174 menu command constructs a circle interior 8 for each selected circle.

| Selected Objects: | Resulting Construction: |
|---|---|
| Circle | Circle interior |
| Three circles | Three circle interiors |

*See also:*
  *Polygon Interior command* 187
  *Arc Sector command* 189
  *Arc Segment command* 189

### 3.4.12.3  Arc Sector

*Selection prerequisites: One or more arcs* [7]

This Construct [174] menu command constructs an arc sector interior [8] for each selected arc.

The arc sector is bounded by the arc and by the radii to the two endpoints of the arc.

The keyboard shortcut for **Arc Sector Interior** is Ctrl+P (Windows) or ⌘P (Mac).

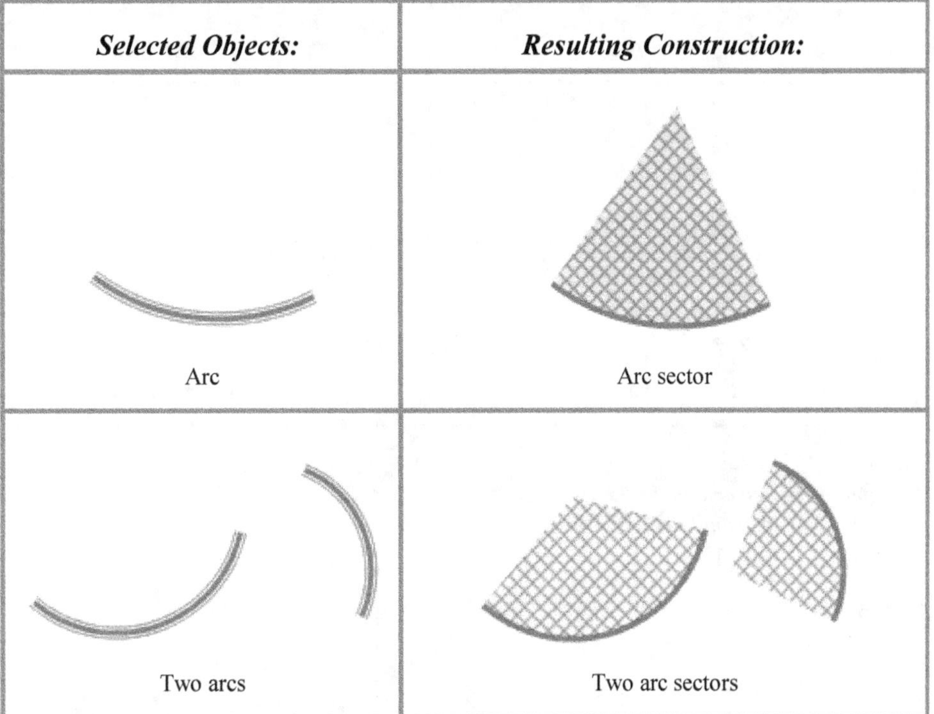

| *Selected Objects:* | *Resulting Construction:* |
|---|---|
| Arc | Arc sector |
| Two arcs | Two arc sectors |

*See also:*
  *Polygon Interior command* [187]
  *Circle Interior command* [188]
  *Arc Segment command* [189]

### 3.4.12.4  Arc Segment

*Selection prerequisites: One or more arcs* [7]

This Construct [174] menu command constructs an arc segment interior [8] for each selected arc.

The arc segment is bounded by the arc and by the chord connecting the two endpoints of the arc.

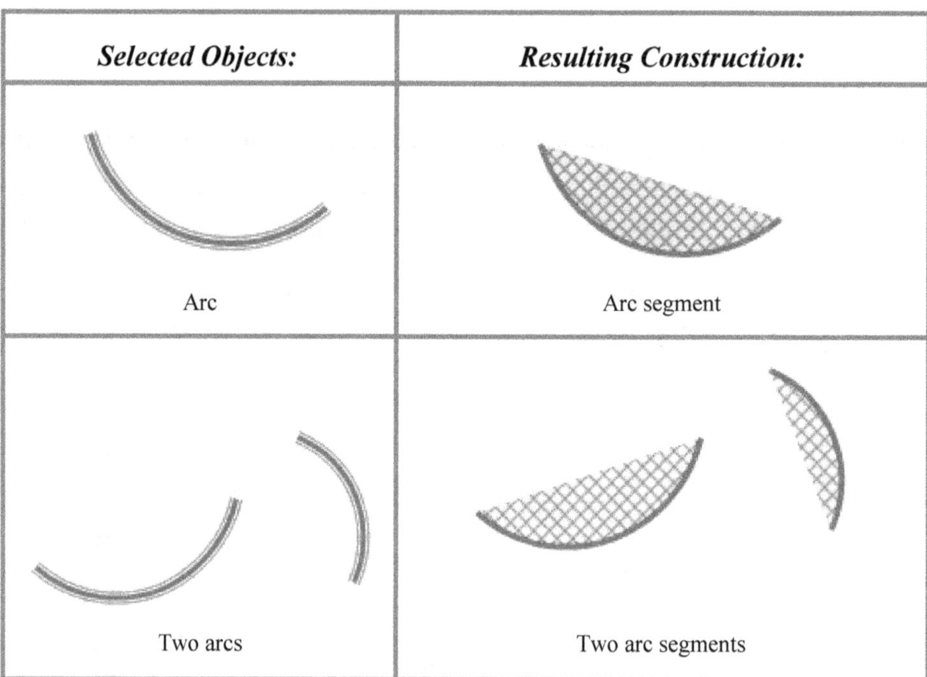

| *Selected Objects:* | *Resulting Construction:* |
| --- | --- |
| Arc | Arc segment |
| Two arcs | Two arc segments |

*See also:*
Polygon Interior command 187
Circle Interior command 188
Arc Sector command 189

## 3.4.13  Locus

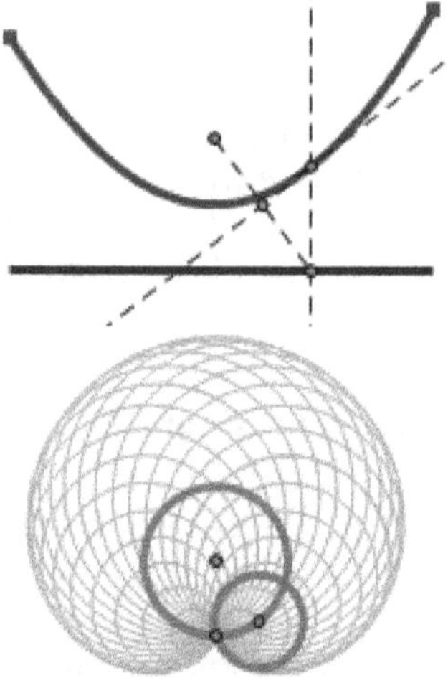

***Selection prerequisites:*** *One of the following combinations:*

- One point 2 (the driver) on a path 89 (the domain) and one object (the driven object) that depends on the point

- One independent point (the driver), one path (the domain) that does not depend on the point, and one object (the driven object) that depends on the point

- One parameter 36 (the driver) and one object (the driven object) that depends on the parameter (the domain is the numeric domain of the parameter)

This Construct [174] menu command constructs the locus of the driven object as the driver moves or varies along the domain.

> The driven object can be a point, a straight object, a circle, an arc, an interior, a picture, a function plot, or another locus.

If the driven object is a function plot, the command becomes **Family of Functions.**

If the driven object is a point locus, the command becomes **Family of Curves.**

If the driven object is a non-point locus, the command becomes **Family of Loci.**

 Note

> One way to think of these prerequisites is that they specify three things: a driver, a domain, and a driven object. The driver is a point or parameter. For a point driver, the domain specifies the path on which it moves; for a parameter driver, the domain specifies the numeric values over which the parameter varies. The driven object is the object controlled by the driver; it's the object that determines the appearance of the locus.

To construct a locus, follow these steps:

1. Select the driver and the driven object. The position of the driven object must depend on the driver. (The order in which you select the objects does not matter.)

2. If the driver is an independent point, select the domain of the driver — a path object that does not depend on the driver.

3. Choose **Construct** [174] | **Locus.**

If the driver is a parameter, the Plot Properties [95] panel appears, allowing you to set the domain of the driver.

See Anatomy of a Locus [14] for detailed examples of the many different types of loci you can construct.

*See also:*
> *Plot Properties for a Locus* [95]
> *Sampling Preferences* [283]

# 3.5  Transform Menu

The Transform menu applies geometric transformations to figures in your sketches, allowing you to create translations, rotations, dilations, reflections, tessellations, scale models, kaleidoscopes, fractals, and much more.

**Basic transformations (Translate** 200**, Rotate** 203**, Dilate** 206**, and Reflect** 207**)** are called *similarity transformations* because they create images that are similar to the pre-images: the image has the same angles as the pre-image, and all distances are changed by the same ratio. To use these transformations, you can either specify fixed distances, angles, and ratios, or use the various **Mark** commands to mark dynamic distances, angles, and ratios in your sketch.

**Iterations** allow you to perform the same transformation repeatedly, transforming a pre-image to form a first image, transforming that first image to create a second image, transforming the second image to create a third image, and so on. You can transform both geometric objects and numeric values, and you can specify the depth of iteration.

**Custom transformations** allow you to specify an arbitrary transformation in terms of two points: an independent point and some other point that depends on the independent point. Once you specify such a transformation, you can apply it to a wide variety of objects in your sketch, including other points, path objects, interiors, and even pictures.

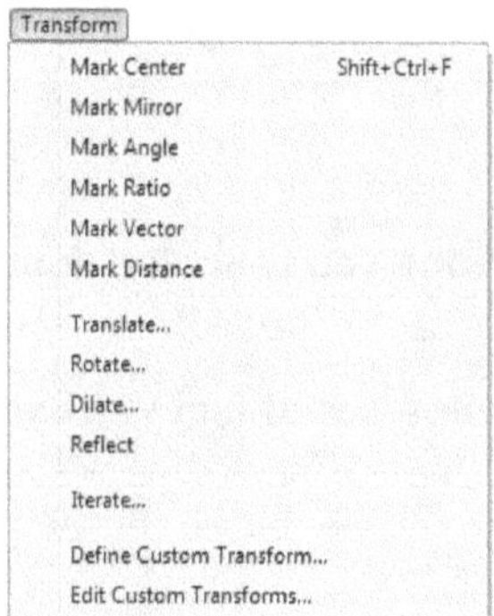


Detailed information is available for each command:
Mark Commands 194
Mark Center 194
Mark Mirror 195
Mark Angle 195
Mark Ratio/Mark Segment Ratio/Mark Scale Factor 197
Mark Vector 198
Mark Distance 199
Translate 200
Rotate 203
Dilate 206
Reflect 207
Iterate 208
Define Custom Transform 211
Edit Custom Transforms 214

## ▼ Basic Transformations

When you specify a basic transformation (**Translate** 200**, Rotate** 203**, Dilate** 206**, and Reflect** 207**),** you specify not only the object to be transformed (the transformational pre-image) but also certain objects or values upon which the transformation is based. For instance, a rotation requires both an angle and a center about which to rotate. The **Mark** 194 commands — **Mark Center** 194**, Mark Mirror** 195**, Mark Angle** 195**, Mark Ratio** 197**, Mark Vector** 198**, and Mark Distance** 199 — let you specify objects in your sketch to be used for your transformations.

These marked objects may be either geometric objects or values. For example, a rotation is defined by a center and an angle. The center is a geometric point about which objects rotate. If the center point moves, then the rotated image changes accordingly. The angle determines how far objects rotate about the center. Rotate by a fixed angle (such as 45°) or by a marked angle (which can be defined by three points, by a marked angle value ⌐92⌐, or by an angle marker ⌐60⌐). If a marked angle changes, then the rotated image also changes dynamically.

In general, follow these steps to construct a basic transformed image of one or more objects:

1. Mark any objects that determine dynamic aspects of your transformation by selecting the objects and choosing the appropriate **Mark** ⌐194⌐ commands (for example, **Mark Center** ⌐194⌐ to mark a selected point as the center of rotation). Most objects can also be marked when the transformation dialog box is open, after you choose the transformation command.

2. Select the object(s) you want to transform. Mathematically, this is the pre-image of your transformation. (Loci, function plots, iterated images, angle markers and tick marks, and text objects cannot be transformed.)

3. Choose the appropriate transformation command from the Transform menu (for example, **Rotate** ⌐203⌐).

4. In the resulting dialog box, specify any objects or values you want to use in your transformation (for example, click a new center point, or specify the angle by typing 45° or clicking an angle measurement or marker in the sketch).

5. A faint preview image appears in the sketch. Check to make sure it's what you expect. Then confirm the dialog box to construct the transformed image.

Once you mark a parameter, Sketchpad remembers that mark even after you change the selections. If you've already marked a parameter for one transformation, you don't need to mark it again to use the same parameter for a second transformation.

## ▼ Iterations

The **Iterate** ⌐208⌐ command allows you to create in a single step multiple transformed images such as this spiral, or any figure resulting from a transformation or construction repeated many times. Use iteration to create complex images such as tessellations and fractals. You can iterate not only geometric objects but also numeric values.

### ▼ Custom Transformations

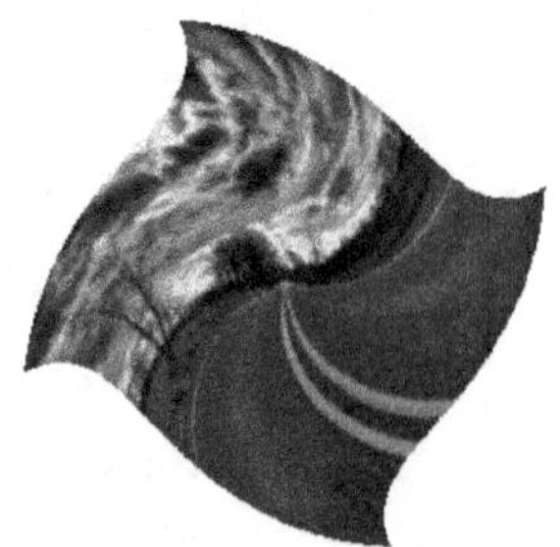

You can define 211 and edit 214 custom transformations. These transformations allow you to define as a transformation any construction that involves two points, one of which depends on the other. Your custom transformations are added to the bottom of the Transform menu, making them very easy to use. Custom transformations are very flexible, because they can incorporate any sequence of construction steps that begin with a point and end with another point.

## 3.5.1    Mark Commands

The first six commands on the Transform 192 menu allow you to mark the objects or values to be used in subsequent basic transformations (**Translate** 200, **Rotate** 203, **Dilate** 206, **and Reflect** 207).

> Any marked object remains marked until you mark a new object of the same type. For example, you don't need to mark the same center more than once, no matter how many times you use it as a marked center.

 Note

> You can use most of the Mark commands whenever your selections include the required object (s) for that mark, even if you have other objects selected as well. Thus, if you've selected five polygons you want to dilate, and realize you forgot to mark the center, then you can select your desired center point, without deselecting the polygons, and choose **Mark Center** 194. The most recently selected point is marked as the center and removed from your selections, and you can go ahead and choose **Dilate** 206 for your polygons.

Most objects can also be marked when the transformation dialog box is open, after you choose the transformation command. Click the object itself to mark it, or click in a caption on a Hot Text 57 link to the object.

*See also:*
  *Mark Center command* 194
  *Mark Mirror command* 195
  *Mark Angle command* 195
  *Mark Ratio command* 197
  *Mark Vector command* 198
  *Mark Distance command* 199

## 3.5.2    Mark Center

***Selection prerequisites:*** *Any number of objects including at least one point*

This Transform 192 menu command marks 194 the most recently selected point as the center point about which future rotations 203 and dilations 206 will take place.

> If you haven't yet marked a center when you choose **Rotate** 203 or **Dilate** 206, Sketchpad

will automatically choose a point and mark it for you.

The keyboard shortcut for **Mark Center** is Shift+Ctrl+F (Windows) or Shift-⌘F (Mac).

To mark a center:

- Select a point and choose **Transform** |192| | **Mark Center,** or

- Double-click a point with the **Selection Arrow** |104| tool.

The marked point will flash briefly to indicate it's been marked as the center for future rotations and dilations. Once marked, this center point will be used for all future rotations and dilations until you mark a different center point.

You can also mark a center after choosing **Rotate** |203| or **Dilate** |206|. With the dialog box open, click the desired center point in your sketch.

## 3.5.3  Mark Mirror

***Selection prerequisites:*** *Any number of objects including at least one straight object*

This Transform |192| menu command marks |194| the most recently selected straight object as the mirror across which future reflections |207| will take place.

To mark a mirror:

- Select a straight object |5| and choose **Transform** |192| | **Mark Mirror,** or

- Double-click a straight object with the **Selection Arrow** |104| tool.

    If you haven't yet marked a mirror when you choose **Reflect** |207|, Sketchpad will automatically choose a straight object and mark it for you.

The marked straight object will flash briefly to indicate it has been marked as the mirror for future reflections. Once marked, this mirror will be used for all future reflections until you mark a different mirror.

## 3.5.4  Mark Angle

***Selection prerequisites:*** *Any number of objects including an angle marker* |60|*, two segments or rays* |5| *with a common endpoint, three points, or an angle value* |92|

    The angle value can be an angle parameter |36|, measurement |36| or calculation |39|.

This Transform |192| menu command marks |194| the most recently selected object(s) to determine the angle to be used for future rotations |203| and polar translations |200|.

    Use a marked angle for a transformation when you want your transformed image to move as the marked angle changes. When you rotate by 45°, the angle of rotation never changes; but when you rotate by ∠ABC, the rotation changes as the positions of points *A, B,* and *C* change.

#### ▼ Mark an Angle Using an Angle Marker

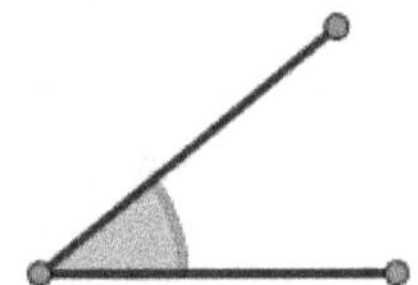

***Selection prerequisites:*** *Any number of objects including an angle marker*

1. Select or create the desired angle marker 60 .

2. Choose **Transform** 192 | **Mark Angle.**

   A brief animation confirms that your angle has been marked.

You can also mark an angle after choosing **Rotate** 203 or **Translate** 200. With the dialog box open, click an angle marker in your sketch.

### ▼ Mark an Angle Using Two Segments or Rays

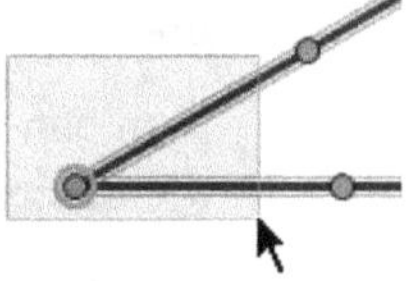

*Selection prerequisites: Any number of objects including two segments or rays with a common endpoint*

1. Select two segments or rays with a common endpoint to define your angle. You may also select the common endpoint (the vertex of the angle).

   A selection rectangle 106 is an easy way to select these objects.

2. Choose **Transform** 192 | **Mark Angle.**

   A brief animation confirms that your angle has been marked.

### ▼ Mark an Angle Using Three Points

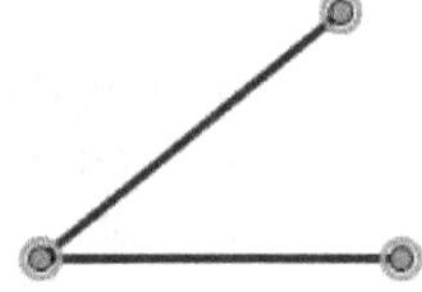

*Selection prerequisites: Any number of objects including three points*

1. Select three points to define your angle. To define an angle by three points, select a point on the initial side, then the vertex, and then a point on the terminal side. For instance, if you select three points shown here in the order *A, B, C,* the result will be a marked angle that specifies the counter-clockwise rotation shown by the arrow.

2. Choose **Transform** 192 | **Mark Angle.**

   A brief animation confirms that your angle has been marked.

### ▼ Mark an Angle Using an Angle Value

m∠ABC = 38.52°

*Selection prerequisites: Any number of objects including an angle value*

1. Select an angle measurement 36 , parameter 36 , or calculation 39 to define your angle. The units must be either degrees or radians.

2. Choose **Transform** 192 | **Mark Angle.**

The selection will flash briefly to confirm that your angle measurement, parameter, or calculation has been marked.

You can also mark an angle after choosing **Rotate**[203] or **Translate**[200]. With the dialog box open, click an angle value in your sketch.

## 3.5.5  Mark Ratio/Mark Segment Ratio/Mark Scale Factor

***Selection prerequisites:*** *Any number of objects including two segments, one value without units, or three collinear points*

This Transform[192] menu command has three forms: **Mark Segment Ratio, Mark Scale Factor,** and **Mark Ratio.**

The command marks[194] the most recently selected ratio or scale factor as the ratio for future dilations [206]. The form of the command depends on your selections:

| *Select:* | *For this command:* | *To use this ratio or scale factor:* |
|---|---|---|
| Two segments ($j$, $k$) | **Mark Segment Ratio** | $j / k$ |
| One value with no units | **Mark Scale Factor** | The selected value |
| Three collinear points ($A$, $B$, $C$) | **Mark Ratio** | $AC / AB$ |

Dilations can use either a fixed ratio (such as ½) or a marked ratio (such as $AC / AB$). When you dilate by ½, the ratio of dilation never changes; but when you dilate by $AC / AB$, the dilation changes as the positions of points $A$, $B$, and $C$ change.

> Use a marked ratio or scale factor when you want your dilated image to change as the marked item changes.

### ▼ Mark Segment Ratio (Two Segments)

***Selection prerequisites:*** *Any number of objects including two segments*

1. Select a segment [5] whose length[217] will be used as the numerator of your ratio.

2. Select a second segment whose length will be used as the denominator.

3. Choose **Transform**[192] | **Mark Segment Ratio.**

You will see a brief animation confirming that the ratio has been marked.

If the first segment is shorter than the second, the ratio is less than 1 and the dilation will shrink objects toward the marked center. If the first segment is longer than the second, the ratio is greater than 1 and the dilation will stretch objects away from the marked center.

You can also mark a segment ratio after choosing **Dilate**[206]. With the Dilate dialog box open, click the two segments in your sketch, one after the other.

### ▼ Mark Scale Factor (A Value with No Units)

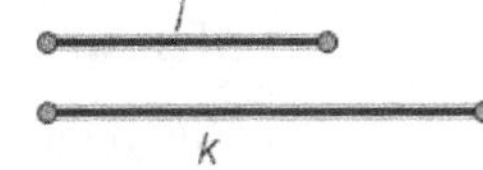

*Selection prerequisites: Any number of objects including an undimensioned value* [92] *(a measurement* [36]*, parameter* [36]*, or calculation* [39]*)*

1. Select an undimensioned value (a pure value, with no units such as inches or degrees) to be the scale factor.

2. Choose **Transform** [192] | **Mark Scale Factor.**

The selection will flash briefly to confirm that your value has been marked.

A convenient way to create a scale factor is to construct a point on a straight object and measure the point's value [221].

If the magnitude of the value you marked as a scale factor is less than 1, the dilation will shrink objects toward the marked center. If the value's magnitude is greater than 1, the dilation will expand objects away from the marked center. If the scale factor is less than 0, image objects will be dilated "through the center," appearing on the opposite side of the center as their pre-images.

You can also mark a scale factor after choosing **Dilate** [206]. With the Dilate dialog box open, click an undimensioned value in your sketch.

## ▼ Mark Ratio (Three Collinear Points)

*Selection prerequisites: Any number of objects including three collinear points*

1. Select three collinear points in your sketch. The points must be in a straight line.

2. Choose **Transform** [192] | **Mark Ratio.**

You will see a brief animation confirming that the ratio has been marked.

When you mark three collinear points, *A, B,* and *C,* as a ratio, Sketchpad dilates [206] by the ratio of signed distances *AC / AB*. If *B* is on the same side of *A* as *C,* the signed ratio of distances is positive; if *B* is on the opposite side of *A* as *C,* the ratio is negative. You may want to use the **Measure | Ratio** [221] command to display this ratio numerically.

One handy way to remember the roles of the three selected points is to think of them as determining a ratio that, if the first selected point was the marked center, would dilate the second selected point to the location of the third selected point.

Another way to remember the roles of the points is to think of them forming a number line, with point *A* representing the origin, *B* the unit point, and *C* the point whose value is used as the ratio.

## 3.5.6    Mark Vector

*Selection prerequisites: Any number of objects including two points*

This Transform [192] menu command marks [194] for future translations [200] the vector determined by the two most recently selected points. The first of these two points is the initial point of the vector, and the second is the terminal point of the vector.

When you choose this command, you will see a brief animation from the initial to the terminal point confirming that the vector has been marked.

After marking a vector, future translations for which you specify the **Translation Vector: Marked** option will be based on this vector. Such translations will translate objects by a distance equal to the distance from the initial to the terminal point and in the same direction as the direction from the initial to the terminal point.

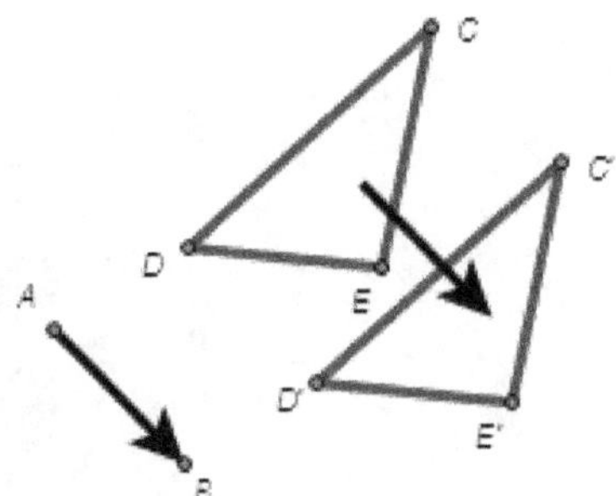

For instance, if you mark the vector from *A* to *B* as in the figure, and then translate $\triangle CDE$, the result will be a second triangle translated by the distance from *A* to *B* and in the direction from *A* to *B*. As you adjust the vector by dragging *A* or *B*, the translated image of $\triangle CDE$ adjusts accordingly.

You can also mark a vector after choosing **Translate** 200. With the Translate dialog box open, click two points in your sketch to designate the initial and terminal points of the desired vector.

### 3.5.7   Mark Distance

***Selection prerequisites:*** *Any number of objects including a distance value*

This Transform 192 menu command marks 194 the one or two most recently selected distance values 92 (measurements 36, parameters 36, or calculations 39) to be used for future polar and rectangular translations 200.

To mark a distance:

1. Select 106 one or two distance measurements, parameters, or calculations (ones with distance units, such as centimeters).

2. Choose **Transform** 192 | **Mark Distance.**

The selected distance value(s) will flash briefly to confirm that your distance has been marked.

If you choose **Mark Distance** with a single selected distance value, this new distance will become the marked distance for polar translation.

If you choose **Mark Distance** with two selected distance values, the first selected value will become the horizontal distance for rectangular translations, and the second will become the vertical distance.

You can also mark a distance after choosing **Translate** 200. With the Translate dialog box open, click a distance value in your sketch.

## 3.5.8   Translate

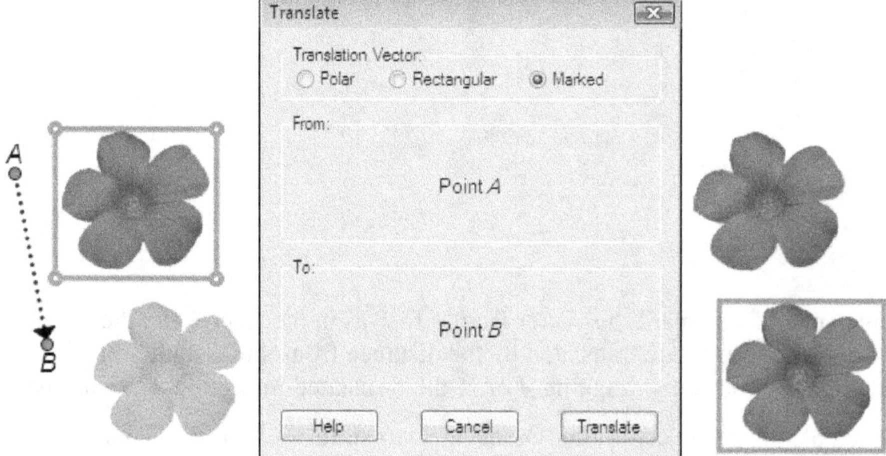


This Transform [192] menu command constructs a translated image of the selected geometric object(s).

Sketchpad has both a **Translate Arrow** [104] tool and a **Translate** command. When you use the tool, you translate the original object. When you use the command, you create a new object — a translated image of the original object.

 Note

You can translate points [2], straight objects [90], circles [6], arcs [7], interiors [8], and pictures [18]. You cannot translate iterated images [24], and you cannot apply the **Translate** command to loci [10], function plots [52], and other sampled objects [95]. (However, you can use a custom transformation [211] to translate point loci [10], function plots [52], parametric plots [54], and sampled transformed paths [89].)

To construct a translated image:

1. Select the object(s) you want to translate.

2. Choose **Transform** [192] | **Translate.** The Translate dialog box appears, and a translated image of your selections appears in the sketch.

3. Choose one of the **Translation Vector** options (**Polar** [201], **Rectangular** [201], or **Marked** [202]), depending on how you want to specify the translation. The **Marked** option is available only if you've already marked a vector [198]. (You can mark a vector now by clicking in the sketch on the two points that define the vector.)

4. Specify any required values for the type of translation you've chosen. You can type values into edit boxes, or you can click objects in the sketch. (Click a distance value to changed the marked distance, an angle value to change the marked angle, or two points to change the marked vector.) While you're choosing your values, you can see in the sketch a copy of the translated image that will be created.

5. When you're satisfied with the translation you've specified, click **Translate.**

*See also:*
> *How to Construct a Segment of Fixed Length* [182]

### 3.5.8.1  Polar Translation Vector

Choose this option to define a translation by specifying a distance and an angle.

> When you specify a polar vector, the initial direction (an angle of 0° or 0 radians) is to the right. A positive angle indicates a counter-clockwise rotation from 0°, and a negative angle indicates a clockwise rotation from 0°.

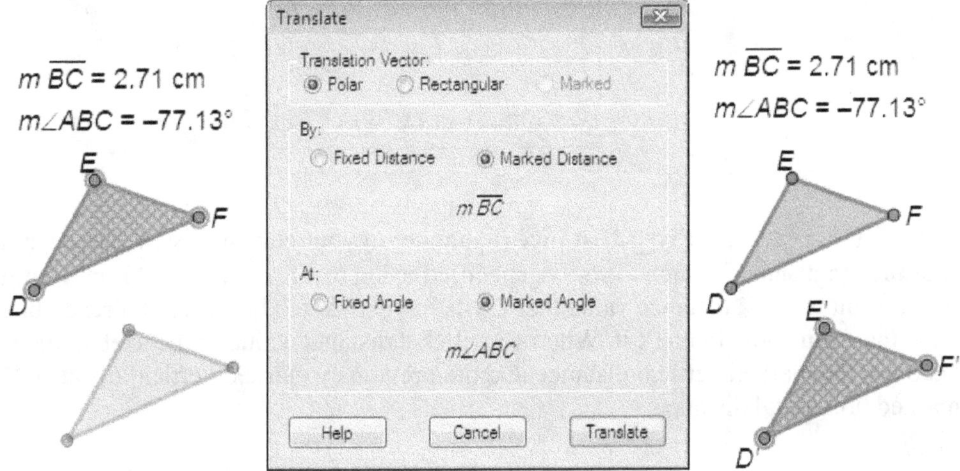


The distance can be a **Fixed Distance** (a number in your current distance unit), or it can be a **Marked Distance** [199] (a distance value [92] you've specified using the **Transform | Mark Distance** [199] command). If you want to use a distance value that exists in your sketch but has not already been marked, click that value now to mark it.

The angle can be a **Fixed Angle** (a number in your current angle unit), or it can be a **Marked Angle** [195] (an angle value [92] you've specified using the **Transform | Mark Angle** [195] command). If you want to use an angle value that exists in your sketch but has not already been marked, you can click that value now to mark it.

*See also:*
> *Rectangular Translation Vector* [201]
> *Marked Translation Vector* [202]

### 3.5.8.2  Rectangular Translation Vector

Choose this option to define the translation by specifying a horizontal distance and a vertical distance.

> When you specify a rectangular vector, positive distances indicate translation to the right or up, and negative distances indicate translation to the left or down.

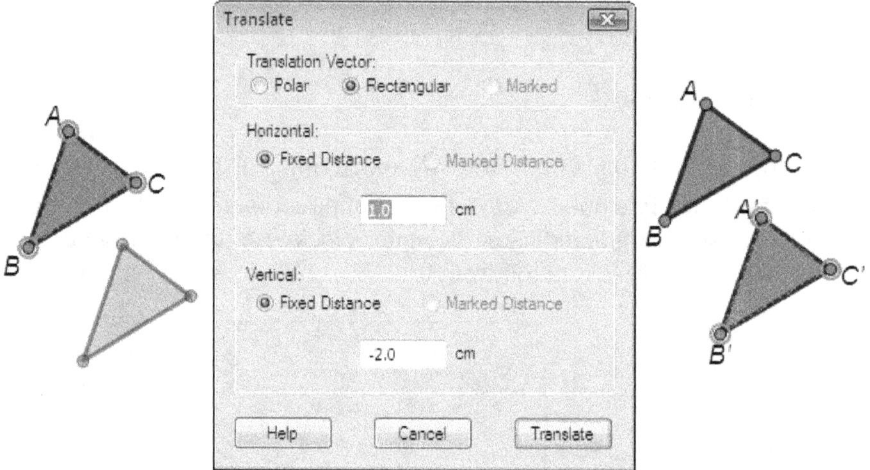

Each distance can be a **Fixed Distance** (a number in your current distance unit), or a **Marked Distance** (a distance value [92] you've specified using the **Transform | Mark Distance** [199] command). If you want to use a distance value that exists in your sketch but has not already been marked, you can click that value now to mark it. When you click a distance value in the sketch, the value you click becomes the marked vertical distance and the previously marked vertical distance becomes the marked horizontal distance.

*See also:*
   *Polar Translation Vector* [201]
   *Marked Translation Vector* [202]

### 3.5.8.3   Marked Translation Vector

Choose this option to define the translation by specifying a vector's initial and terminal points.

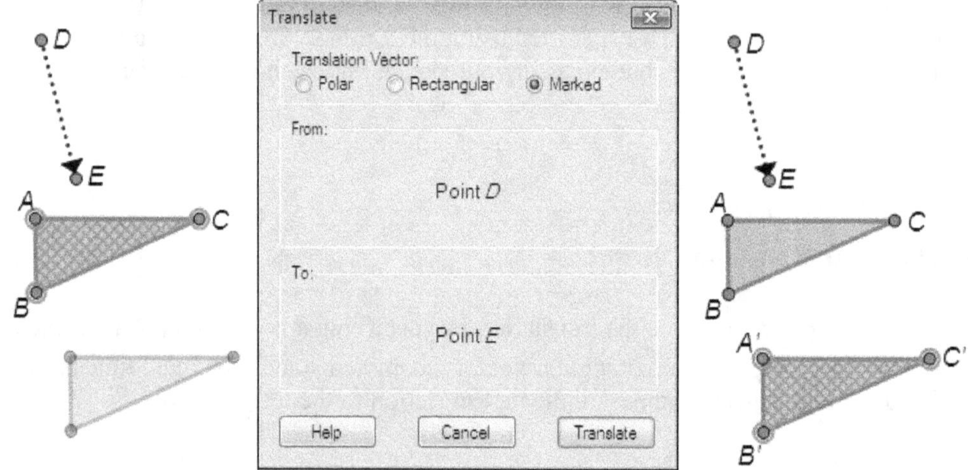

This option is enabled only if you've already marked a vector in your sketch. If you haven't already used the **Transform | Mark Vector** [198] command to mark a vector, you can mark one now by clicking two points in your sketch — first the initial point, and then the terminal point.

*See also:*

*Polar Translation Vector* [201]
*Rectangular Translation Vector* [201]

## 3.5.9    Rotate

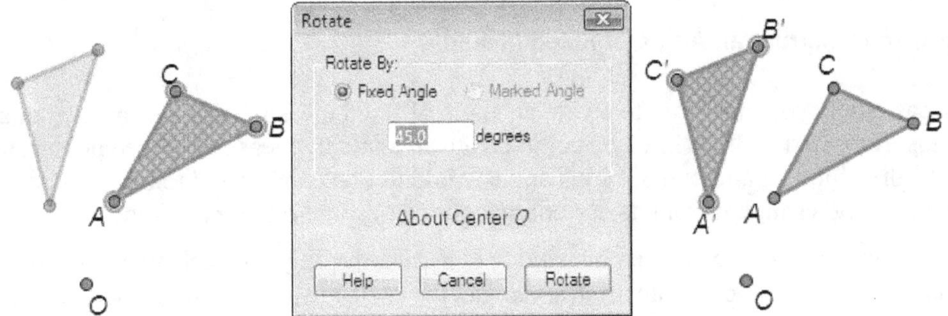


This Transform [192] menu command constructs a rotated image of the selected geometric object(s).

Sketchpad has both a **Rotate Arrow** [104] tool and a **Rotate** command. When you use the tool, you rotate the original object. When you use the command, you create a new object — a rotated image of the original object.

 Note

You can rotate points [2], straight objects [90], circles [6], arcs [7], interiors [8], and pictures [18]. You cannot rotate iterated images [24], and you cannot apply the **Rotate** command to loci [10], function plots [52], and other sampled objects [95]. (However, you can use a custom transformation [211] to rotate point loci [10], function plots [52], parametric plots [54], and sampled transformed paths [89].)

Before choosing this command, you may want to mark a center point [194] and mark an angle [195].

Mark a center point by double-clicking with the **Arrow** [104] tool. If you rotate without first marking a center, Sketchpad will mark one for you.

To rotate objects:

1. Select the object(s) you want to rotate.

2. Choose **Transform** [192] | **Rotate.** The Rotate dialog box appears, and a rotated image of your selections appears in the sketch.

3. Choose either **Fixed Angle** or **Marked Angle,** as described below.

4. You can click a point in the sketch to change the marked center, or you can click an angle marker [60] or angle value [92] to change the marked angle.

5. When you have chosen the options you want and entered any required values, click **Rotate.**

The rotated image appears.

### Rotate by Fixed Angle

Choose **Fixed Angle** to enter a fixed (numeric) angle of rotation using your sketch's current angle units.

Positive angles result in counter-clockwise rotations, and negative angles result in clockwise rotations.

## Rotate by Marked Angle

Choose **Marked Angle** to rotate your selection based on an angle you've specified using the **Transform | Mark Angle** |195| command. This choice is disabled if you haven't already marked an angle. However, if you want to use an  angle marker or angle value that exists in your sketch but has not already been marked, click that value now to mark it.

### 3.5.9.1    How to Construct an Angle of Fixed Measure

Occasionally you may want to create an angle of fixed measure — for example, an angle that measures exactly 33°. While you could create an angle by measuring three points and dragging them until they form an angle of 33°, this angle would not be constructed to be fixed at 33°. Dragging it again would change it from its current magnitude to some other magnitude.

To fix an angle in Sketchpad, you need to construct the angle in such a way that dragging cannot change its magnitude. For an arbitrary angle, the easiest way to do this is with the Transform |192| menu.

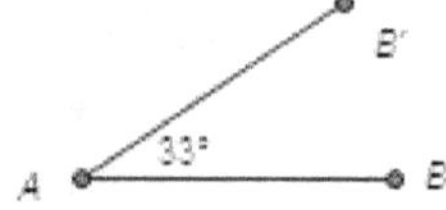


1. Use the **Point** |112| tool to to construct two points, *A* and *B,* in your sketch.

2. Select point *A* and choose **Transform | Mark Center** |194|. Point *A* is marked as the center of future rotations and dilations.

3. Select point *B* and choose **Transform | Rotate** |203|.

4. In the dialog box, enter the fixed angle by which you want to rotate. Click the **Rotate** button to confirm your choices.

5. Sketchpad constructs point *B'* as the rotated image of *B* by your requested angle. Even if you drag *A, B,* or *B',* Sketchpad will maintain this angle's magnitude, because you've defined *B'* to be the rotated image of *B* by this angle.

You can now construct rays or segments connecting *A* to *B* and to *B'* to incorporate the fixed angle into your sketch.

*See also:*
*How to Construct a Segment of Fixed Length* |182|
*How to Construct Congruent Angles* |204|

### 3.5.9.2    How to Construct Congruent Angles

Occasionally you may want to create an angle that's congruent to an existing angle. You can also use these methods to help make a triangle or other polygon that's congruent to an existing triangle or polygon, but can be positioned and rotated as you please while always remaining congruent to the original.

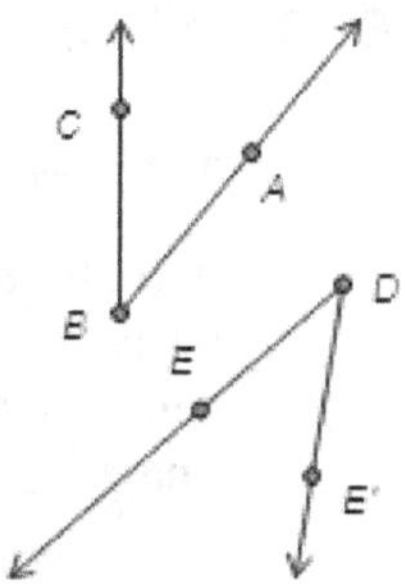

For this construction you'll use the **Transform | Rotate** 203 command to make a new angle congruent to an original angle ∠*ABC*.

1. Use the **Ray** 115 tool to construct a starting ray *DE*. This will be the initial side of the new angle. (If you have an existing straight object, you can use it instead of the ray.)

2. Mark endpoint *D* of the ray or other straight object as the center of rotation by double-clicking it with the **Arrow** 104 tool. This point will be the vertex of the new angle.

3. Select points *A, B,* and *C* that define the original angle, and choose **Transform | Mark Angle** 195.

4. Select point *E,* the other point that determines the initial side of the angle.

5. Choose **Transform | Rotate** 203 to rotate point *E* by the marked angle. The rotated image is point *E'.*

6. Use the **Ray** 115 tool to construct the ray from *D* through the new point *E'.*

7. Test your construction by changing the original angle, and by dragging points *D* and *E* that determine the initial side of the new angle.

> If you like, use an existing straight object in place of ray *DE.* You could even use a segment that you've constructed to be congruent to another segment 184.

## Construct a Congruent Triangle

To use ASA (Angle - Side - Angle) to construct a new triangle congruent to an existing triangle, begin with a congruent segment 184. You'll use this as the initial side of the angles at the two endpoints of the segment. Then use the congruent angle construction twice (steps 2 through 6) to construct two angles, one on each end of the segment, that are congruent to the corresponding angles in the original triangle.

Similarly, you can use SAS to construct a new triangle congruent to an existing triangle. In this case, begin with the congruent angle construction to make an angle congruent to an angle from the original triangle. Then use **Construct | Circle by Center+Point** 182 to make two circles centered at the vertex of the new angle and with radii equal to the lengths of the two sides of the original triangle. The points where the rays intersect the circles are the other two vertices of the  new triangle.

How would you use SSS for the same purpose?

And what happens if you try using SSA?

*See also:*
   *How to Construct a Segment of Fixed Length* 182
   *How to Construct an Angle of Fixed Measure* 204
   *How to Construct Congruent Segments* 184

## 3.5.10  Dilate

This Transform [192] menu command constructs a dilated image of the selected geometric object(s).

In a dilated image, a designated ratio is used to move every point of the original closer to or farther away from the center point. If the ratio is greater than 1, the image points are farther away from the center than the originals and the image is larger than the original image. If the ratio is less than 1, the image points are nearer to the center and the image is smaller.

> Sketchpad has both a **Dilate Arrow** [104] tool and a **Dilate** command. When you use the tool, you dilate the original object. When you use the command, you create a new object — a dilated image of the original object.

 Note

> You can dilate points [2], straight objects [90], circles [6], arcs [7], interiors [8], and pictures [18]. You cannot dilate iterated images [24], and you cannot apply the **Dilate** command to loci [10], function plots [52], and other sampled objects [95]. (However, you can use a custom transformation [211] to dilate point loci [10], function plots [52], parametric plots [54], and sampled transformed paths [89].)

Before choosing this command, you may want to mark a center point [194] and mark a ratio or scale factor [197].

> Mark a center point by double-clicking with the **Arrow** [104] tool. If you dilate without first marking a center, Sketchpad will mark one for you.

1. Select the object(s) you want to dilate.

2. Choose **Transform** [192] | **Dilate.** The Dilate dialog box appears, and a dilated image of your selections appears in the sketch.

3. Choose either **Fixed Ratio** or **Marked Ratio,** as described below.

4. Click a point in the sketch to change the marked center, click a measurement [36] with no units to set the marked scale factor [197], or click two segments to set the marked segment ratio.

5. When you have chosen the options you want and entered any required values, click **Dilate.**

The dilated image appears.

### ▼ Dilate by Fixed Ratio

> Choose **Fixed Ratio** to designate a fixed ratio by entering both a numerator and a denominator. A ratio smaller than 1 results in an image that's smaller than the original, and a ratio greater than 1 results in an image that's larger.

### ▼ Dilate by Marked Ratio

Choose **Marked Ratio** to dilate your selection based on a marked ratio or scale factor you've specified using the **Transform | Mark Ratio** [197] command. This choice is disabled if you haven't already marked a ratio. However, if you want to use a ratio of two segments that exist in your sketch, click the segments now to mark them. Similarly, if you want to use a scale factor that exists in your sketch, click it now to mark it.

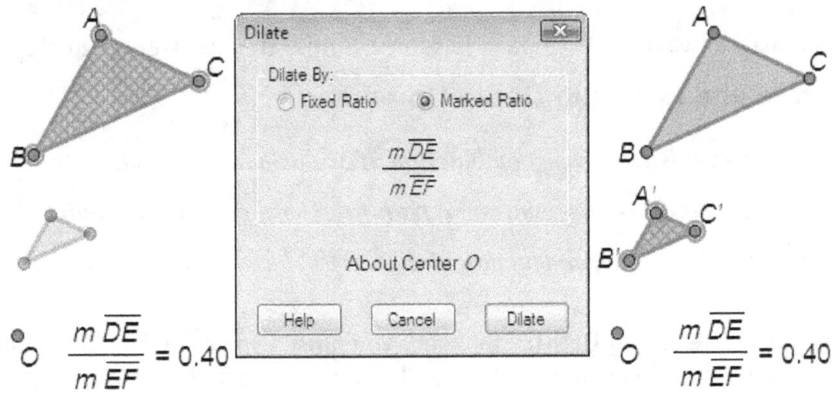

## 3.5.11  Reflect

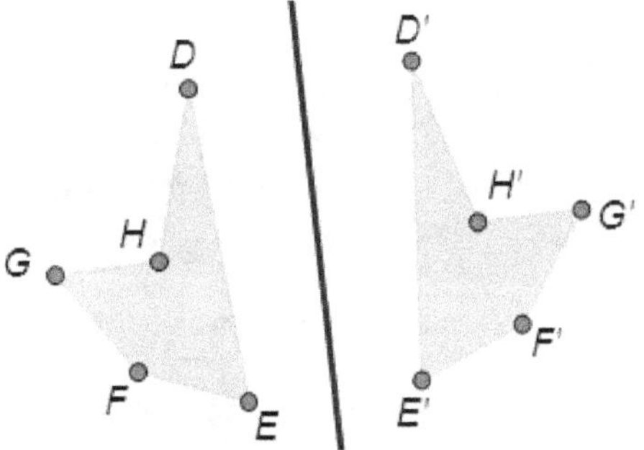

This Transform [192] menu command constructs a mirror image of the selected geometric object(s) across a marked mirror.

 Note

You can reflect points [2], straight objects [90], circles [6], arcs [7], interiors [8], and pictures [18]. You cannot reflect iterated images [24], and you cannot apply the **Reflect** command to loci [10], function plots [52], and other sampled objects [95]. (However, you can use a custom transformation [211] to reflect point loci [10], function plots [52], parametric plots [54], and sampled transformed paths [89].)

Before choosing this command, you may want to mark a mirror by selecting a straight object and choosing **Transform | Mark Mirror** [195].

You can also mark a mirror by double-clicking a straight object with the **Arrow** [104] tool. If you reflect without first marking a mirror, Sketchpad will mark one for you.

A brief animation indicates that the straight object has been marked as a mirror for subsequent reflections.

To reflect objects:

1. Select the object(s) you want to reflect.

2. Choose **Transform** [192] | **Reflect.**

The reflected image appears.

## 3.5.12  Iterate

This Transform [192] menu command has three forms: **Iterate, Iterate to Depth,** and **Terminal Point.**

### ▼ Iterate and Iterate to Depth Commands

*Selection prerequisites: Some combination of one or more pre-image "seed" objects:*

- *Independent points [2] or points on a path [89]. These points must define other points in your sketch.*

- *Parameters [36] or independent calculations [39]. These values must define other values in your sketch.*

To enable the **Iterate to Depth** command, you must also select a single value that determines the depth of iteration. This value must be the last object selected, and you must hold the Shift key when choosing the command.

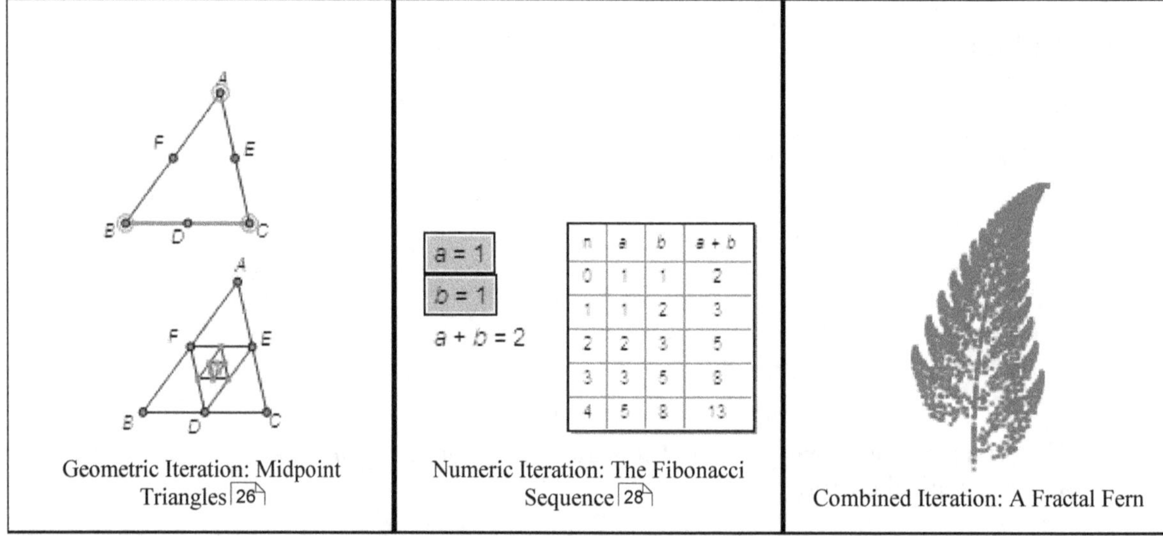

| Geometric Iteration: Midpoint Triangles [26] | Numeric Iteration: The Fibonacci Sequence [28] | Combined Iteration: A Fractal Fern |

When you choose either **Iterate** or **Iterate to Depth,** the Iterate dialog box [209] appears so that you can specify the mapping that defines the iteration.

When you choose **Iterate to Depth,** the last value selected dynamically controls the depth of the iteration [30].

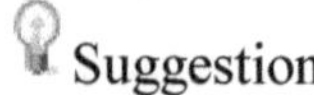 Suggestion

> Sometimes you may want to create an iteration that begins with a constructed point or a calculated value, even though the **Iterate** command does not allow you to use such points or values as its prerequisites. You can often get around this limitation. First construct the iteration separately, creating independent points and parameters to serve as the seed objects. After you've created the iteration, merge the independent points to your desired constructed points, and edit your parameters to change them into calculations based on other values in the sketch.

## ▼ Terminal Point Command

*Selection prerequisites: A single iterated point image*

This command constructs a point at the position corresponding to the very last step of the iteration. If the depth of the iteration $\boxed{30}$ changes, the terminal point $\boxed{30}$ moves accordingly.

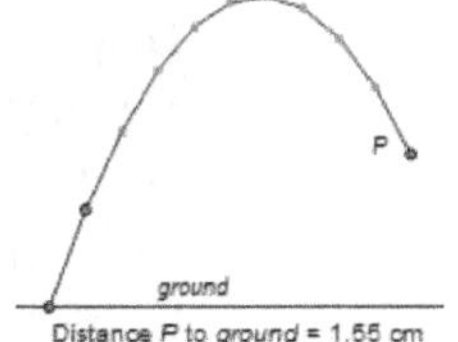


*Subtopics:*


*See also:*


### 3.5.12.1  Using the Iterate Dialog Box

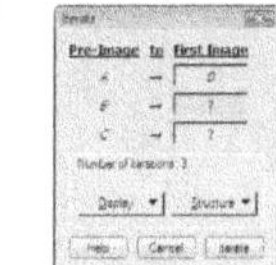

When you choose **Transform** $\boxed{192}$ | **Iterate** $\boxed{208}$, the Iterate dialog box appears.
Drag the Iterate dialog box out of the way if it's hiding the destination images you want to click.

Use the top part of the dialog box to specify the mapping used at each step to create the next iteration image. Fill in each box of the right-hand column by clicking in the sketch on the object to which each pre-image object should be mapped.

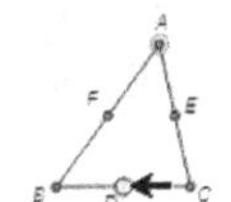

Suggestion

To make this easier, Sketchpad highlights in the sketch the pre-image object currently being matched. For instance, in this example, Sketchpad highlights point $A$ in the sketch when it's time to click the object that point $A$ maps to. Sketchpad then highlights point $B$ in the sketch when it's time to click the object that $B$ maps to, and so forth.

## ▼ Use the Display Pop-up Menu to Control the Appearance of the Iteration.

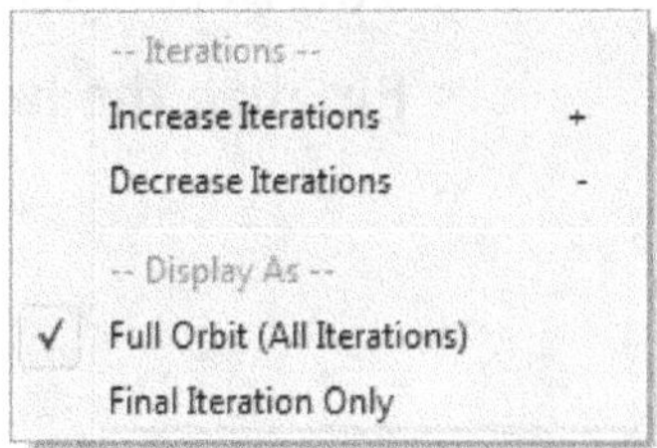


**Increase or Decrease Iterations:** Increase or decrease the number of times your rule is iterated. You can also use the + or − keys, without opening the **Display** pop-up menu, to increase or decrease the number of iterations.

When you're creating an iteration using parametric depth [30], you cannot change the number of iterations by hand.

**Display the Full Orbit or Final Iteration Only:** Display all iterations, or only the final iteration. When you display all iterations, the set of all images of an iterated object is sometimes called the *orbit* of that object.

It's often convenient to display all iterations when you're iterating using a single map, and to display only the final iteration when you're iterating using multiple iteration maps [211].

The **Transform** [192] | **Terminal Point** [30] command allows you to construct a final iterated point image as an actual point that you can use in further constructions.

### ▼ Use the Structure Pop-up Menu to Control the Structure of the Iteration.

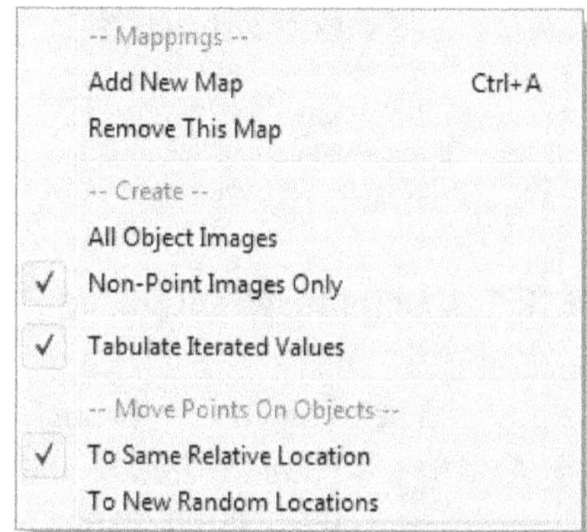


**Add/Remove Map:** Add a new iteration map [211] or remove the current map.

**Create Images:** Create images of all objects that depend on [80] the selections or create images of only the nonpoint objects.

 Suggestion

Often — especially when working with multiple maps — you won't want to see the iterated images of points, but only of segments, polygons, and so forth. Sketchpad automatically switches this option — to only create nonpoint object images — when you start working with multiple maps, though you can switch it back if you prefer to create point images as well as nonpoint images.

**Tabulate Values:** Determine whether to create a table of all iterated measurements [30].

**Move Points on Objects:** Set iterated points on objects [176] to stay at the same relative location as the original or set them to use new random locations [30] each time they're iterated.

### ▼ Finalize the Iteration

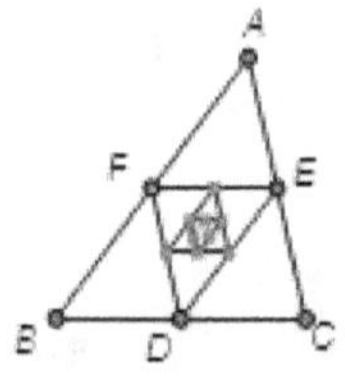

Once you've specified the destination for each selected pre-image, click **Iterate** to finalize the iteration.

The constructed iteration [24] appears. Use Iteration Properties [35] to change the depth or other characteristics of the iteration.

You can see step-by-step examples of using this dialog box to create a geometric iteration, [26] a numeric iteration, [28] and a  Sierpiński gasket [33] (using multiple iteration maps).

*See also:*
  *Iterations and Iterated Images* [24]

### 3.5.12.2  Multiple Iteration Maps

By specifying a destination point for each of the iterated independent points, you create an iteration map. This map describes how to transform the pre-image to create a transformed copy of the original objects. For most iterations, you create a single map, so that each iteration step produces a single copy of the original objects.

For some iterations, one step of the iteration produces two or more copies of the original objects. Each such copy of the original objects requires its own map, so such iterations require multiple maps.

<table>
<tr><td>

For example, a binary tree splits in two at each junction, so the iteration rule has two maps: one to produce the right branch (Map #1) and one to produce the left branch (Map #2).

</td><td>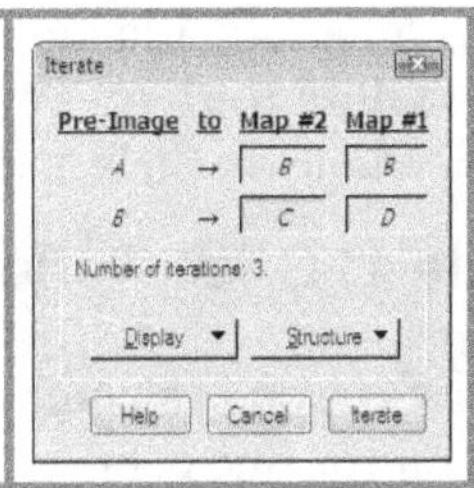</td><td>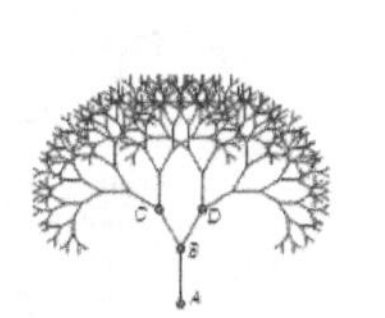</td></tr>
<tr><td>

Similarly, a parallelogram tessellation requires you to iterate the original parallelogram both horizontally (Map #1) and vertically (Map #2).

To create such an iteration, select the seed points and choose **Transform | Iterate** [208]. In the Iterate dialog box, [209] define the first map normally. Then choose **Add New Map** from the Structure pop-up menu and define the second map. After you define all the maps that make up your iteration rule, click **Iterate**.

</td><td>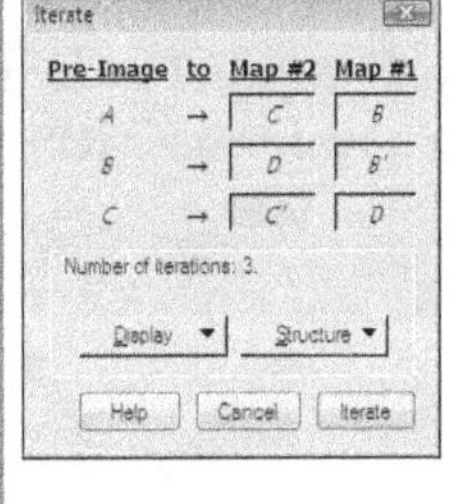</td><td>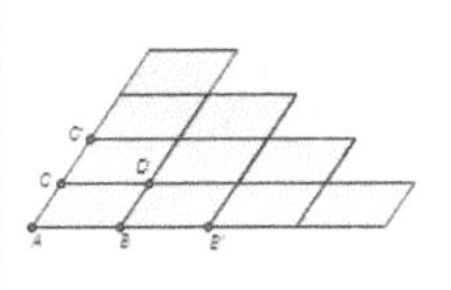</td></tr>
</table>

Fractals and tessellations are the most common geometric constructions for which the iteration rule requires multiple iteration maps.

See How to Construct a Sierpiński Gasket [33] for step-by-step directions to create a famous fractal.

*See also:*
  *Using the Iterate Dialog Box* [209]
  *Iterations and Iterated Images* [24]

## 3.5.13  Define Custom Transform

*Selection prerequisites: Two points, one of which depends on the other*

This Transform [192] menu command defines a custom transformation.

A custom transformation is a general transformation, defined by an example that involves two points: a pre-image point and an image point that depends on the pre-image point. When you apply such a transformation to another object, every point of the object undergoes the same transformation that relates the defining points.

The relationship between the pre-image and image points can involve any combination of geometric operations (constructions and transformations) and numeric and algebraic operations (measurements, calculations, functions, and plotting). This generality makes it possible to define an extremely wide variety of custom transformations.

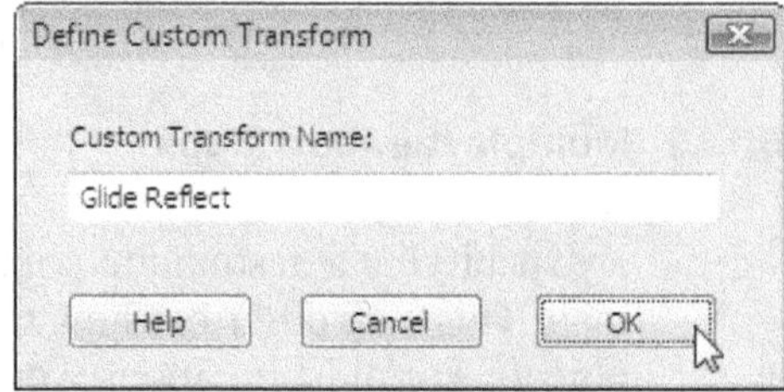

To create the transformation, select the pre-image and image points and choose **Transform** |192| | **Define Custom Transform.** In the dialog box that appears, give the transformation a name, such as **Glide Reflect.** This new transformation appears as a command at the bottom of the Transform menu. To apply this transformation to another object in your sketch, select that object and choose **Transform | Glide Reflect.** You can also apply a custom transformations using numbered keyboard shortcuts (in Windows Ctrl+1, Ctrl+2, and so forth; on Mac, ⌘1, ⌘2, and so forth).

### ▼ Example 1: Glide Reflect

A glide reflection is a combination of two basic transformations built into Sketchpad: a reflection across a mirror line combined with a translation parallel to the mirror. To define a glide reflection as a custom transformation, reflect point $A$ across a mirror, and then translate it by the vector from $B$ to $C$, parallel to the mirror. Use points $A$ and $A''$ to define the custom transformation. Then apply the transformation repeatedly to a pre-image picture to produce a glide reflection.

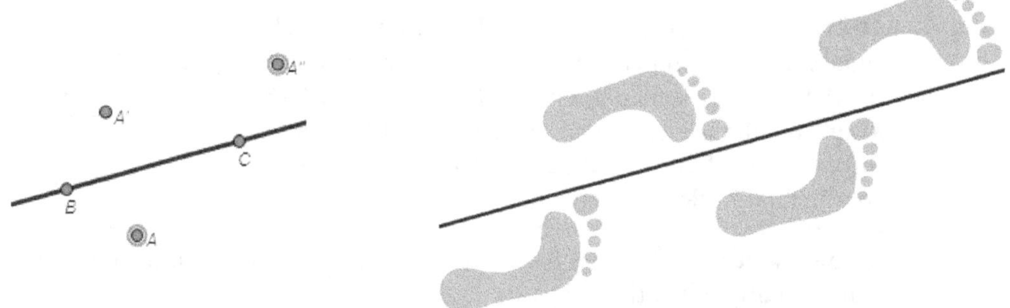

### ▼ Example 2: Scale Horizontally

Dilating |206| an object scales it both horizontally and vertically. To scale an object only horizontally but not vertically, you must define a custom transformation. One way of defining such a transformation is to measure the coordinates of the point to be scaled, multiply the $x$-value by a scale factor (leaving the $y$-value unchanged), and plot the scaled point $A'$ defined by the new coordinates. Apply the resulting transformation to a circle (to create an ellipse) or to a picture (to stretch it horizontally but not vertically).

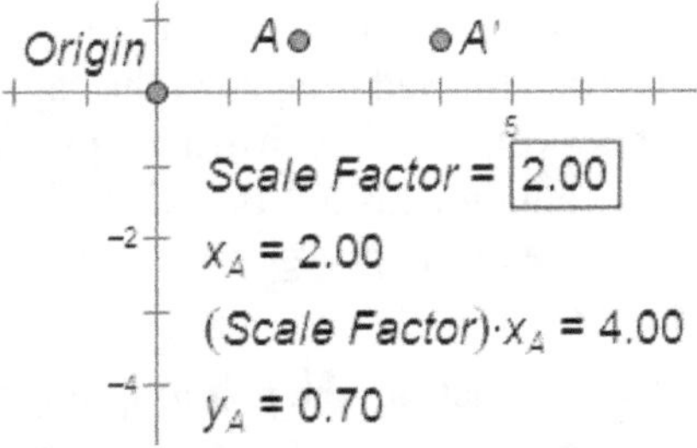

## ▼ Example 3: Swirl

One way to create a swirling effect is to start with an original point and rotate it about the origin, rotating by an angle that gets larger as the point gets farther from the origin. To define a swirl as a custom transformation, measure the polar coordinates $(r, \theta)$ of point $A$ (with $\theta$ in radians) and calculate a rotation $\alpha$ by multiplying $r$ by 0.3. Construct a rotated point by plotting point $A'$ at $(r, \theta + \alpha)$. Then use points $A$ and $A'$ to define the custom transformation. Applying this transformation to the pre-image on the left produces the image on the right. (The pre-image also shows the defining points $A$ and $A'$ and the origin point $B$ in the center of the picture.)

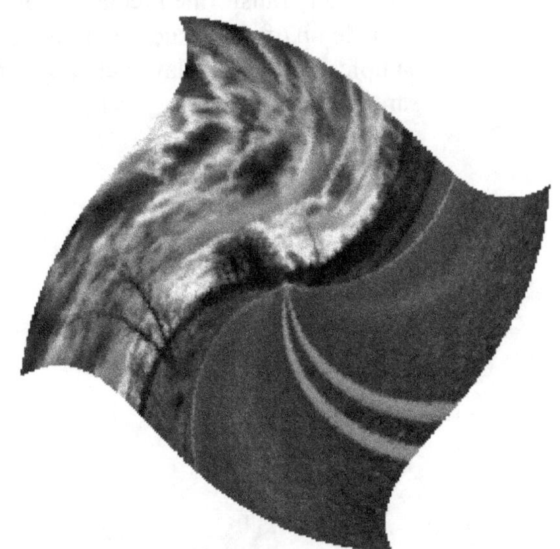

## ▼ Custom Transformed Images

When a custom transformation (such as a glide reflection) is made up of one or more of Sketchpad's basic transformations (translations, rotations, dilations, and/or reflections), images created by the custom transformation are always similar to their pre-images. Angles are the same, and distances are always in the same ratio. Transformed images of straight objects are remain straight objects, transformed images of circles remain circles, and transformed polygons are similar to their pre-image polygons. Consequently, you can use these transformed images in the same way as their pre-images: you can find the midpoint of a transformed segment, you can find the radius of a transformed circle, and you can measure the perimeter and area of a transformed polygon.

But when a custom transformation (such as horizontal scaling or swirling) includes steps different from the basic transformations, there is no guarantee that the resulting images are similar to their pre-images. When a straight object (like the yellow line on the road in the example) is transformed, the result may be curved. When a circle is transformed, the result may be an ellipse, or may be some entirely different shape. The image of a polygon may not have straight edges. With such a transformation, you cannot use the transformed images in the same way as the pre-images. The image of a segment is not straight, so you cannot construct a parallel or perpendicular to it. The image of a circle isn't circular, so you cannot measure its radius. The image of a polygon does not have straight edges, and the image of the polygon's interior may even be outside of its frame.

When a path object is transformed by such a transformation, the resulting image is a sampled transformed path 90 such as the stretched circle in Example 2 above. Sketchpad calculates the shape of the transformed path by using points on the pre-image path and transforming them to locate the image path. Each such transformed point produces one sample on the transformed path; the full image path is made up of many such samples. Choose **Edit | Properties | Plot** 95 to set the number of samples used to display such an object and to determine whether it's displayed continuously or discretely.

When a picture is transformed by such a transformation, the resulting image is a sampled transformed picture 23; Sketchpad calculates the transformed picture by using small rectangles from the pre-image picture and transforming them to display the resulting transformed picture. Each such transformed rectangle produces one sample of the transformed picture; the full picture is made up of many such samples. Choose **Edit | Properties | Plot** 98 to set the number of samples used to display such a picture and to determine how much distortion is allowed in the sample rectangles.

## 3.5.14   Edit Custom Transforms

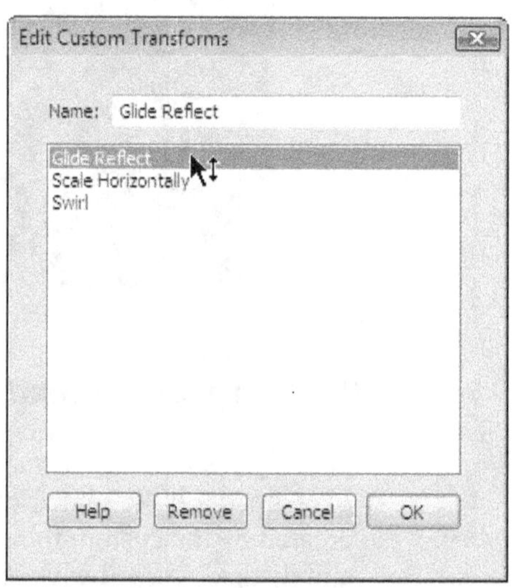


This Transform 192 menu command enables you to rename, reorder, or remove a custom transformation.

Rename an existing transformation by typing a new name in the **Name** field (Windows) or by double-clicking the name in the list (Mac).

Remove a custom transformation by selecting it and pressing **Remove.**

Reorder custom transformations by dragging them in the list.

If you delete either of the points that define your transformation, or if you change the construction so that the image point no longer depends on the pre-image point, the transformation becomes invalid. In this case you can either remove the custom transformation, or you can change the sketch (by editing it or using undo) to restore the connection between the two points.

You cannot undo the removal of a custom transformation. Instead you must select the two points and recreate the transformation.

## 3.6    Measure Menu

The Measure menu allows you to measure numeric properties [36] of selected objects. The commands in the top portion of the menu measure objects' geometric properties; the commands in the bottom portion measure analytic properties.

> If you want to measure a specific property and its command is not available, make sure you have only that command's prerequisites selected. (You may have too many or too few objects selected.)

Detailed information is available for each command:

Length [217]
Distance [218]
Perimeter [218]
Circumference [218]
Angle [218]
Area [219]
Arc Angle [219]
Arc Length [220]
Radius [221]
Ratio [221]
Value of Point [221]
Coordinates [223]
Abscissa (x) [223]
Ordinate (y) [224]
Coordinate Distance [224]
Slope [224]
Equation [225]

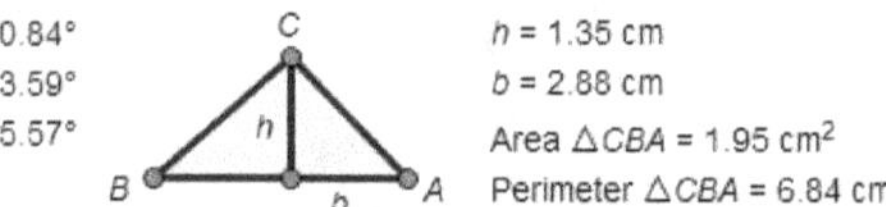


To measure an object's properties, select the object and choose from the available commands in the Measure menu. Sketchpad produces a measurement — a named numeric value [36] in the proper units — as the result. When you drag or change an object that you've measured, the measured value changes accordingly.

To use any of these commands, you must select the appropriate objects (the *prerequisites*) in the sketch.

| Command: | Prerequisites: |
| --- | --- |
| **Length** [217] | One or more segments [5] |
| **Distance** [218] | Two points [2], or one point and one straight object [5] |
| **Perimeter** [218] | One or more polygon, arc sector, or arc segment interiors [8] |

| | |
|---|---|
| **Circumference** [218] | One or more circles [6] or circle interiors [8] |
| **Angle** [218] | Three points [2] (select the vertex as the second point) |
| **Area** [219] | One or more interiors [8] or circles [6] |
| **Arc Angle** [219] | One or more arcs [7], or a circle [6] and two or three points [2] on the circle |
| **Arc Length** [220] | One or more arcs [7], or a circle [6] and two or three points [2] on the circle |
| **Radius** [221] | One or more circles [6], circle interiors [8], arcs [7], or arc interiors [8] |
| **Ratio** [221] | Two segments [5] or three collinear points [2] |
| **Value of Point** [221] | One or more points [2] on paths [89], or one path and one point not on the path |
| **Coordinates** [223] | One or more points [2] |
| **Abscissa (x) or** [223] **Polar Distance (d)** [223] | One or more points [2] |
| **Ordinate (y) or** [224] **Polar Angle)** [224] | One or more points [2] |
| **Coordinate Distance** [224] | Two points [2] |
| **Slope** [224] | One or more straight objects [5] |
| **Equation** [225] | One or more lines [5] or circles [6] |

## 3.6.1 Length

$A$ •———• $B$

$m\ \overline{AB} = 2.65$ cm

*Selection prerequisites: One or more segments* [5]

This Measure [216] menu command measures the length of each selected segment, using the distance units chosen on the Units panel [276] of the Preferences [275] dialog box.

The length of a segment is equal to the distance [218] between its endpoints.

## 3.6.2  Distance

$A$ ∘    ∘$B$

$AB$ = 2.12 cm

∘$C$

$D$ ∘————————∘$E$

Distance $C$ to $\overline{DE}$ = 0.92 cm

*Selection prerequisites: One of the following combinations of point(s) and straight object(s):*

- *Two points* 2

- *One or more points* 2 *and one straight object* 90

- *One point* 2 *and one or more straight objects* 90

This Measure 216 menu command measures the distance between two points, or the distance from a point to a straight object, using the distance units chosen on the Units panel 276 of the Preferences 275 dialog box.

Select several points and one straight object to measure the distance from each of the points to the straight object.

Select one point and several straight objects to measure the distance from the point to each of the straight objects.

The distance from a point to a line is the shortest distance from the point to the line and is measured along the perpendicular. The distance from a point to a ray or segment is defined to be the same as the distance from the point to the straight line that contains the ray or segment.

>  To measure the distance between two points on a coordinate system in grid units, use the Coordinate Distance 224 measurement instead.

## 3.6.3  Perimeter

*Selection prerequisites: One or more polygons, arc sectors, or arc segment interiors* 8

This Measure 216 menu command measures the perimeter of each selected polygon or arc interior, using the distance units chosen on the Units panel 276 of the Preferences 275 dialog box.

The perimeter of an arc sector is the sum of the arc length and the lengths of the two radii bounding the arc sector. The perimeter of an arc segment is the sum of the arc length and the length of the chord bounding the arc segment.

## 3.6.4  Circumference

*Selection prerequisites: One or more circles* 6 *or circle interiors* 8

This Measure 216 menu command measures the circumference of each selected circle or circle interior, using the distance units chosen on the Units panel 276 of the Preferences 275 dialog box.

## 3.6.5  Angle

*Selection prerequisites: One of the following:*

- *Three points* 2

- *Two straight objects* 90 *with a common endpoint, optionally including the common endpoint*

- *One or more angle markers* [60]

This Measure [216] menu command measures an angle using the angle units chosen on the Units panel [276] of the Preferences [275] dialog box.

If you select three points, the first selected point defines the initial side of the angle, the second point defines the vertex, and the third point defines the terminal side of the angle.

> Be sure to select the vertex point as the second point.

If you select two straight objects that share an endpoint, the first straight object defines the initial side and the second defines the terminal side. You can select the common vertex or not, as you prefer.

> It's often convenient to use a selection rectangle to select both straight objects and their common vertex.

If you select an angle marker, the initial and final sides are determined by the angle marker.

If angle units in Preferences [276] are set to **directed degrees** or **radians,** the value of the measurement can be either positive or negative. A counter-clockwise angle results in a positive measurement, and a clockwise angle results in a negative measurement. Possible values for angles defined by three points or by two straight objects range from $-180°$ to $+180°$, or from $-\pi$ radians to $\pi$ radians.

If the angle units are set to **degrees,** all angle measurements are positive, and possible values for angles defined by three points or by two straight objects range from $0°$ to $180°$ .

Possible values for an angle defined by an angle marker depend on angle unit preferences and on whether the angle marker is simple, reflex, counter-clockwise, or clockwise [60].

## 3.6.6  Area

*Selection prerequisites:* One or more interiors [8] or circles [6]

This Measure [216] menu command measures the area of each selected polygon, circle, circle interior, arc segment interior, and arc sector interior, using the distance units chosen on the Units panel [276] of the Preferences [275] dialog box.

## 3.6.7  Arc Angle

*Selection prerequisites:* One of the following:

- *One or more arcs* [7]

- *A circle* [6] *and two points on the circle*

- *A circle* [6] *and three points on the circle*

This Measure [216] menu command measures the angle of each selected arc using the angle units chosen on the Units panel [276] of the Preferences [275] dialog box.

If one or more arcs [7] are selected, this command measures the angle of each selected arc. If the angle units in Preferences are set to **degrees** or **directed degrees**, the value ranges from $0°$ to $360°$. If the angle units are set to **radians,** possible values range from 0 radians to $2\pi$ radians.

> *Arc angle* (sometimes also called arc measure) refers to the central angle — that is, the angle formed by the radii connecting the circle's center to the endpoints of the arc, as shown here.

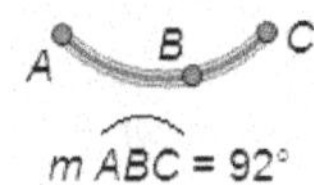

If a circle and two points are selected, this command measures the angle of the minor arc on the circle defined by the two selected endpoints. If the angle units in Preferences are set to **degrees**, possible values range from 0° to 180°. If the angle units in Preferences are set to **directed degrees**, values range from −180° to 180°; clockwise arcs result in negative values, and counter-clockwise arcs result in positive values. If the angle units are set to **radians**, possible values range from −$\pi$ radians to $\pi$ radians.

$m\ \overset{\frown}{CD}$ on $\odot AB$ = 76°

$m\ \overset{\frown}{ED}$ on $\odot AB$ = −95°

If a circle and three points are selected, this command measures the angle of the minor or major arc that starts at the first selected point, passes through the second point, and ends at the third point. If the angle units in Preferences are set to **degrees**, possible values range from 0° to 360°. If the angle units in Preferences are set to **directed degrees**, possible values range from −360° to 360°; clockwise arcs result in negative values, and counter-clockwise arcs result in positive values. If the angle units are set to **radians**, possible values range from −$2\pi$ radians to $2\pi$ radians.

$m\ \overset{\frown}{CED}$ on $\odot AB$ = −284°

*See also:*
  *Arc Length command* 220

## 3.6.8   Arc Length

***Selection prerequisites:*** *One of the following:*

- *One or more arcs* 7

- *A circle* 6 *and two points on the circle*

- *A circle* 6 *and three points on the circle*

This Measure 216 menu command measures the length of each selected arc using the distance units chosen on the Units panel 276 of the Preferences 275 dialog box.

If one or more arcs are selected, this command measures the length of each selected arc.

Length $\overset{\frown}{ABC}$ = 3.47 cm

If a circle and two points are selected, this command measures the length of the minor arc on the circle defined by the two selected endpoints. .

Length $\overset{\frown}{CD}$ on $\odot AB$ = 2.18 cm

If a circle and three points are selected, this command measures the length of the minor or major arc on the circle that starts at the first selected point, passes through the second point, and ends at the third point.

Length $\overset{\frown}{CED}$ on $\odot AB$ = 8.09 cm

*See also:*
  *Arc Angle command* 219

### 3.6.9　Radius

*Selection prerequisites: One or more circles* 6 *, circle interiors* 8 *, arcs* 7 *, or arc interiors* 8

This Measure 216 menu command measures the radius of each selected object using the distance units chosen on the Units panel 276 of the Preferences 275 dialog box.

The radius of an arc is the same as the radius of the circle on which the arc falls.

### 3.6.10　Ratio

*Selection prerequisites: Two segments* 5 *or three collinear points* 2

Collinear points are points that lie on the same straight line.

This Measure 216 menu command measures the ratio of two lengths or distances.

- If two segments are selected, this command measures the ratio of the length of the first selected segment to the length of the second.

- If three collinear points $A$, $B$, and $C$ are selected in order, this command measures the ratio of the distance $AC$ to the distance $AB$.

  If points $B$ and $C$ are on the same side of $A$, the ratio is positive; if $B$ and $C$ are on opposite sides of $A$, the ratio is negative.

Another way to think of the ratio defined by three collinear points is to think of a number line with its origin at point $A$ and its unit point at $B$. The position of point $C$ on this number line determines the value of the measurement.

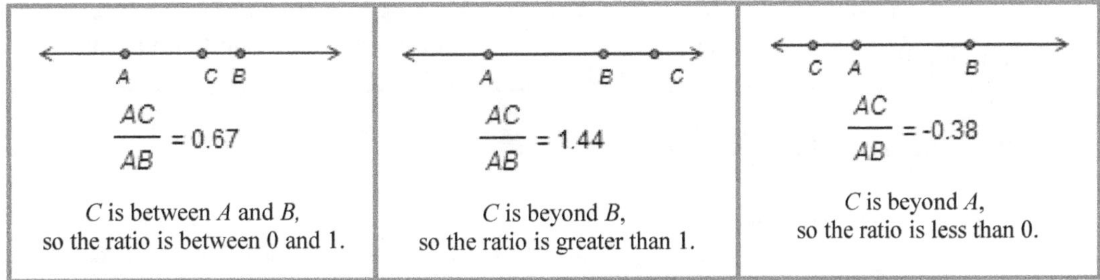


If point $C$ is constructed on straight object $AB$, the ratio measurement is the same as the value of point $C$ on its path. In this case it's easier to select only point $C$ and choose **Measure** 216 | **Value of Point** 221.

### 3.6.11　Value of Point

*Selection prerequisites: One or more points* 2 *constructed on path objects* 89 *, or one path object and one point not on the path*

If the selection is a path and a point not on the path, you must hold the Shift key to enable the command.

This Measure 216 menu command measures the selected point's relative position on the path on which it is constructed. The measured value depends on the type of path.

If a path and a point not on the path are selected, the command measures the value of the location on the path that's closest to the selected point.

For some path objects, including axes, function plots, and parametric loci, the value of a point is determined explicitly by the numbering of the axis or by the value of the parameter. For other path objects, the value of a point is determined relative to the defining points of the path. For instance, points on segments, rays, lines, or arcs are measured relative to the two defining points, which are assigned values of 0 and 1. Points on closed paths (including circles and interiors) are measured relative to a defined starting point that is assigned a value of 0; these values go from 0 up to (but not including) 1.

| Object | Possible Values |
|---|---|
| Segment | $0 \le t \le 1$ |
| Ray | $0 \le t$ |
| Line or Axis | all real numbers |
| Circle | $0 \le t < 1$ |
| Arc | $0 \le t \le 1$ |
| Interior (Circle Interior, Polygon, or Arc Interior) | $0 \le t < 1$ |
| Point Locus defined by a point driver | $0 \le t \le 1$ |
| Point Locus defined by a parameter driver | domain of driving parameter |
| Function Plot | domain of independent variable |

*Subtopic:*
  *How to Construct a Slider* 222

*See also:*
  *Plot Value on Object command* 237

## 3.6.11.1  How to Construct a Slider

Often it is useful to construct a slider for controlling a numeric value. For example, if you wanted to create a graph 233 of the line $y = mx + b$ in which you can slide a point back and forth to control the values of $m$ and $b,$ you might want to set up sliders for $m$ and $b$. To make a slider for $m$:

1. Use the **Line** 115 tool to construct a line. Label the two resulting points $A$ and $B$.

2. Use the **Point** 112 tool to construct point $C$ on the line.

3. Select point $C$ and choose **Measure** 216 | **Value of Point** 221.

4. Select 106 and hide 167 the line and point $B$, leaving only two points and the measured ratio.

5. Construct a segment 115 connecting $A$ and $C$.

6. Double-click the measured ratio with the **Text** 120 tool, and set its label to $m$. Hide point $A$, and change the label of point $C$ to $m$ to match the label of the value.

Your basic slider is complete. As you drag point *m,* the value changes accordingly.

You can vary this construction for a number of purposes.

- For positive values, use a ray instead of a line in step 1.

- For values between 0 and 1, use a segment instead of a line in step 1.

- For a different scale, change the distance from *A* to *B,* or multiply the slider value by some scaling factor.

- For integer values, use the Calculator 257 to round the value of the slider.

    You can also create a numeric value (a parameter 36 ) directly, using the **Graph** 233 | **New Parameter** 226 command. The advantage of the slider described here is that sliding the control point back and forth provides a powerful, visual way to change the value.

## 3.6.12  Coordinates

*Selection prerequisites: One or more points* 2

This Measure 216 menu command measures the coordinates of the selected points. The measurement is with respect to the marked coordinate system 41 .

    If there is no marked coordinate system, Sketchpad marks an existing coordinate system 234 or creates a new one 233 .

If the marked coordinate system has a square or rectangular grid form 235 , the coordinates are measured as an $(x, y)$ pair.

If the marked coordinate system has a polar grid form, the coordinates are measured as an $(r, \theta)$ pair.

If you hold the Shift key while choosing this command, it becomes **Abscissa & Ordinate** or **Polar Distance & Direction,** depending on whether the coordinate system is rectangular or polar. In these cases, the command measures the coordinates ($x$ and $y$ or $r$ and $\theta$) as separate values, allowing you to use them as values in calculations and elsewhere.

The Shift key is a one-step shortcut to choosing **Abscissa** 223 and **Ordinate** 224 (or **Polar Distance** 223 and **Polar Direction** 224 ) as two separate commands.

*See also:*
    *Abscissa (x) command* 223
    *Ordinate (y) command* 224
    *Define Coordinate System* 233

## 3.6.13  Abscissa (x) / Polar Distance (r)

*Selection prerequisites: One or more points* 2

This Measure 216 menu command measures the abscissa ($x$) or polar distance ($r$) of the selected points. The measurement is with respect to the marked coordinate system 41 .

    If there is no marked coordinate system, Sketchpad marks an existing coordinate system 234 or creates a new one 233 .

If the marked coordinate system has a square or rectangular grid form 235 , the abscissa ($x$) is measured; if it has a polar grid form, the polar distance ($r$) is measured.

To measure both $x$ and $y$ (or $r$ and $\theta$ if the grid form is polar) at the same time, hold the Shift key and choose **Measure | Abscissa & Ordinate** 223 .

*See also:*
   *Ordinate (y) command* 224
   *Coordinates command* 223

## 3.6.14  Ordinate (y) / Polar Direction (θ)

***Selection prerequisites:*** *One or more points* 2

This Measure 216 menu command measures the ordinate ($y$) or polar direction ($\theta$) of the selected points. The measurement is with respect to the marked coordinate system 41.

   If there is no marked coordinate system, Sketchpad marks an existing coordinate system 234 or creates a new one 233.

If the marked coordinate system has a square or rectangular grid form 235, the ordinate ($y$) is measured; if it has a polar grid form, the polar direction ($\theta$) is measured.

To measure both $x$ and $y$ (or $r$ and $\theta$ if the grid form is polar) at the same time, hold the Shift key and choose **Measure | Abscissa & Ordinate** 223.

*See also:*
   *Abscissa (x) command* 223
   *Coordinates command* 223

## 3.6.15  Coordinate Distance

***Selection prerequisites:*** *Two points* 2

This Measure 216 menu command measures the distance between two selected points based on the marked coordinate system 41.

   If there is no marked coordinate system, Sketchpad marks an existing coordinate system 234 or creates a new one 233.

This command differs from the **Distance** 218 command because the coordinate distance is based, not on physical distance (that is, not on inches or centimeters), but on the unit size of the marked coordinate system. A coordinate distance measurement has no units.

*See also:*
   *Abscissa (x) command* 223
   *Ordinate (y) command* 224
   *Coordinates command* 223
   *Distance command* 218

## 3.6.16  Slope

***Selection prerequisites:*** *One or more straight objects* 5

This Measure 216 menu command measures the slope of each selected line with respect to the marked coordinate system 41.

   If there is no marked coordinate system, Sketchpad marks an existing coordinate system 234 or creates a new one 233.

## 3.6.17  Equation

*Selection prerequisites: One or more lines* [5] *or circles* [6]

This Measure [216] menu command measures the equation of each selected object with respect to the marked coordinate system [41].

If there is no marked coordinate system, Sketchpad marks an existing coordinate system [234] or creates a new one [233].

The equation of a line [5] is expressed in one of the following three forms:

| | |
|---|---|
| $y = c$ | for a horizontal line |
| $x = c$ | for a vertical line |
| $y = m \cdot x + b$ | for any other line |

For a circle [6], the equation is expressed in one of the following two forms:

| | |
|---|---|
| $(x - h)^2 + (y - k)^2 = r^2$ | if the coordinate system is square [235] |
| $\dfrac{(x - h)^2}{a^2} + \dfrac{(y - k)^2}{b^2} = 1$ | if the coordinate system is not square [235] |

In other words, a Euclidean circle described by an equation in square coordinates has a circle's equation. The same shape, described in nonsquare coordinates, has the equation of an ellipse.

# 3.7    Number Menu

The Number menu provides commands that create, calculate, and tabulate numeric values 92 and that define functions 45 in several different ways.

Create parameters 36 that you can set to different values, adjust, and animate within a domain you specify.

Use the Calculator 257 to derive new properties by calculating relationships between existing measurements. For example, choose **Calculate** to sum the measured interior angles of a triangle or compute the ratio of a circle's measured circumference to its radius.

Tabulate values 39 (parameters, measurements, and calculations) by hand, or automatically as the values change.

Define a new function 45 by entering its expression. The expression can include values in your sketch, an extensive set of built-in functions, and other functions that you define yourself.

Define a function that's the derivative of another function.

Define a function based on the shape 51 of a curve you draw with the **Marker** 122 tool.

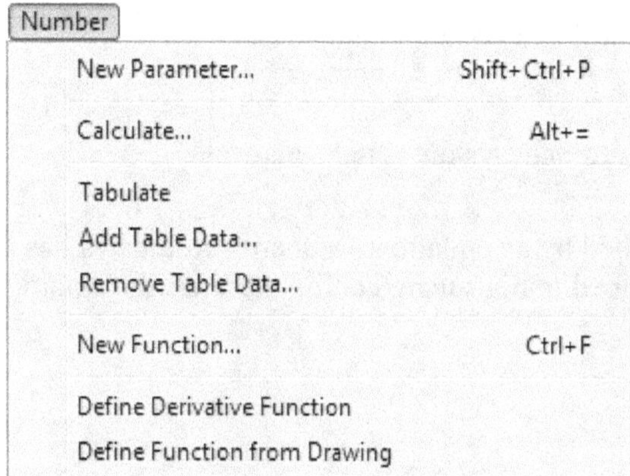

Detailed information is available for each command:
New Parameter 226
Calculate 227
Tabulate 228
Add Table Data 228
Remove Table Data 229
New Function 230
Define Derivative Function 231
Define Function from Drawing or Picture 232

## 3.7.1    New Parameter

$$t_1 = \boxed{1.0}$$
$$t_2 = 5°$$
$$t_3 = \boxed{3.00}\ cm$$

This Number 226 menu command creates a new parameter 36 in your sketch. A parameter is a number that can easily be changed. It's convenient to use parameters in places where you need to have a number but want to be able to change that number easily.

The keyboard shortcut for **New Parameter** is Shift+Ctrl+P (Windows) or Shift-⌘P (Mac).

Use parameters in calculations 39, in functions 45, and as values by which to transform 192 objects. For example, you might create two parameters, named $m$ and $b$, and use them in plotting 241 the function $y = mx + b$. Or you might create a parameter that varies from 0° to 360° and use it as a marked angle to rotate 203 a polygon 8.

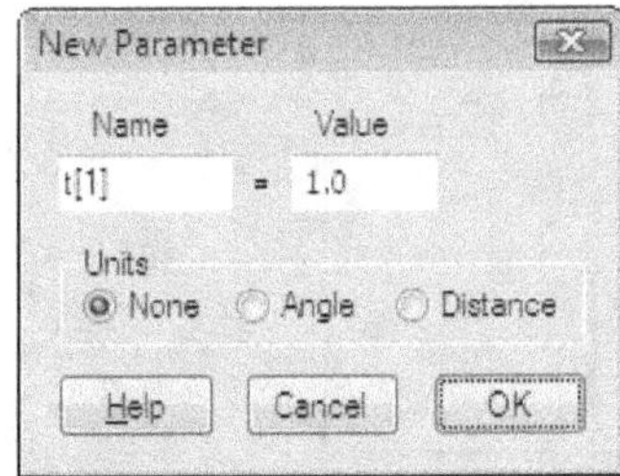

When you type the initial value, the number of decimal places you type determines both the precision and the keyboard adjustment, as shown in these examples:

| Initial Value | Decimal Places | Displayed Precision | Keyboard Adjustment |
|---|---|---|---|
| 5 | 0 | units | 1 |
| 5.0 | 1 | tenths | 0.1 |
| 5.00 | 2 | hundredths | 0.01 |

The keyboard adjustment is the value by which the parameter changes when you select it and press the + or − key.

To change an existing parameter's keyboard adjustment, choose **Edit | Properties | Parameter** 38.

To show or hide an existing parameter's edit box, or to change its displayed precision, choose **Edit | Properties | Value** 94.

When you choose **New Parameter,** a dialog box appears.

> **Name:** Type a new name for the parameter. If you want the name to include a subscript, enter the subscript within square brackets. For example, the default name t[1] appears as $t_1$ in the sketch.
>
> > To change the value of a parameter, you can double-click it with the **Arrow** 104 tool, use the **Display | Animate** 171 or **Edit | Edit Parameter** 157 commands, or create an Animation action button 149.
>
> **Value:** Set the initial value of the parameter by typing a value in this field.
>
> **Units:** Choose whether the parameter's value uses no units, uses the current angle unit 276, or uses the current distance unit 276.

 Parameters and Sliders

> To create an easily modified numeric value, use either a parameter or a slider 222. A parameter is easier to create, and you can adjust it more precisely and easily animate it over a specific numeric range. A slider is harder to create, but you can manipulate it easily (by dragging the point at the tip) and control it in a continuous and satisfying way.

## 3.7.2   Calculate

This Number 226 menu command displays the Calculator 257 and allows you to create a calculation 39 in the sketch.

> The keyboard shortcut for **Calculate** is Alt+= (Windows) or ⌘= (Mac).

A calculation can use constants and mathematical operations, and it can use measurements, calculations, and parameters 92 that already exist in the sketch. The calculation can also use

Sketchpad's standard functions as well as any user-defined functions that already exist in the sketch.

### 3.7.3 Tabulate

*Selection prerequisites: One or more values* [92]*, equations* [36]*, or text objects* [99] *that contain values, coordinates, or equations*

| Area $\odot CD$ | (Radius $\odot CD)^2$ | Area $\odot CD$ / (Radius $\odot CD)^2$ |
|---|---|---|
| 2.54 cm$^2$ | 0.81 cm$^2$ | 3.1416 |
| 14.79 cm$^2$ | 4.71 cm$^2$ | 3.1416 |
| 36.46 cm$^2$ | 11.61 cm$^2$ | 3.1416 |
| 51.13 cm$^2$ | 16.27 cm$^2$ | 3.1416 |
| 92.80 cm$^2$ | 29.54 cm$^2$ | 3.1416 |

This Number [226] menu command creates a new table [39] of the selected objects' values. Sketchpad's tables contain entries describing a set of measurements' [36] values at different points in time. In addition to measured values, you can tabulate parameters [36], calculations [39], measured coordinate pairs [223] and equations [225], functions [45], and Hot Text captions [56].

Once you've created a table, you can add data [228] to it manually or automatically, and you can remove its data [229] in order to collect new data.

By default, new tables contain one row that dynamically tracks the current values of all tabulated measurements. Values in this row change as measured objects are dragged or animated. Turn off dynamically-changing values by choosing **Edit | Properties | Table** [41].

### 3.7.4 Add Table Data

This Number [226] menu command displays a dialog box allowing you to add one or more rows of data to a table [39]. Each row of data captures the values of the tabulated measurements at the moment the row is added. **Add Table Data** is available when a single table created by the **Number | Tabulate** [228] command is selected.

As a shortcut, you can add a single row to a table by double-clicking[109] it with the **Arrow**[104] tool.

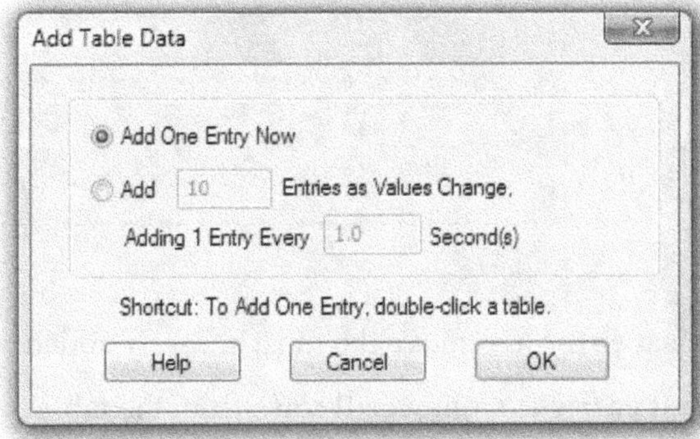

In the dialog box:

- Choose **Add one entry now** to add a single row to your table, capturing the current values of tabulated measurements.

- Choose **Add entries as values change** to add one or more rows automatically when the tabulated measurements next change. In the edit boxes, enter how many rows you want to add to the table, and the rate at which to add them.

Sketchpad lets you collect up to 25 rows of new data at a time. If you want to collect more than that, collect them in groups of 25 rows at a time.

 Note

You cannot use the **Add Table Data** command for a table of iterated measurements[30] created using the **Transform | Iterate**[208] command. If you want to add data to a table of iterated measurements, you must increase the level of iteration using Iteration Properties[35] (or by selecting the iterated measurement table and pressing the + key).

To remove data from a table, choose **Number | Remove Table Data**[229].

To remove only the last row from a table, hold the Shift key and double-click[109] the table with the **Arrow**[104] tool.

## 3.7.5 Remove Table Data

This Number[226] menu command displays a dialog box allowing you to remove data from a table[39]. **Remove Table Data** is available when a single table created by the **Number | Tabulate**[228] command is selected.

As a shortcut, you can remove the last row from a table by holding the Shift key and double-clicking[109] the table with the **Arrow**[104] tool.

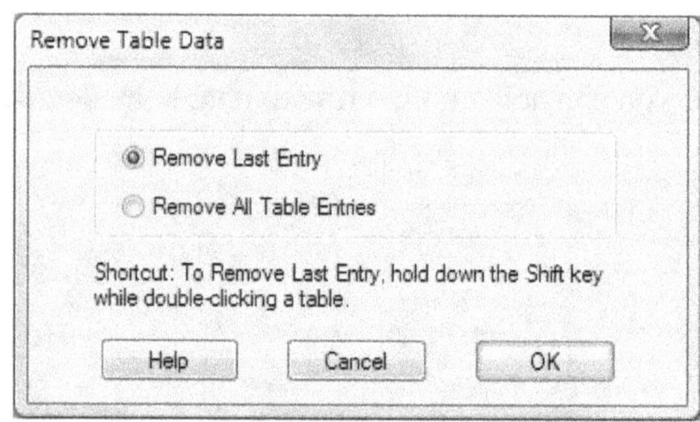

In the dialog box:

- Choose **Remove last entry** to remove the most-recently added row.

- Choose **Remove all entries** to remove all rows from the table.

    When you remove all previously added rows, Sketchpad adds a new first row containing the tabulated measurements' current values.

 Note

You cannot use the **Remove Table Data** command for a table of iterated measurements [30] created using the **Iterate** [208] command. If you want to remove data from a table of iterated measurements, you must decrease the level of iteration using Iteration Properties [35] (or by selecting the table and pressing the – key).

To add data to a table, choose **Number | Add Table Data** [228].

To add a single row containing the current values of the data, double-click [109] the table with the **Arrow** [104] tool.

## 3.7.6 New Function

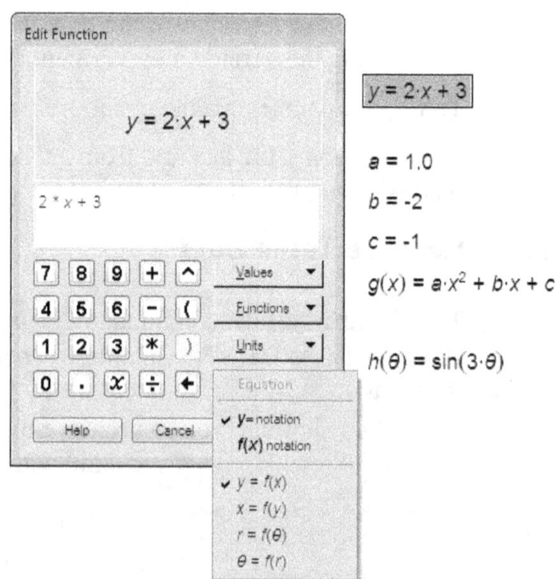

This Number [226] menu command displays the Calculator [257] and allows you to create a new function [45] in the sketch.

The keyboard shortcut for **New Function** is Ctrl+F (Windows) or ⌘F (Mac).

- Use the Calculator's keypad to enter a number, an operation, or the independent variable.

  If the coordinate system is rectangular or square, the independent variable is *x;* if the coordinate system is polar, the independent variable is $\theta$.

- Use the Values pop-up menu to enter a selected value, a new parameter, $\pi$, or e. You can also create a new parameter using the keyboard shortcut: Shift+Ctrl+P (Windows) or Shift-⌘P (Mac).

- Use the Functions pop-up menu to enter standard functions such as absolute value, square root, and trigonometric functions.

- Use the Units pop-up menu to enter angle or distance units.

- Use the Equation pop-up menu to set the function's display to either **y= notation** or **f(x) notation** and to determine the form in which it's graphed.

- Click a measurement, calculation, or parameter 92 in the sketch to enter its value into the expression.

- Click an already-defined function 45 in the sketch to use it in the expression.

Once you've created a function, you can plot it by choosing **Graph | Plot Function** 241.

To create a function and plot it with one command, choose **Graph | Plot New Function** 241.

## 3.7.7    Define Derivative Function

$$f(x) = x{\cdot}\sin(x)$$
$$f'(x) = x{\cdot}\cos(x) + \sin(x)$$

This Number 226 menu command creates a new function 45 that is the derivative of the selected function with respect to that function's independent variable.

- Derivative functions update automatically when the function they differentiate is edited.

- In most ways derivative functions behave just like other functions you create. You can calculate with derivative functions, evaluate derivative functions for specific arguments, and plot derivative functions. You can even create the derivative of a derivative. The only thing you can't do with derivative functions is to edit them directly; you must edit the original function instead.

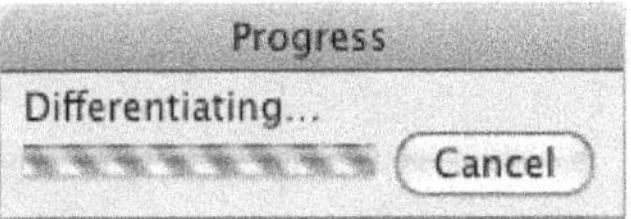


- Differentiation of complicated functions can be time-consuming. In these cases, a dialog box like the one at right appears. Click **Cancel** if you want to interrupt computation of an exact derivative. If you cancel the computation, Sketchpad will provide you with an approximate derivative, in the form

$$f'(x) = \frac{f(x+0.005)-f(x-0.005)}{0.01}$$

This approximate derivative can still be plotted, and can even be differentiated itself. Although there is some loss in accuracy, in most cases the approximation resembles the exact derivative.

The approximate derivative is always used to find the derivative of a function defined by a drawing or picture [51].

See Sketchpad's Internal Mathematics [325] for a more detailed discussion of derivatives.

### 3.7.8 Define Function from Drawing or Picture

***Selection prerequisites:*** *One drawing or picture*

This Number [226] menu command creates a new function from the selected drawing or picture based on the marked coordinate system.

If there is no marked coordinate system, Sketchpad marks an existing coordinate system [234] or creates a new one [233].

#### ▼ Define Function from Drawing

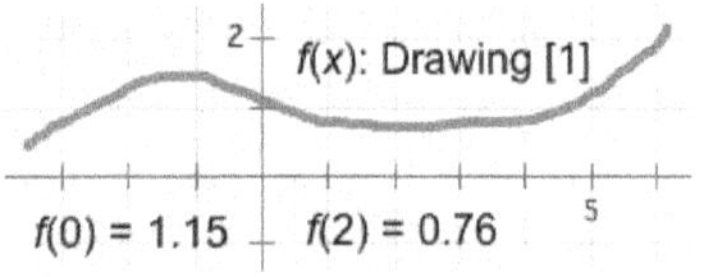


To create a function using this command, use the **Marker** [122] tool to draw a curve on your coordinate system. With the drawing [18] selected, choose **Define Function from Drawing.**

The selected drawing defines the value of the function, and the width of the drawing defines its domain. If the drawing crosses itself or doubles back horizontally so that more than one part of the drawing corresponds to a given horizontal position, the highest part of the drawing determines the function value for that horizontal position.

#### ▼ Define Function from Picture

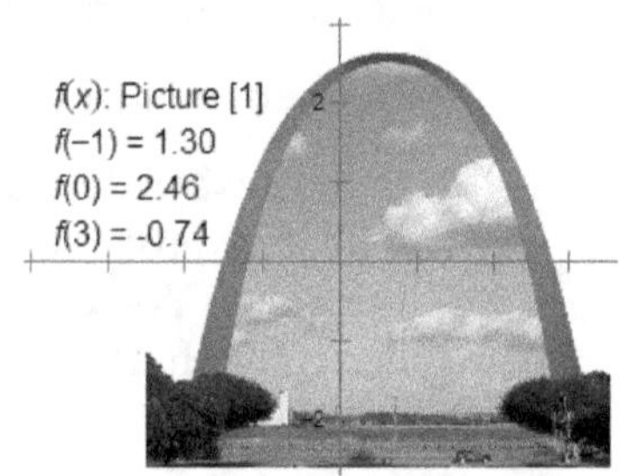


Import a picture, either by dragging and dropping into your sketch or by copying from some other document and pasting into Sketchpad. With the picture [18] selected, choose **Define Function from Picture.**

The top edge of the picture defines the value of the function, and the width of the picture defines its domain.

## 3.8    Graph Menu

The Graph menu allows you to create and manipulate coordinate systems[41], to show or hide the grid of a coordinate system, to make points snap to integer coordinates, to plot a single value on an axis or other path, to plot points or table data on a coordinate system, and to construct a function plot or parametric plot on a coordinate system.

This menu offers a variety of commands that serve as a foundation for investigations and activities in analytic geometry and algebra.

Use commands on the Measure[216] menu to measure various quantities on the coordinate plane, including coordinates[223] of points, coordinate distances[224] between points, slopes[224] of straight objects, and equations[225] of lines and circles.

Use commands on the Number[226] menu to create a parameter[226], calculation[227], or table[228], and to define a function symbolically[230], as a derivative[231], or graphically[232].

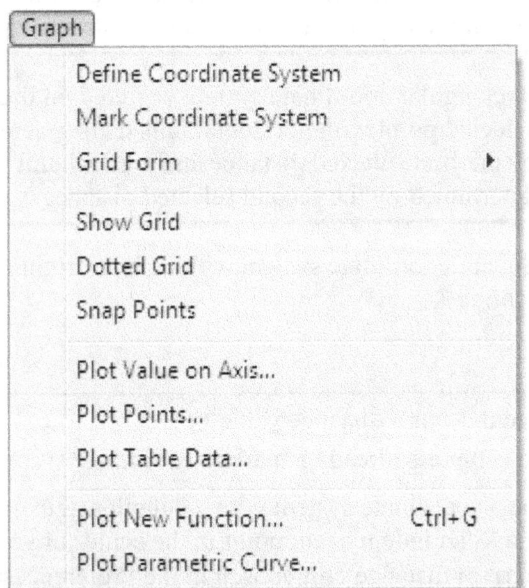


Detailed information is available for each command:
Define Coordinate System[233]
Mark Coordinate System[234]
Grid Form[235]
Show Grid[236]
Dotted Grid[236]
Snap Points[237]
Plot Value on Axis[237]
Plot Points/Plot as (x, y)/Plot as (r, theta)[239]
Plot Table Data[240]
Plot Function/Plot New Function[241]
Plot Parametric Curve[241]

### 3.8.1    Define Coordinate System

This Graph[233] menu command creates a new coordinate system[41] and marks it as the active coordinate system. The type and scale of the created coordinate system depends on your selections, as described in this table.

(If you choose a command, such as **Measure | Coordinates**[223], that requires a coordinate system but you haven't created one in your sketch, Sketchpad defines and marks a coordinate system for you.)

| *Selections* | *Command* | *Result* |
| --- | --- | --- |
| One point | **Define Origin** | Square coordinate system centered on the selected point with default unit scale |

| One circle | **Define Unit Circle** | Square coordinate system centered on the selected circle with unit scale determined by the circle's radius |
|---|---|---|
| One defining distance* | **Define Unit Distance** | Square coordinate system centered on a default origin with unit scaling determined by the defining distance |
| One point and one defining distance* | **Define Unit Distance** | Square coordinate system centered on the selected point with unit scaling determined by the defining distance |
| Two defining distances* | **Define Unit Distances** | Rectangular coordinate system centered on a default origin, with horizontal unit scaling determined by the first selected distance and vertical unit scaling determined by the second selected distance |
| One point and two defining distances* | **Define Unit Distances** | Rectangular coordinate system centered on the selected point with horizontal unit scaling determined by the first selected distance and vertical unit scaling determined by the second selected distance |
| Nothing or anything other than the above | **Define Coordinate System**** | Square coordinate system with default origin and unit scaling |

* A defining distance can be either a segment $\boxed{5}$ or a distance value $\boxed{92}$.

** **Define Coordinate System** is disabled if there's already a marked coordinate system.

As described in the table, some commands define a coordinate system with a default origin or default unit scaling. Sketchpad constructs a default origin as an independent point in the center of your window. The default unit scale is based on the current distance unit chosen in the Preferences $\boxed{276}$ dialog box, but can be adjusted by dragging the coordinate system's unit point or axis tick labels.

Since most activities require only a single coordinate system, the **Define Coordinate System** choice (which defines an entirely default coordinate system) is disabled if your sketch already has a coordinate system. If you really want a second coordinate system, you must select the appropriate objects as listed in the first column of the table.

*See also:*
*Mark Coordinate System command* $\boxed{234}$
*Multiple Coordinate Systems* $\boxed{43}$

## 3.8.2   Mark Coordinate System

***Selection prerequisites:*** *A coordinate system's axis, origin point, unit point, unit circle, or grid*

This Graph $\boxed{233}$ menu command marks the coordinate system $\boxed{41}$ associated with the selected object as the coordinate system on which to measure or plot new objects.

When you have multiple coordinate systems $\boxed{43}$, one of those coordinate systems is the marked coordinate system. All of the Graph $\boxed{233}$ menu commands apply to the marked coordinate system, as do those Measure 216 menu commands that require a coordinate system.

The marked coordinate system is normally the most recently created coordinate system; you use this command to mark a different coordinate system.

Normally you'll have only a single coordinate system, and you won't have to worry about which coordinate system is marked.

*See also:*
   *Define Coordinate System command* 233

## 3.8.3   Grid Form

This Graph 233 menu command changes the grid appearance and scaling of the marked 234 coordinate system 41.

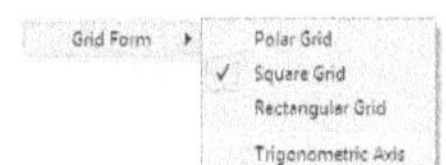

Sketchpad's coordinate systems can have several possible forms:

| | |
|---|---|
| **Polar Grid:** A polar coordinate system has a set of grid lines that are circles (at a constant distance from the origin, or $r$ value) and a set of grid lines that pass through the origin (at a constant angle from the origin, or $\theta$ value). Any coordinate system can be made polar.<br><br>If the horizontal and vertical axes of a polar coordinate system have different scaling, the constant-distance grid lines appear as ellipses rather than circles. | Polar |
| **Square Grid:** A square coordinate system has the same scaling on the horizontal and vertical axes and has grid lines that are horizontal (at a constant $y$ value) and vertical (at a constant $x$ value). Any coordinate system, except one defined in terms of two different distances, can be made square. | Square |
| **Rectangular Grid:** A rectangular coordinate system has independent scaling for the horizontal and vertical axes and has grid lines that are horizontal (constant $y$ value) and vertical (constant $x$ value). Any coordinate system, except one defined in terms of a unit circle, can be made rectangular. | Rectangular |
| **Trigonometric Axis:** The horizontal axis is numbered with fractions and multiples of $\pi$. You can use a trigonometric axis on any kind of coordinate system.<br><br>When you choose trigonometric axes for a polar grid, both axes become trigonometric.<br><br>You can turn trigonometric numbering on or off for any axis by selecting it and choosing **Edit | Properties | Axis** 44. | Square with Trigonometric Axis |

If you choose one of these commands with a sketch that doesn't yet have a coordinate system, Sketchpad creates a default coordinate system with the chosen form.

*See also:*
*Show Grid command* 236
*Define Coordinate System command* 233
*Multiple Coordinate Systems* 43

### 3.8.4　Show Grid

This Graph 233 menu command shows or hides the grid lines of the marked 234 coordinate system 41.

When the grid of the marked coordinate system is showing, the command changes to **Hide Grid.**

If you choose **Show Grid** with a sketch that doesn't yet have a coordinate system, Sketchpad creates a default coordinate system.

If you hold the Shift key, the command changes to **Show Coordinate System** or **Hide Coordinate System,** and has the effect of showing or hiding the grid, the axes, and the origin of the marked coordinate system.

*See also:*
*Define Coordinate System command* 233
*Grid Form command* 235
*Snap Points command* 237

### 3.8.5　Dotted Grid

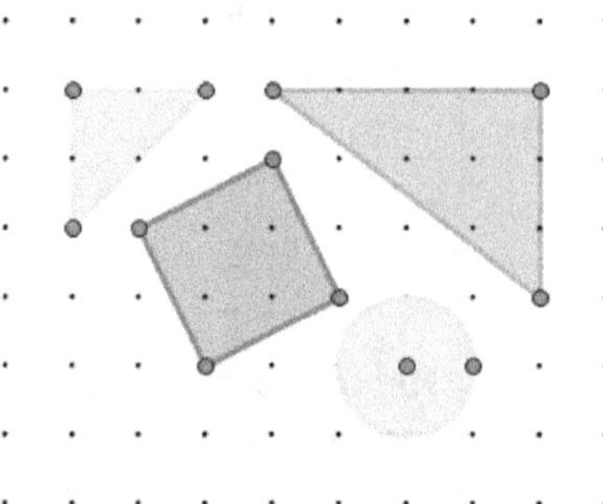

This Graph 233 menu command changes the display of the coordinate system between grid lines and grid dots.

Choose **Dotted Grid** once to change the display to grid dots. Choose it again to change the display back to grid lines.

The position of grid lines or grid dots depend on the scale of the coordinate system. If a coordinate system has a unit distance of 1 cm, the lines or dots appear at every unit (1, 2, 3, ...) If you reduce the scale so that the unit distance is 0.5 cm, they appear at every two units (2, 4, 6, ...). If you increase the scale so the unit distance is 2 cm, they appear at every 0.5 unit (0.5, 1.0, 1.5, ...).

To make points stick to the grid dots:

1. Choose **Graph | Snap Points** 237. When point snapping is turned on, independent points snap to integer positions.

2. If necessary, adjust the scale of the coordinate system so that the grid dots also appear at integer positions.

Use this "snappy grid" for geoboard 44 activities and other activities, such as Pick's Theorem, that make use of the grid dots.

## 3.8.6   Snap Points

This Graph 233 menu command causes independent points 80 to snap to nearby locations when you drag them. Choose this command once to activate it. When snapping is active, a check-mark appears next to the command. Choose the command a second time to deactivate snapping.

Use **Snap Points** when you want to work with "nice" whole-number coordinates. Combine it with a **Dotted Grid** 236 to make a geoboard 44.

The locations to which points snap when this command is active depend on the grid form 235 of the marked 234 coordinate system 41. For rectangular and square grids, points snap to locations with whole number coordinates (that is, to the integer lattice). For polar grids, points snap to locations whose distance from the origin is a whole number and whose angle with respect to the horizontal axis is a multiple of 15°. In either case, points you drag when **Snap Points** is active only snap to locations that are relatively close by. In other words, if you have rescaled your coordinate system 109 so that the distance between integer coordinates is great, points you drag will snap only to integer locations that are close to them.

*See also:*
   *Define Coordinate System command* 233

## 3.8.7   Plot Value on Axis

This Graph 233 menu command constructs, on one or more paths, one or more plotted points determined by value(s) you select or type.

If you select the value(s) or path(s) before choosing the command, Sketchpad constructs the plotted point(s) immediately.

Otherwise, a dialog box appears so that you can designate a value to be plotted and the path on which to plot it. (If you choose **Graph | Plot Value** in a sketch with no visible axes or paths, Sketchpad will offer to create new coordinate axes to permit plotting values.)

The position of a plotted point on a path depends on the path. For instance, you can plot values between 0 and 1 on a segment, you can plot values greater than or equal to zero on a ray, and you can plot any value on a line. See the **Measure | Value of Point** 221 command for details of how values are defined on various path objects.

The **Plot Value on Axis** command takes several forms, depending on the selected objects.

| Selection(s) | Command | Result |
| --- | --- | --- |
| One or more values and one path | **Plot Value on Path*** | Immediately plots the selected value on the selected path. |
| One value and multiple paths | **Plot Value on Paths*** | Immediately plots the selected value on each selected path. |

| One path | **Plot Value on Path...*** | Opens the Plot Value dialog box using the selected path. |
|---|---|---|
| Other | **Plot Value on Axis...** | Opens the Plot Value dialog box. |

*With these selections, the command name changes to describe the selected path(s).

When the Plot Value dialog box appears, enter the appropriate value and path.

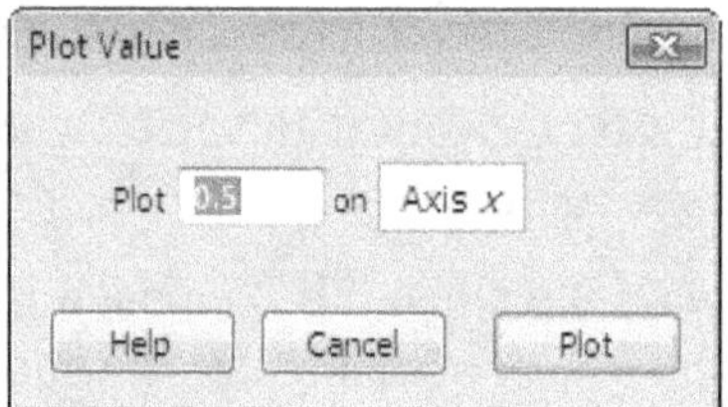

Type a number in the entry field on the left to plot a fixed value, or click in the sketch on a value (or a Hot Text 57 link to a value) to plot that value.

The initial object in the entry field on the right is either the path you selected or *Axis x* if you didn't select a path. Click a path object in the sketch to plot the value on that object.

When you've filled in the entry fields properly, click **Plot** to plot the point on the path.

### ▼ Examples

In the example below, only the quadrilateral *ABCD* is selected, so the Plot Value dialog box appears to allow you to enter a value. As a result of typing the value 0.75, the resulting plotted point is 0.75 of the way around the perimeter of the quadrilateral.(The perimeter starts at the first vertex, point *A,* and continues around the remaining vertices in the order in which they were selected to construct the quadrilateral.)

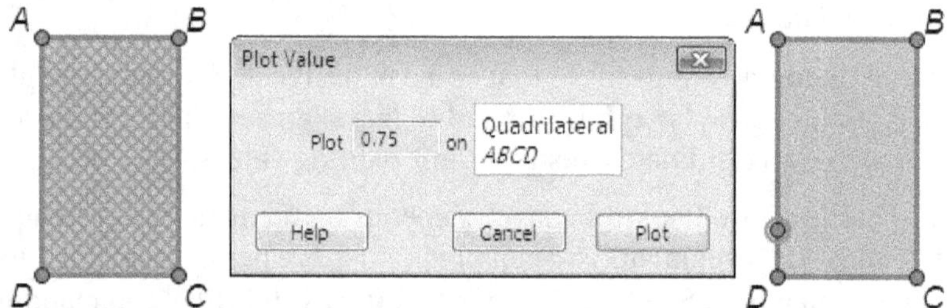

In the example below, a circle and four values are selected. The result of the command is to immediately plot all four values on the circumference of the circle, with a value of 1 corresponding to the entire circumference. (The value of a point on a circle measures counter-clockwise revolutions around the circle starting from the point directly to the right of the center.)

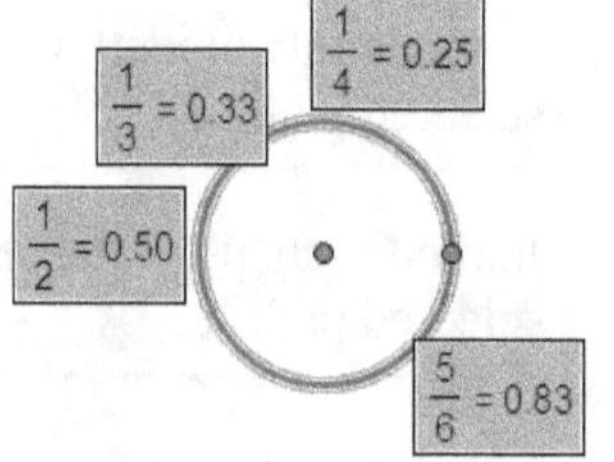

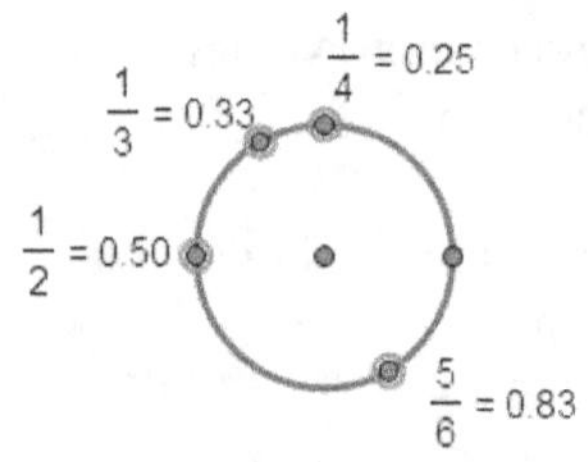

Selections: One circle and four values

Result: Four points plotted on the circle

In the example below, three segments and one value are selected. The result of the command is to immediately plot the value on all three segments.

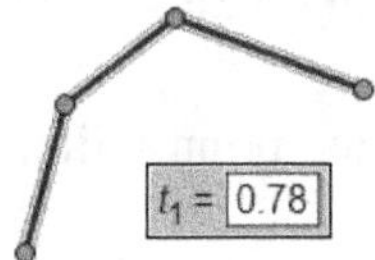

$t_1 = \boxed{0.78}$

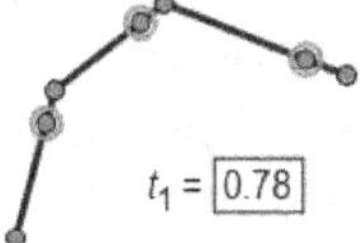

$t_1 = \boxed{0.78}$

Selections: Three segments and one value

Result: Three plotted points, one on each segment

*Other plotting commands:*
*Plot Points/Plot as (x, y)/Plot as (r, theta) command* 239
*Plot Table Data command* 240
*Plot Function/Plot New Function command* 241
*Plot Parametric Curve command* 241

## 3.8.8 Plot Points/Plot as (x, y)/Plot as (r, θ)

This Graph 233 menu command plots one or more points on the marked coordinate system 234 at the specified coordinate location. If there is no marked coordinate system, Sketchpad creates a default coordinate system 41. Depending on the selected objects in your sketch, the command can appear as **Plot as (x,y)**, **Plot as (r, θ)**, or **Plot Points**.

### ▼ Plot as (x,y) / Plot as (r, theta)

***Selection prerequisites:*** *Two values* 92

If you have two selected parameters 36, measurements 36 or calculations 39, the command appears as **Plot as (x, y)** or (if there's a marked polar coordinate system 234) as **Plot as (r, θ)**.

Use this command to graph measurements or calculations that change over time.

When you choose the command, Sketchpad plots the point determined by the two selected values, with the first measurement used as the $x$ or $r$ coordinate and the second as the $y$ or $\theta$ coordinate. As the measurements change, the position of the plotted point changes to match.

### ▼ Plot Points

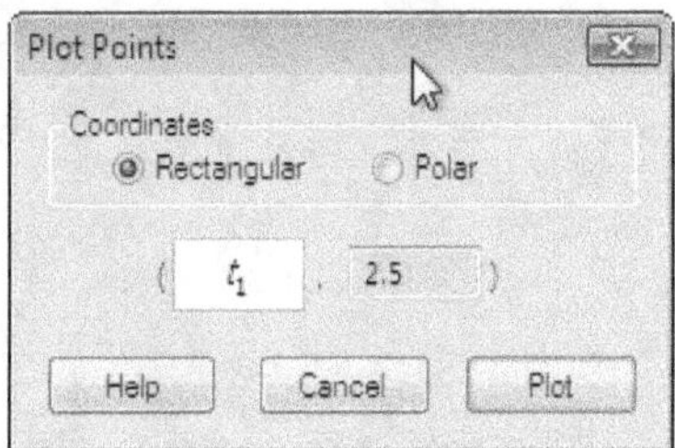

If you don't have two selected values, the command is **Plot Points.**

This command displays a dialog box that allows you to plot one or more points on the marked coordinate system 234 by entering the coordinates of each point to plot.

In the dialog box:

- Choose **Rectangular** to plot points using $(x, y)$ values or choose **Polar** to plot them using $(r, \theta)$

values.

- Click in either entry field and type numbers with decimal points, or type expressions such as 2 +3 and $\pi/4$.

- Click in either entry field and then click in the sketch on a value 92 (or a Hot Text 57 link to a value) to enter that value into the field.

  When you use a value from the sketch, the plotted point will move accordingly when the value changes.

- Click **Plot** to plot the current values; the dialog box remains open to allow you to plot additional points.

- After plotting your last point, click **Done** to close the dialog box.

*Other plotting commands:*
*Plot Value on Axis command* 237
*Plot Table Data command* 240
*Plot Function/Plot New Function command* 241
*Plot Parametric Curve command* 241

## 3.8.9  Plot Table Data

***Selection prerequisites:*** *One table* 39

This Graph 233 menu command plots a point on the marked coordinate system 234 for each row of the selected table. If there is no marked coordinate system, Sketchpad creates a default coordinate system 41.

This command displays a dialog box that allows you to choose whether to plot the table data using rectangular or polar coordinates, and to choose which column of the table to plot as the $x$ (or $r$) coordinate and which column to plot as the $y$ (or $\theta$) coordinate.

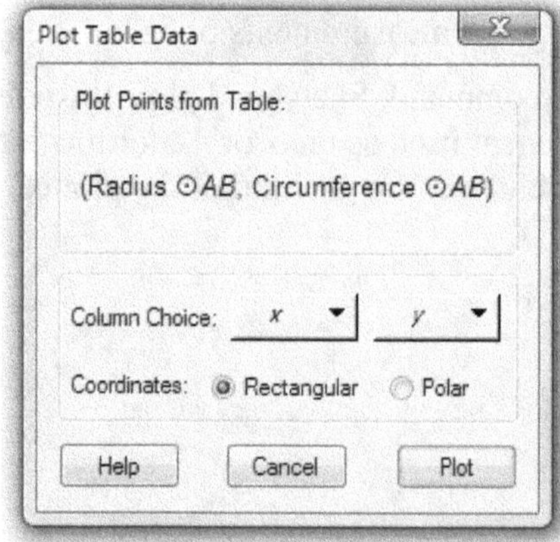


*Other plotting commands:*
*Plot Value on Axis command* 237
*Plot Points/Plot as (x, y)/Plot as (r, theta) command* 239
*Plot Function/Plot New Function command* 241
*Plot Parametric Curve command* 241

## 3.8.10   Plot Function/Plot New Function

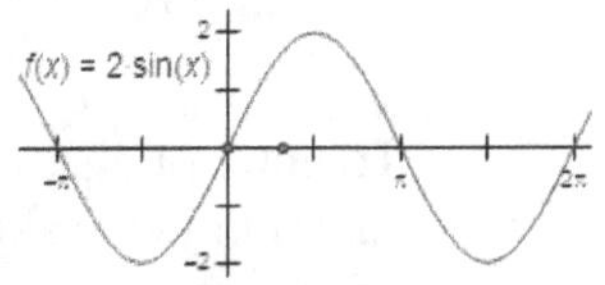


***Selection prerequisites:*** *One or more functions* 45, *or no objects*

This Graph 233 menu command plots one or more selected functions 45, or creates and plots a new function if no objects are selected.

The keyboard shortcut for **Plot New Function** is Ctrl+G (Windows) or ⌘G (Mac).

If you're plotting a new function, the Calculator 257 appears, just as it does for the **Number | New Function** 230 command, to allow you to define the function to be plotted.

Once you've plotted a function, you can change the domain or the number of samples by using the Plot Properties 97 dialog box. You can also change the domain by dragging the arrows at the ends of the function plot.

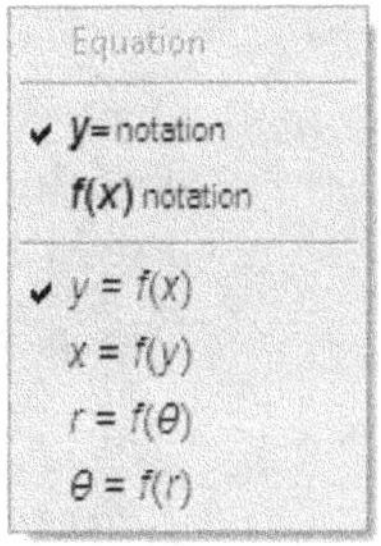

You can change the equation of the function to plot it as a polar function $r = f(\theta)$, or to plot it with axes reversed, as $x = f(y)$. To do so, double-click the function, or select it and choose **Edit | Edit Function** 157. When the Calculator appears, choose the equation you want from the Equation pop-up menu. You can also set the function's display to either **y= notation** or ***f*(x) notation.**

*Other plotting commands:*
*Plot Value on Axis command* 237
*Plot Points/Plot as (x, y)/Plot as (r, theta) command* 239
*Plot Table Data command* 240
*Plot Parametric Curve command* 241

## 3.8.11   Plot Parametric Curve

This Graph 233 menu command constructs the parametric plot 54 defined by two functions 45. (If your sketch doesn't contain two functions, a warning dialog box appears.)

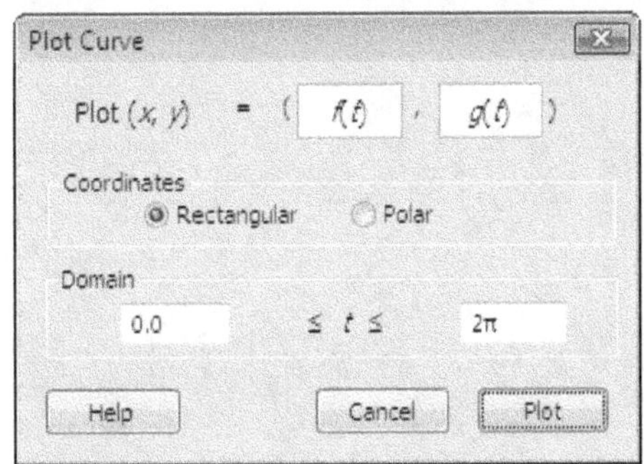

If one or two functions are selected when you choose this command, the selected functions appear in the edit boxes at the top of the dialog box. The first function is used for $x$ (or $r$), and the second for $y$ (or $\theta$).

To enter or change the function in either edit box, highlight that edit box and then click the desired function in the sketch. (If the dialog box hides the function you want to click, move the dialog box out of the way.)

**Coordinates:** Use these radio buttons to choose whether the curve is plotted using **Rectangular** ($x$, $y$) or **Polar** ($r$, $\theta$) coordinates.

**Domain:** Use these edit boxes to determine the domain of the independent variable used to evaluate the functions. You can type a number or an expression in either box.

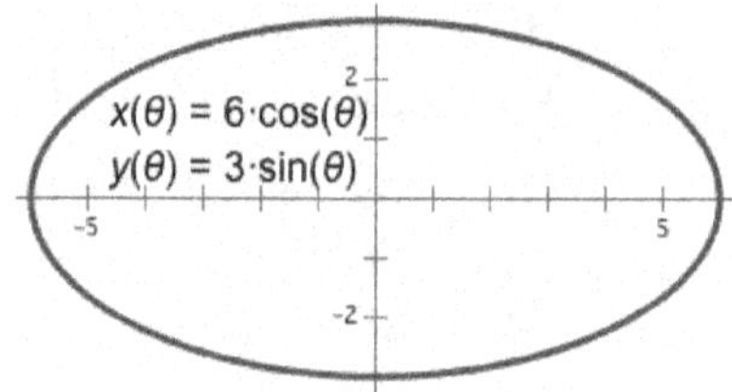


Once you've constructed a parametric plot, use the Plot Properties 97 dialog box to change the domain, the number of samples, or the display of arrowheads at the ends of the curve. (Drag these arrowheads as another way to change the domain.)

*Other plotting commands:*
  *Plot Value on Axis command* 237
  *Plot Points/Plot as (x, y)/Plot as (r, theta) command* 239
  *Plot Table Data command* 240
  *Plot Function/Plot New Function command* 241

## 3.9    Window Menu

This menu contains commands for arranging your open document windows on the screen and for navigating among those documents.

This menu contains different commands depending on the operating system.

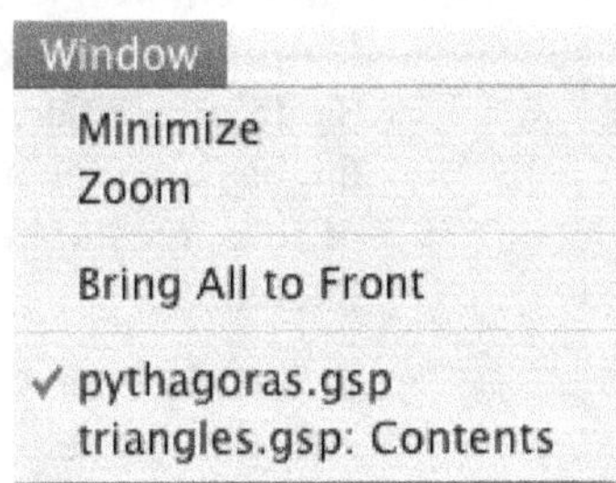
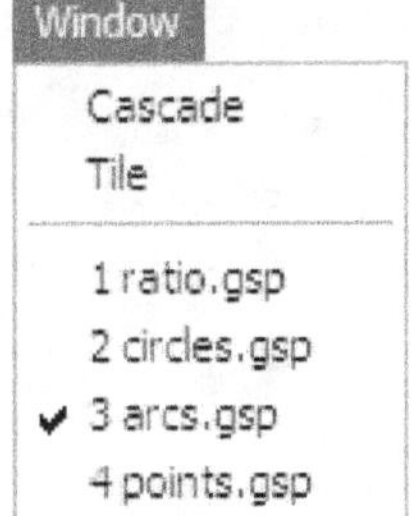

### *Minimize, Zoom (Mac OS X)*

These commands minimize or zoom the active sketch window.

### *Bring All to Front (Mac OS X)*

This command brings all Sketchpad windows in front of all non-Sketchpad windows.

### *Cascade, Tile (Microsoft Windows)*

These commands arrange all open document windows in either an overlapping or a tiled fashion, starting from the upper-left corner of the main Sketchpad window.

### *Window List*

Choose a document from this list to bring that document's window in front of all other document windows. This list contains up to ten documents, with a checkmark indicating the document whose window is in front of all other document windows.

### *More Windows (Microsoft Windows)*

This command appears at the bottom of the window list when you have more than ten open document windows. Use this command to bring to the front a window that doesn't appear among the first ten in the list.

## 3.10 Help Menu

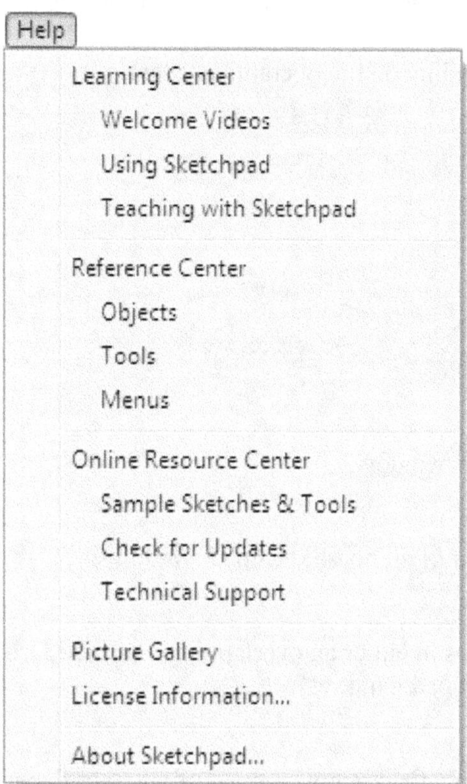

The Help menu contains commands that you can use to get help on using Sketchpad.

The **Learning Center** is the place to begin learning about Sketchpad. View the Welcome to Sketchpad videos, use interactive tutorials to learn the basics of using Sketchpad, and take advantage of resources to help you teach effectively with Sketchpad.

The **Reference Center** provides detailed reference information about every aspect of Sketchpad, including chapters covering Sketchpad's objects, tools, and menus. Use the Table of Contents, the index, or the full-text search to find information easily. (You are using the Reference Center now.)

Visit the **Online Resource Center** to download materials for classroom activities and projects; find links to sample sketches, reports and presentations, bibliographic materials, and other resources; find out about professional development opportunities; check for program updates; or contact our technical support team.

The **Picture Gallery** contains a collection of pictures that you can use in your sketches and activities as you experiment with transformations and other mathematical investigations, or just to dress up your Sketchpad presentations.

**About Sketchpad** appears here in the Windows version only; it appears on the Apple menu in the Macintosh version. This command displays a dialog box identifying the registered user name, the specific version number of the program, and assorted program information and credits.

In addition to the Help menu, most of Sketchpad's dialog boxes have **Help** buttons that you can click to get specific information about using that dialog box.

## 3.11  Context Menu

Context menus provide a convenient shortcut to many of the menu commands that apply to a specific window or object. (Context menus do not appear in the main menu; they appear only when you request them.)

To display a Context menu:

***Windows:*** Right-click in a sketch window, sketch object, tool object, or information balloon.

***Mac:*** Hold down the Ctrl key while you click in a sketch window, sketch object, tool object, or information balloon.

There are four different windows or objects that have Context menus:

- **The Sketch:** Click empty space in the sketch window to change preferences 159 or to use commands that apply to the entire sketch.

- **A Sketch Object:** Click a sketch object to change the object's properties 158 or attributes or to use commands that apply to that object.

- **A Tool Object:** Click a given object or a step in the Script View 286 to change the object's properties 158 or attributes.

- **An Information Balloon:** Click an information balloon 124 to display the object's Properties 158, to change the size of the balloon's text, or to control whether the balloon displays parents, children, or both.

Context menus show only available commands.

For information on any of the commands in the Context menu, refer to that specific topic in the Menu Reference 136.

### ▼ Sketch Context Menu

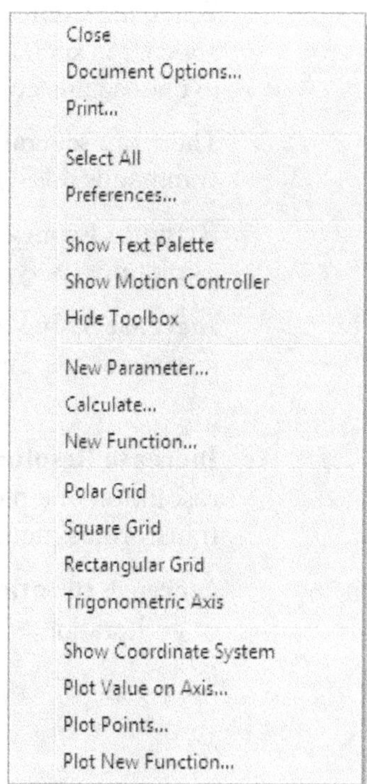

Right-click (or Ctrl-click) empty space in the sketch to change the sketch preferences [159] or to use commands that apply to the entire sketch.

When the Context menu for a sketch appears, if there's room on the screen, the menu appears in such a position that Preferences [159] is directly under the mouse, making it especially easy to choose this command.

## ▼ Sketch Object Context Menu

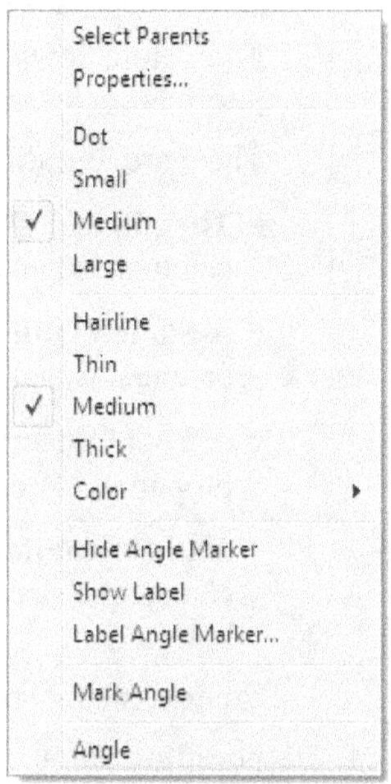


Right-click (or Ctrl-click) a sketch object to change the object's properties [158] or attributes or to use commands that apply to that object.

When the Context menu for a sketch object appears, if there's room on the screen, the menu appears in such a position that Properties [158] is directly under the mouse, making it especially easy to choose this command.

There are several commands that appear only in the Context menu for certain sketch objects; these commands don't appear in any other menu.

**Bring to Front** [247] and **Send to Back** [247] allow you to change the layer of layered objects [91] such as pictures, polygons and other interiors, and loci and iterations of these objects.

**Increase Value** [247] and **Decrease Value** [247] allow you to increase or decrease the value of a parameter [36]. The value changes by the Keyboard Adjustment amount set in Parameter Properties [38].

**Increase Resolution** [247] and **Decrease Resolution** [247] allow you to increase or decrease the resolution of a plotted object [95] (a locus, a function plot, a sampled path, or a sampled transformed picture).

**Increase Iterations** [248] and **Decrease Iterations** [248] allow you to increase or decrease the depth of an iteration [24].

### ▼ Tool Object Context Menu

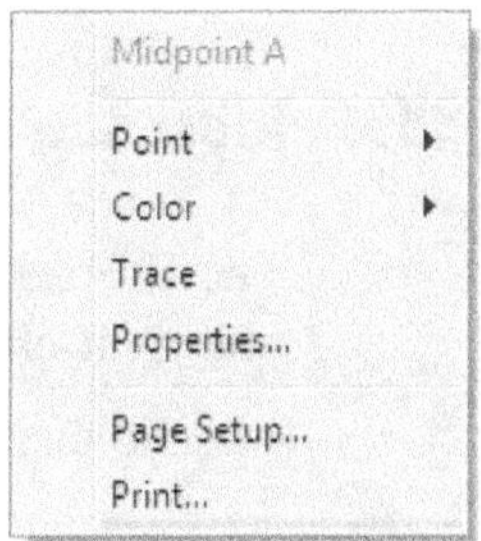


Right-click (or Ctrl-click) a given object or a step in the Script View 286 to change the object's properties 158 or attributes, or to print the script view.

### ▼ Information Balloon Context Menu

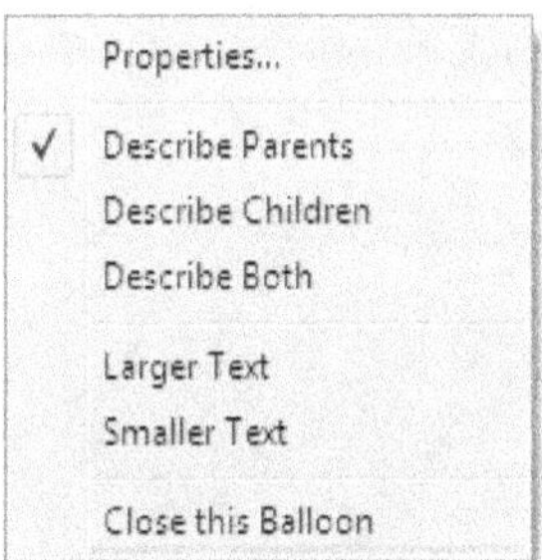


Right-click (or Ctrl-click) an information balloon 124 to display the object's Properties 158, to control whether the balloon displays parents, children, or both, and to change the size of the balloon's text.

## 3.11.1  Bring to Front/Send to Back

The **Bring to Front** and **Send to Back** commands appear in the Context 245 menu when you right-click (windows) or Ctrl-click (Mac) on a layered object 91.

Choose these commands to move the designated layered object in front of other layered objects or behind other layered objects.

In addition to moving a layered object in front of or behind other objects, you can also set its opacity 92, to determine whether other objects can be seen through it.

## 3.11.2  Increase Value/Decrease Value

The **Increase Value** and **Decrease Value** commands appear in the Context 245 menu when you right-click (windows) or Ctrl-click (Mac) on a parameter 36.

Choose these commands to increase or decrease the value of the parameter. The value changes by the **Keyboard adjustment** amount set in Parameter Properties 38.

## 3.11.3  Increase Resolution/Decrease Resolution

The **Increase Resolution** and **Decrease Resolution** commands appear in the Context 245 menu when you right-click (windows) or Ctrl-click (Mac) on a plotted object 95.

Choose these commands to increase or decrease the resolution of the plotted object (a locus, a function plot, a sampled path, or a sampled transformed picture).

## 3.11.4 Increase Iterations/Decrease Iterations

The **Increase Iterations** and **Decrease Iterations** commands appear in the Context 245 menu when you right-click (windows) or Ctrl-click (Mac) on an iteration or iterated image 24 .

Choose these commands to increase or decrease the depth of the iteration.

# Windows

# 4 Windows

In addition to the geometric, algebraic, and other objects that comprise Sketchpad constructions (and the tools and menu commands you use to create them) there are several windows, dialog boxes, and palettes that allow you to control and manipulate your experience with Sketchpad.

| | |
|---|---|
| Documents 250 | Learn about the document window 253 and how to manage the pages and tools 254 that a document includes. |
| Calculator 257 | Perform calculations using measurements from your sketch, and create and edit functions. |
| Motion Controller 263 | Start, stop, or pause objects in motion, speed them up, slow them down, or reverse their direction. |
| Text Palette 270 | Change the font, size, style, and color of text 271, and use mathematical formatting 272 in text objects. |
| Properties 273 | Find out how the selected object was constructed, or change how it appears or behaves. |
| Preferences 275 | Change Sketchpad's defaults for units 276, color 277, text 279, and several tools 280. |
| Advanced Preferences 281 | Change advanced defaults for exporting 282, sampling 283, and system settings 284. |
| Script View 286 | View a script of the active custom tool 125, change the order of its given objects, and format the resulting objects. |
| Color Picker 290 | Choose specific colors for objects, and change the colors in the color menu. |

## 4.1 Documents

A document in Sketchpad contains one or more *sketches*—that is, one or more collections of related mathematical objects. To create these objects, you'll use Sketchpad's Toolbox 102 and menus 136.

A document appears on your screen in a document window 253, and can be saved to your computer's hard drive or to your network. Once saved, the document can later be reopened from the same place.

When a document contains more than a single sketch, each individual sketch appears on its own page.

To change to a different page, do one of the following:

- Click a page tab at the bottom left of the window.

- Press the Page Down key or the Page Up key on your keyboard.

- Press a Link action button 75 that links to a different page.

- Open the Document Options 254 dialog box and double-click a page in the list of pages.

A document can also contain any number of custom tools 125, which extend the functionality of the fundamental **Point** 112, **Compass** 113, **Straightedge** 115, and **Polygon** 118 tools in the Toolbox.

Use **File | Document Options** 142 or choose **Tool Options** 254 from the Custom Tools 126 menu to manage the pages and tools contained in a Sketchpad document.

### ▼ Document Pages

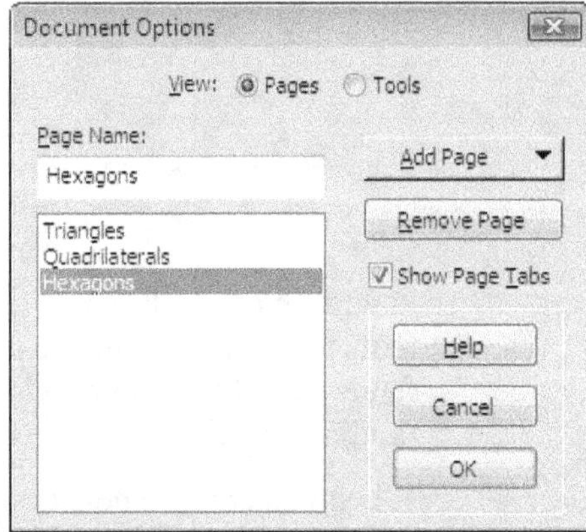

A new Sketchpad document always begins with a single page or sketch — a view of the geometric plane. Over time, you may want to add additional pages to a document. For example, you may want to organize a series of sketches that develop an argument; you may want to present an activity that has several parts; or you may want to explore a conjecture in more depth than would be possible in a single sketch.

In each of these cases, it's convenient to store several sketches together as pages of the same document.

To add pages to a document — and to name, copy, reorder, or remove existing pages — choose **File | Document Options** 142.

When a document has more than one page, page tabs normally appear at the bottom left of the document window, and a page name or number appears in the window's title bar. To switch from one page of your document to another, click a page tab or press the Page Up or Page Down key on your keyboard. You can also create Link buttons 151 to jump between pages.

Use **File | Document Options** 142 to hide or show the page tabs.

## ▼ Document Tools

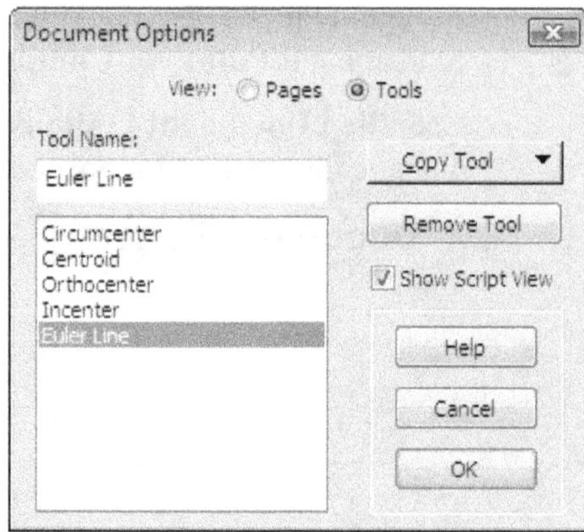

In addition to multiple pages, a Sketchpad document may contain one or more custom tools 125 — tools that you or someone else has created. Custom tools extend the functionality of Sketchpad's fundamental tools to provide new mathematical objects or to provide new ways to construct familiar mathematical objects.

When you create a new custom tool, it becomes part of your document. Use **File | Document Options** 142 to rename, reorder and remove custom tools from the active document, and to copy tools from a different open document into the active document.

Organize custom tools you use frequently by storing each collection of related tools in its own document. If you use a document's pages to describe the tools it contains — and give examples of their use — your document becomes a handy "toolkit" that you can share with other people. For example, you might want to make a kit of tools for constructing different centers of a triangle or one for constructing various regular polygons.

When you're working in a sketch, you can use not only the custom tools located in the document on which you're working, but also any custom tools contained in other open documents. To work with custom tools contained in a document on your hard drive, open the document so its custom tools become available. Alternatively, if you store any documents in the Tool Folder 133, tools from these documents are available whenever you start Sketchpad — even if the documents that contain them are not open.

View the details of a tool in written form using the Script View 286. This view shows you a list of each given object and each constructed object of the tool, allows you to modify the tool in various ways, and allows you to watch the tool step by step as it operates. Use the **Show Script View** command in the Custom Tools menu 126, or the **Show Script View** checkbox in the Tool Options 254 dialog box, to see the Script View of the active custom tool.

*Subtopics:*

## 4.1.1    Document Windows

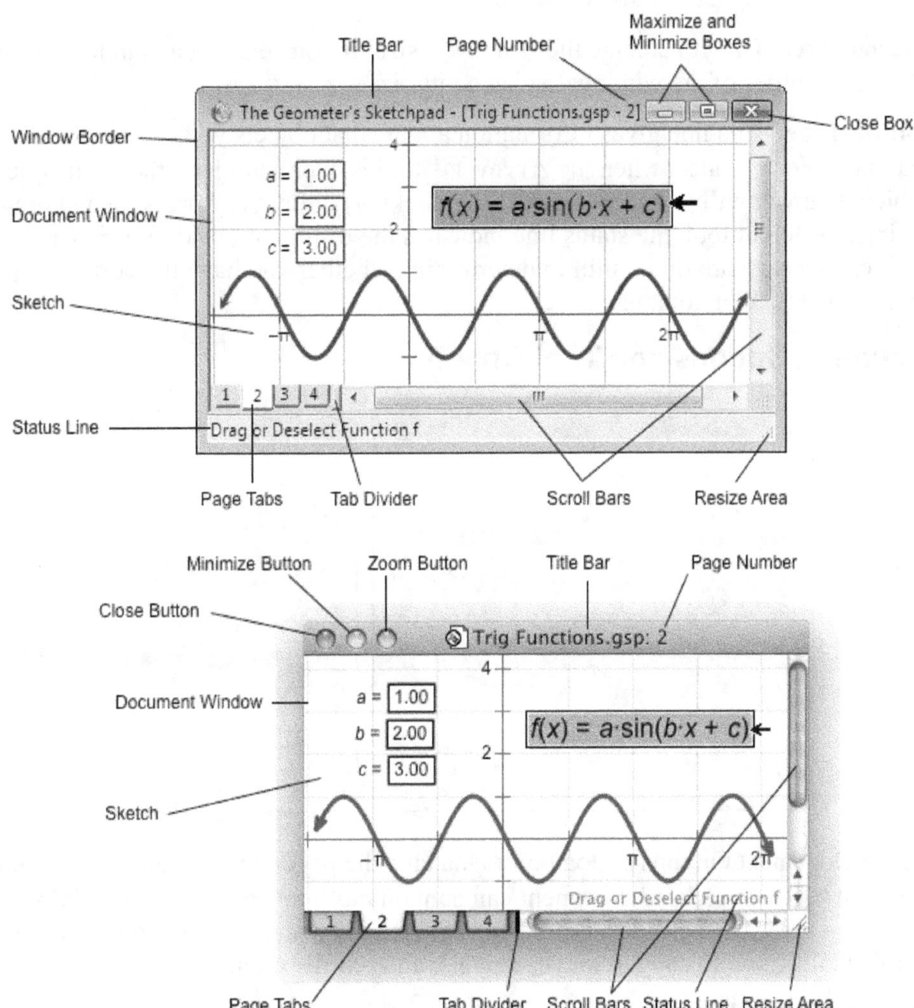


A Sketchpad document window is shown above. The window contains a sketch area within which you construct mathematical figures, and various controls with which you can manipulate the window and the sketch.

- **Title Bar:** Drag to reposition the window on the screen.

- **Close Box/Button:** Click to close the window.

- **Zoom, Maximize and Minimize Boxes/Buttons:** Click to expand the window to full size or to shrink it to an icon.

- **Page Tabs:** Click to change pages.

  Tabs are present only if your document has more than one page. Choose **File | Document Options** 142 to add or remove pages, or to hide or show the page tabs.

- **Tab Divider:** Drag to provide more or less space for page tabs.

- **Scroll Bars:** Click or drag to scroll the window.

  Because some sketch objects, like lines and rays, extend beyond the normal scrollable area, you can always press a scroll bar's buttons even when the scroll bar itself is at its limit.

You can also scroll the window by pressing the Alt key (Windows) or Option key (Mac) 295 and dragging in the window.

- **Resize Area:** Drag to change the window's size. If you're using a Windows computer, you can drag any border of the window to change its size.

- **Status Line:** This line gives information about what objects you've selected and what actions you can take. For instance, when the **Arrow** 104 tool is over an object, the status line identifies the object that you will select or deselect (by clicking) or drag (by pressing and dragging). When you're using a different tool, the status line indicates the results of clicking or releasing the pointer. This line can be particularly helpful in determining whether you have the correct objects selected to enable a particular command.

## 4.1.2   Document Options and Tool Options

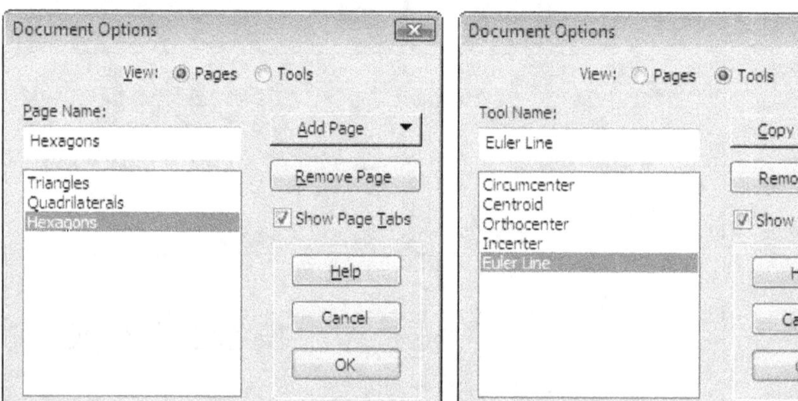

Use the Document Options dialog box to manage the pages and custom tools 125 contained in a document 250. A Sketchpad document can contain multiple pages; use this dialog box to add, remove, rename, and reorder the pages. Similarly, a Sketchpad document can contain multiple custom tools; use this dialog box to copy, remove, rename, and reorder them.

When you choose **File | Document Options** 142 this dialog box appears showing the pages in your sketch.

When you choose **Tool Options** from the Custom Tools 125 menu this dialog box appears showing the custom tools in your sketch.

Click **Pages** or **Tools** at the top of the dialog box to switch between the two views.

### ▼ List of Pages/Tools

This list shows all the pages or tools in the current document. You can perform the following actions directly on the list items:

- Click a page or tool in the list and then change its name.

- Press and drag a page or tool in the list to change the position of that page or tool in your document.

- When you're viewing pages, click the name of a page in the list to display that page in the document.

- Double-click a page name to display that page and close the dialog box.

### ▼ Page Name/Tool Name

To rename a page or tool, type a new name here for the page or tool that is chosen in the list box.

### ▼ Add Page

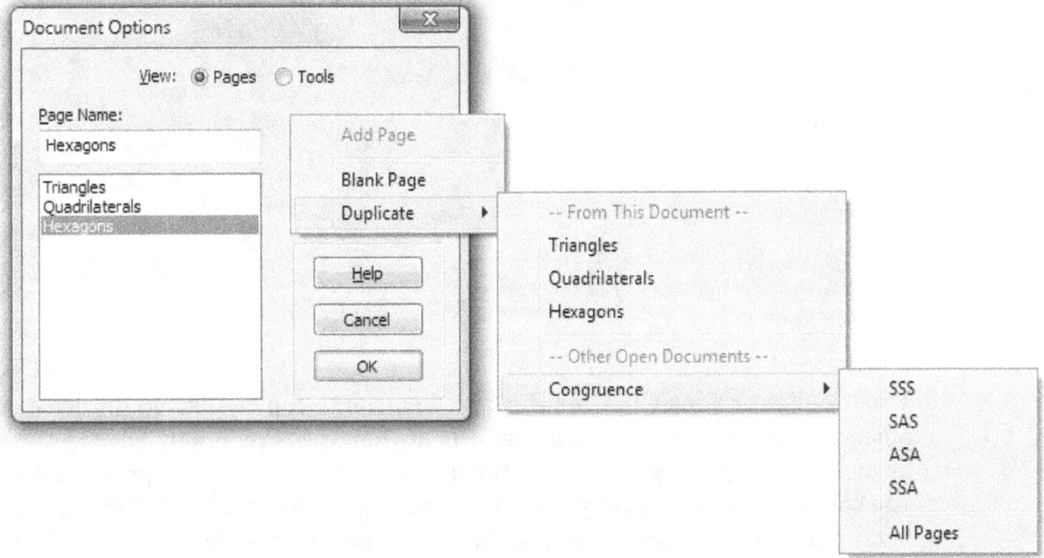


When you're viewing pages, use this pop-up menu to add new pages to your document. Choose **Blank Page** to add a new blank page to your document. Choose from the **Duplicate** submenu to add a duplicate copy of a page from any open document. The top part of the **Duplicate** submenu lists pages from the current document. Listed below a divider are any other open documents; these serve as submenus from which you can choose pages to duplicate and add to the current document.

### ▼ Show Page Tabs

This checkbox appears only when you view page options, not when you view tool options. When **Show Page Tabs** is checked, tabs appear along the bottom left of the document, showing the names or numbers of the pages. Use these tabs to navigate among the pages of a document.

> When page tabs aren't showing, use the Page Up and Page Down keys on your keyboard, Link 151 buttons, or the Document Options dialog box to move from page to page in your document.

Moving the divider at the left edge of a document's horizontal scroll bar all the way to the left has the same effect as unchecking **Show Page Tabs.** Moving the divider away from the left edge of the window has the same effect as checking **Show Page Tabs.**

### ▼ Show Script View

This checkbox appears only when you view tool options, not when you view page options. Check or uncheck this box to show or hide the script view 286 of the most-recently chosen custom tool 125. The script view allows you to see the given objects and the steps 288 of the tool, to change the properties of the steps, and to observe and control the tool as it functions.

### ▼ Copy Tool

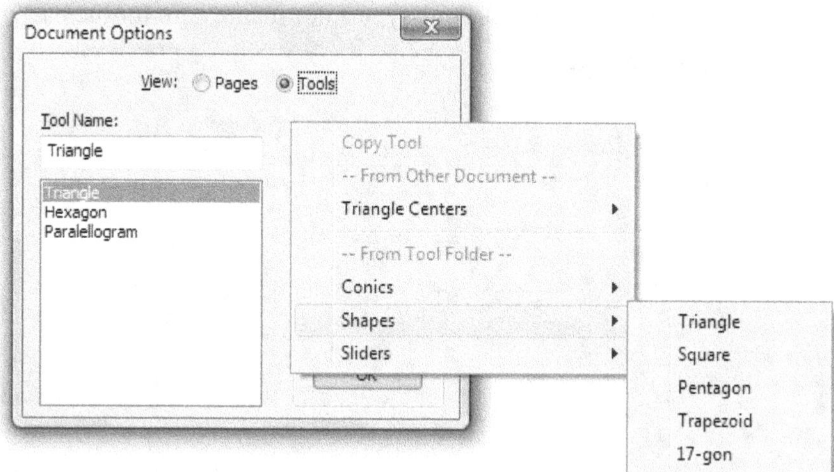

When you're viewing tools, use this pop-up menu to copy tools into the active document from other open documents or from documents in your Tool Folder[133]. The top part of the Copy Tool menu lists any other open documents that contain tools; each item provides a submenu from which you can choose a tool to copy and add to the current document. The bottom part of the Copy Tool menu allows you to copy tools from Sketchpad documents located in your Tool Folder[133].

> Tools that you use frequently and not just in one document can be stored in the Tool Folder[133]. These tools are always available whenever you're using Sketchpad.

### ▼ Remove Page/Tool

Use this button to permanently remove a page or tool from a document. First choose from the list box the page or tool you wish to remove, then click **Remove Page** or **Remove Tool.** Every document must have at least one page, so you cannot remove the only page from a one-page document.

> Before removing a tool or page from your document, you may want to save the document with a different name in order to preserve a copy of that page or tool in case you want it later.

Once you remove a tool or page and click **OK,** it's gone for good — you cannot get it back. If you decide you don't really want to remove that tool or page, you must click **Cancel** rather than **OK** in order to leave your document unchanged.

## 4.2    Calculator

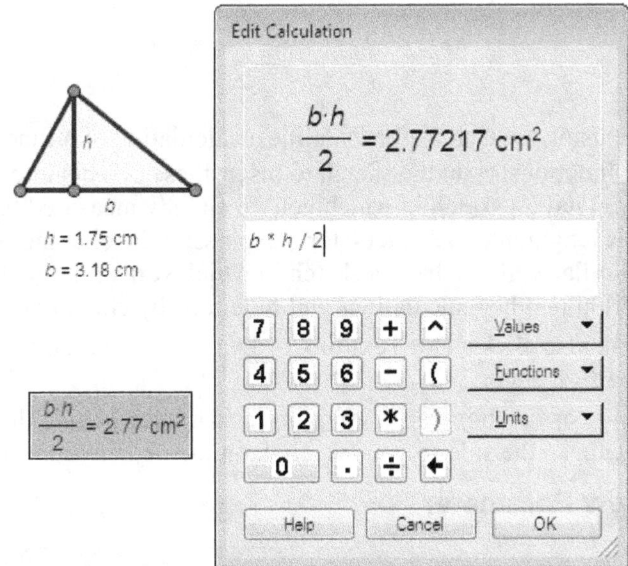

Use the Calculator to create or edit two kinds of Sketchpad objects: calculations [39] and functions [45].

**Calculations:** Calculations are values generated using numbers, mathematical operations, functions [45], measurements [36], or parameters [36] from your sketch. These calculations can be used for a wide variety of purposes: to define distances, angles, and scale factors by which objects are transformed; to plot points; and to determine the results of other calculations and functions.

**Functions:** When you define a function in Sketchpad, you can also use numbers, mathematical operations, measurements, parameters, or other functions from your sketch. Once you've defined a function, use it for various purposes: plot it, evaluate it, differentiate it, or use it to define other calculations or functions.

There are several different purposes for which you'll use the Calculator:

- To create a new calculation, choose **Number | Calculate** [227].

- To create a new function, choose **Number | New Function** [230].

   Choose **Graph | Plot New Function** [241] to create a new function and plot it immediately.

- To edit a calculation, double-click the calculation with the **Selection Arrow** [104] tool, or select the calculation and choose **Edit | Edit Calculation** [157].

- To edit a function definition, double-click the function with the **Selection Arrow** [104] tool, or select the function and choose **Edit | Edit Function** [157].

- To change a parameter [36] into a calculation, select the parameter, choose **Edit | Edit Parameter** [157], and change the expression to something more complicated than just a number.

- To change a calculation into a parameter [36], edit it so that it consists of one number only, with or without distance or angle units.

### ▼ Insert a Value or Function from the Sketch

$$m\,\overline{AB} = 2.61\ \text{cm}$$

$$f(x) = a{\cdot}x^2 + b{\cdot}x + c$$

While you're using the Calculator to define a calculation or a function, you can click existing values [92] or functions [45] in the sketch to insert these objects into your new expression. For instance, if you have a sketch in which you've already measured the length of segment *AB,* you can insert this length into the Calculator's expression by clicking on the existing measurement in the sketch. Similarly, if you have a sketch in which you've defined a function $f(x) = ax^2 + bx + c,$ use this function in other calculations and functions by clicking it in the sketch.

You can also click a Hot Text link to a value or function to insert that value or function into the Calculator.

If a value or function you want to insert is hidden behind the Calculator itself, drag the Calculator to the side so you can click on the object you want to insert.

### ▼ Insert a New Parameter

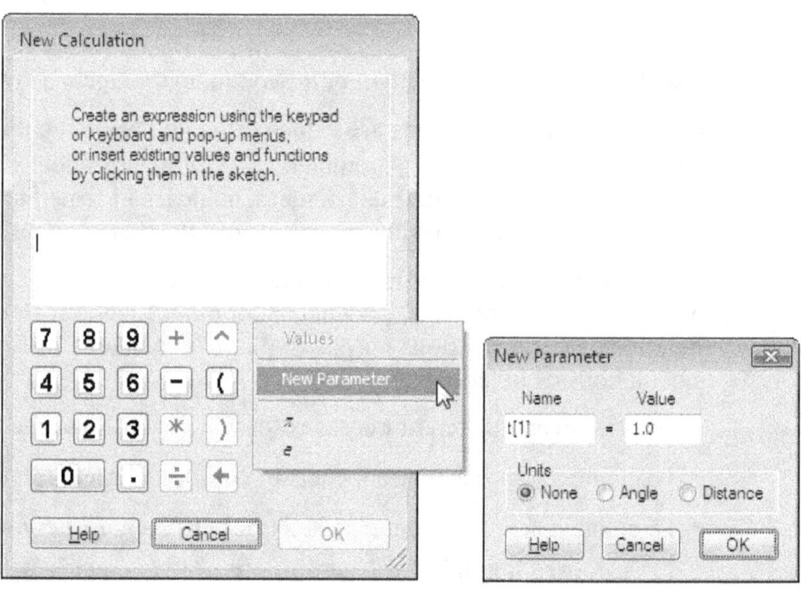


While you're using the Calculator to define a calculation or a function, you can create a new parameter [36] and insert it into your expression.

1. Choose **New Parameter** from the Values pop-up menu [259], or use the keyboard shortcut Shift+Ctrl+P (Windows) or Shift-⌘P (Mac). The New Parameter dialog box appears

2. Type the name you want to use for the new parameter. You can also set the initial value of the parameter.

3. Click **OK.** The parameter is inserted into your expression and added to your sketch.

   The result is the same as if you had used **Number | New Parameter** [226] to create the parameter before using the Calculator.

*See also:*


## 4.2.1    Parts of the Calculator

The Calculator has several parts: a keypad, an input area, a preview area, and pop-up menus for inserting special mathematical elements

### ▼ Keypad

Click the buttons in the Calculator's keypad to insert numbers, decimal points, operators, and parentheses into your calculation or function. Instead of clicking one of these buttons, you can type the corresponding key on your computer's keyboard. (Use the / key — the forward slash — on the keyboard for division.)

Click the left-arrow key on the keypad, or the Backspace key (Windows) or Delete key (Macintosh) on the keyboard, to delete the last item you typed. If you're using the Calculator to define a function, you can click the key labeled $x$ to enter the value of the independent variable.

The $x$ key is present only when you're defining a function; it changes to $y$, $\theta$, or $r$ depending on the form of the equation.

### ▼ Input Area

$$(x + 1) \wedge 2 / (2x)$$

The input area displays each element you insert into the calculation or function as you enter it. Refer to the input area to see exactly what you've entered. You can also click in the input area or press the right- or left-arrow key on the keyboard to change the insertion point and add new numbers, operators, and so forth in the middle of an existing expression.

If the input area does not form a valid mathematical expression, the portion of the input area up to the first error is shown in black. The portion following the first error appears in red.

### ▼ Preview Area

$$\frac{(x + 1)^2}{2 \cdot x} = 2$$

The preview area shows a mathematically formatted preview of the input area when it contains a valid expression. Use the preview to be sure that you've entered the desired calculation or function correctly — that you have the parentheses in the right places and that you have the correct order of operations.

### ▼ Pop-up Menus

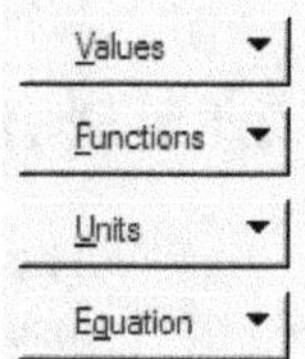

Depending on whether you're defining a calculation or a function, either three or four pop-up menus appear. The last of these, the Equation pop-up menu, appears only when you're defining or editing a function rather than a calculation.

> You can click an existing value $\boxed{92}$ or function $\boxed{45}$ in the sketch to insert it into the Calculator. You can also click a Hot Text link to a value or function to enter it.

### ▼ Values

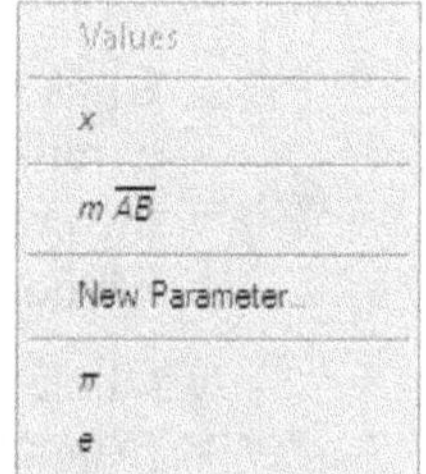

This pop-up menu allows you to enter the value of any selected measurement in the sketch, to insert a new parameter, or to insert the value of the constant $\pi$ or $e$. The values that appear in this menu include measurements or calculations that were selected in the sketch when the Calculator was opened. If you want to use a value from the sketch that doesn't appear in this menu, click the value in the sketch to insert it into your expression. (If the value is hidden behind the Calculator, drag the Calculator by its title bar to move it out of the way and then click the value.)

When you edit a calculation or function, you can insert only values that don't depend on the object you're editing. In other words, if you're editing a calculated value $2 \cdot AB$, and your sketch has another calculation which uses this result to calculate $2 \cdot AB + 2 \cdot CD$, you cannot insert the value of the second calculation into the first.

If you're defining a function, you can also use the Values pop-up menu to insert the value of the independent variable $x$, $y$, $r$, or $\theta$.

## ▼ Functions

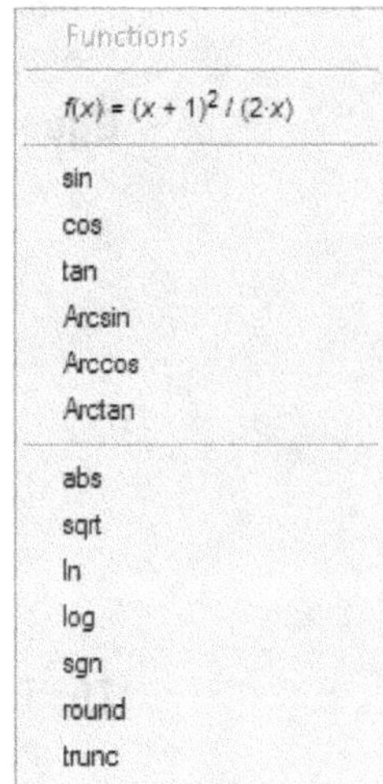

This pop-up menu allows you to use in your expression any selected function you've already defined in the sketch, or to use any of Sketchpad's standard functions. Sketchpad's standard functions include the trigonometric functions, the inverse trig functions, and these additional functions:

| | |
|---|---|
| **abs** | absolute value |
| **sqrt** | square root |
| **ln** | natural logarithm (base $e$) |
| **log** | common logarithm (base 10) |
| **sgn** | signnum (returns +1, 0, or -1 depending on whether its argument is positive, zero, or negative) |
| **round** | round (rounds its argument to the nearest whole number) |
| **trunc** | truncate (truncates its argument by removing the fractional part For example, trunc(2.5) = 2, and trunc(–7.8) = –7.) |

The **sgn** function is especially useful in creating a calculation that makes a decision based on the value of a variable, measurement, or parameter.

You can also click on an existing function that you've defined in the sketch to insert it into the Calculator.

## ▼ Units

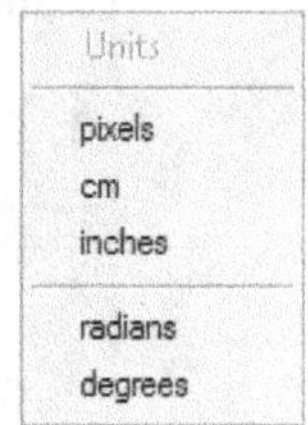

The Units pop-up menu allows you to insert any desired angle or distance unit (pixels, cm, inches, radians, or degrees). A unit must be attached to a number, so the Units menu is enabled whenever you've just inserted a numeric constant — either an ordinary number or one of the constants $\pi$ and $e$.

*See also:*
Plot Function 241
New Parameter 226

▼ **Equation**

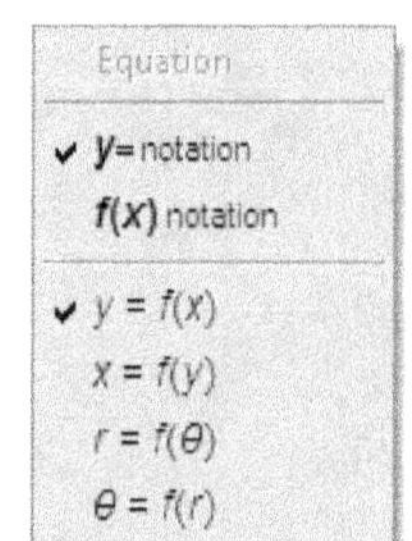


The Equation pop-up menu appears only when you create a function 230 or edit a function 47, and allows you to set the form in which the function plot appears.

Use the first two choices to determine whether your equation appears in *y=* notation (for instance, $y = x^2$) or in *f(x)* notation (for instance, $f(x) = x^2$).

To change the default setting for new functions, choose **Edit | Preferences | Text** 279.

Use the bottom four choices to determine whether the independent variable is $x$, $y$, $\theta$, or $r$. Choose *y = f(x)* or *x = f(y)* for a square or rectangular plot, 235 or choose *r = f(θ)* or *θ = f(r)* for a polar plot. 235

*See also:*
How to Use Signum to Construct a Piecewise Function 262

## 4.2.2    How to Use Signum to Construct a Piecewise Function

Sometimes you may want to construct a function 45 that behaves one way in part of its domain and differently in another part of its domain. Such functions, called *piecewise* functions, are very important in interpolation, in fitting curves to data, and in designing and producing shapes and surfaces in applications such as automobile design. The signum function makes it possible to do this in Sketchpad.

The signum function is useful any time you want a calculation which makes some sort of decision — which performs a different calculation when some value changes.

For instance, you could construct a function whose plot 52 is in the shape of a cosine wave when $x > 0$, but is parabolic when $x < 0$. Here's how to do this with the signum function:

| | |
|---|---|
| **1.** Choose **Graph \| Plot New Function** 241. | |
| **2.** Enter the cosine-wave part of the function, as shown at right. When $x > 0$, the signum function returns $+1$, and the value of the multiplier is 1. But when $x < 0$, the signum function returns $-1$, and the value of the multiplier is 0. This first part of the function will be a cosine wave to the right of the origin, but will always be zero to the left of the origin. | $\cos(x)\left(\frac{1+\text{sgn}(x)}{2}\right)$ |
| **3.** Add the parabola part of the function as shown at right. When $x > 0$, the multiplier will be 0, and when $x < 0$, the multiplier will be 1. | $(x^2+\cdots)\left(\frac{1-\text{sgn}(x)}{2}\right)$ |

Here's the result when you click *OK:*

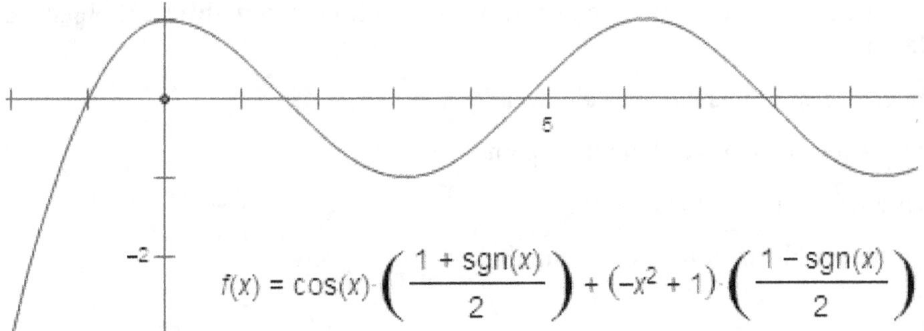

The functions in this example were chosen in such a way that the two functions join continuously and smoothly. Can you construct a different piecewise function in which the join is not continuous, or in which it's continuous but not smooth?

Here's how to use the signum function to solve the general problem of defining a function $h(x)$ whose value is $f(x)$ for all $x < k$ and is $g(x)$ for all $x > k$:

$$h(x) = f(x) \cdot \left( \frac{1 - \mathrm{sgn}(x - k)}{2} \right) + g(x) \cdot \left( \frac{1 + \mathrm{sgn}(x - k)}{2} \right)$$

# 4.3   Motion Controller

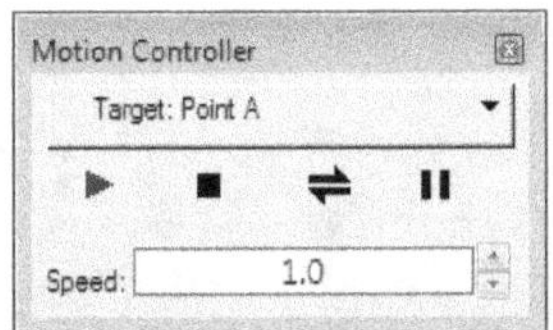

Use the Motion Controller to start objects animating and to control the motion of objects in your sketch.

Movement is at the heart of dynamic geometry. By animating and dragging objects in Sketchpad, you can rapidly explore many different variants of a construction, allowing you to make discoveries and investigate conjectures in a way that would be impossible without motion. Animation also allows you to demonstrate and present your findings in a more interesting and effective way than you could ever do with a static diagram.

Many of the functions of the Motion Controller are available through Sketchpad's menus, action buttons, and so forth. By providing a shortcut to these functions, the Motion Controller makes it quicker and easier for you to use animation in your sketch.

If you want even more control of animation than the Motion Controller provides, create Animation [69] and Movement [71] buttons by choosing **Edit | Action Buttons** [148].

You use the Motion Controller to start objects moving, stop them, reverse their direction, and change their speed. It gives you easy access to the functionality of the **Animate** [171], **Increase Speed** [172], [172] **Decrease** [172] **Speed** [172], [172] and **Stop Animation** [172] commands in the Display [162] menu, while providing better control of the speed and direction of moving objects. It makes it easy to control the motion of a specific moving object without affecting other moving objects.

The Motion Controller appears when you start an animation or when you choose **Display | Show Motion Controller** [173].

When you move objects in Sketchpad, whether by using the Motion Controller, by choosing **Animate** |171|, or by pressing an Animation or Movement action button, different objects move or change in different ways.

- Independent points move freely in the plane.

- Points on paths move along their paths.

- Parameters change their values.

- All other objects move by moving their parent objects.

*Subtopics:*

## 4.3.1    Parts of the Motion Controller

Each part of the Motion Controller affects a different aspect of motion.

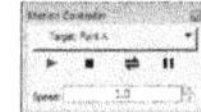

**Target:** This button describes the objects that will be affected by the Motion Controller buttons. If at least one object is moving in your sketch, you can click to display a pop-up menu of moving points and changing parameters. Choose an object from the menu to select it and to make it the new target.

If there are movable selected objects in your sketch, the target matches the selections. If there are selected objects that cannot be moved, there's no target. If the sketch contains moving objects but no selections, the target is all moving objects.

> To set the target to an object that's not moving, select it with the **Selection Arrow** [104] tool.

> To set the target to a moving object, click the Target button and choose the object from the list that appears.

**Animate:** Click this button to animate each target object. This button has the same effect as the **Display | Animate** [171] command.

**Stop:** Click this button to stop each moving target object. This button has the same effect as the **Display | Stop Animation** [172] command.

**Reverse:** Click this button to reverse the direction of each moving target object. This button is enabled when at least one target object is moving in a fixed direction (rather than in random directions).

**Pause:** Click this button to pause all motion. Unlike the other elements of the Motion Controller, this button affects every moving object, not just the target.

**Speed:** Click and type a new speed here to change the speed of each moving target object.

**Increase/Decrease Speed Arrows**: Click one or the other of these arrows to increase or decrease by one step the speed of each moving target. These arrows have the same effect as the Display | Increase Speed [172] and **Display | Decrease Speed** [172] commands.

> Each speed step is a little more than 25%; three speed steps in a row changes the speed by a factor of two.

Drag the Motion Controller by the title bar to reposition it. If you're using a Windows computer, you can also dock it to either the top or bottom of the Sketchpad window.

*See also:*
*Show Motion Controller command* [173]
*Using the Motion Controller* [266]

## 4.3.2   Using the Motion Controller

There are many different animation-related tasks you can accomplish using the Motion Controller 263.

You can also use Animation 149 buttons and Movement 150 buttons to move objects automatically.

### ▼ Start an Animation

1. Select one or more objects to animate. The objects must be geometric objects or parameters 267.

2. Click the Animate button 265 in the Motion Controller. Each selected object begins moving.

   The only objects you can't animate are captions, calculations, functions, action buttons, measurements, and pictures. (Except for action buttons, each of these objects can be attached 153 to a point, and the point can be animated.)

The Animate button has the same effect as the **Display | Animate** 171 command.

### ▼ Select a Moving Object

To modify the motion of a particular moving object, you must select the object to make it the target of the Motion Controller. If the object were not moving, you would simply click it with the **Arrow** 104 tool to select it — but it's not always so easy to click a moving object.

You may be able to select the desired moving object with a mouse click if it's moving slowly or with a selection rectangle 106 if there aren't any other objects near it. If not, follow these steps.

1. Press and hold on the Target menu to show the list of target objects.

2. If the object you want appears on the list, choose it. The object is selected, and you're done.

   If the object is not an independent point, a point on path, or a parameter, the object is moved by moving its parents, and it doesn't appear on the list. In this case, continue with steps 3 through 5.

3. Click the Pause button in the Motion Controller.

   Motion stops.

4. Select the object.

5. Click the Pause button again to release the button and restart the motion.

   The object remains selected and is listed as the target object.

### ▼ Stop a Moving Object

You can stop the motion of a single object while leaving other objects in motion.

1. Select the moving object as described in the previous section.

2. Click the Stop button in the Motion Controller. The selected object stops moving.
   Clicking Stop with nothing selected stops all moving objects.

The Stop button has the same effect as the **Display | Stop Animation** 172 command.

### ▼ Reverse the Direction of a Moving Object

You can reverse the direction of any moving object as long as it's not animating randomly.

1. Select the moving object as described above.

2. Click the Reverse button in the Motion Controller. The selected object reverses its direction.

Clicking Reverse with nothing selected reverses the direction of all moving objects.

### ▼ Set, Increase, or Decrease the Speed of a Moving Object

You can change the speed of any moving object.

1. Select the moving object by clicking the Target button and choosing the object from the pop-up menu.

2. Set a new speed by either typing a new speed into the Motion Controller's speed control or by clicking the up or down arrows of the speed control.

3. The selected object moves at the new speed.

    When you have several objects moving at different speeds and you want to make all of them go faster or slower, use the speed arrows rather than typing a specific speed.

The up and down arrows have the same effect as the **Display | Increase Speed** 172 and **Display | Decrease Speed** 172 commands.

*See also:*
 *Principles of Animation* 267
 *Parts of the Motion Controller* 265

## 4.3.3    Principles of Animation

Just about every object you can create in Sketchpad can be animated. To animate an object, you can choose **Display | Animate** 171, use the Motion Controller 263, or create an Animation action button 149 . Of these three methods, action buttons give you the most control of the details of motion.

    The only objects you can't animate are captions, calculations, functions, action buttons, measurements, and pictures. (Except for action buttons, each of these objects can be attached 153 to a point, and the point can be animated.)

    You can also move an object automatically by creating a Movement action button 150.

### ▼ How Objects Move

Different objects in Sketchpad move in different ways.

- An independent point 2 moves randomly on the plane.

- A point on path 176 moves along its path 89. If the path is an interior (polygon, circle interior, or arc interior), the point moves along the perimeter of the interior.

    For open paths (straight objects, arcs, and some point loci) default direction is bidirectional. For circles and circle interiors, it's counter-clockwise, and for other closed paths it's forward (based on the order of the parent objects). You can change the direction using the Motion Controller or the properties of the Animation button.

- A parameter 36 changes its value.

- Any other object moves by moving its parent objects 80.

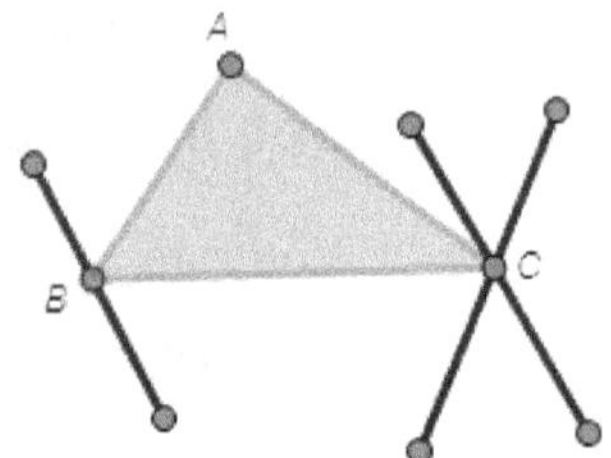

If you animate this triangle interior, it moves its parent objects (vertices *A, B,* and *C*). Vertex *A* (an independent point) moves randomly on the plane, vertex *B* (a point on a segment) moves bidirectionally along its path, and point *C* (an intersection) moves the intersecting segments, which in turn move their endpoints. Thus your ability to animate any geometric object is ultimately based on animating independent points and points constructed on paths. All other objects are animated indirectly — by animating their parents.

Except for parameters, non-geometric objects (action buttons, measurements, calculations, functions, and captions) cannot be animated (unless they are merged |153⟩ to a point). A parameter is like an independent point in the sense that its value, like the position of an independent point, does not depend on other objects. Sketchpad animates a parameter by changing its value.

Because independent points, points on paths, and parameters are the only objects that can be animated independently of their parents, these are the only objects that Sketchpad directly animates, and they are normally the only objects that appear in the Motion Controller's *Target* |265⟩ pop-up menu.

> Other objects can be listed as the Motion Controller target if they are selected. For example, if you select the interior of △*ABC,* the triangle will be listed as the target. Even though the triangle is listed as the target, any motion changes you make will directly affect points *A, B,* and *C* and will affect the triangle indirectly.

## ▼ Animation of an Independent Point

An independent point is animated by moving it randomly on the plane. The animation speed determines how far it's likely to move in each random step. You cannot control the direction for an independent point. To specify the starting animation speed explicitly or to make the motion occur one time only, create an Animation button |149⟩ and use the Animate Properties |69⟩ panel.

> Merge |153⟩ the independent point to a path |89⟩ if you want it to move in a specific direction.

## ▼ Animation of a Point on Path

A point on path is animated by moving along its path. If the path is closed (for example, a circle, a polygon, an arc sector, or an arc segment) the animation proceeds around and around the path. If the path is a segment or an arc, the animation proceeds bidirectionally — back and forth along the path. If the path is infinite, as with a line or a ray, Sketchpad animates the point bidirectionally and tries to use the portion of the path that's visible in the window. To specify the direction or speed explicitly or to make the point travel its path only once, create an Animation button |149⟩ and use the Animate Properties |69⟩ panel.

> To designate a specific portion of a line or ray as the domain for an animating point, construct a segment collinear with the line or ray and merge |153⟩ the point to the segment.

## ▼ Animation of a Parameter

A parameter [36] is animated by changing its value within its domain. The default domain, direction, and speed of a parameter's variation depends on the units of the parameter, as shown in this table.

> The speed listed here is a maximum. A parameter may change more slowly if your computer is busy with many tasks.

| *Units* | *Domain* | *Direction* | *Speed* |
|---------|----------|-------------|---------|
| None | −100 to 100 units | Bidirectional | 1 unit/sec |
| Degrees | 0° to 360° | Increasing | 45°/sec |
| Radians | 0 to $2\pi$ | Increasing | $\pi/4$ radians/sec |
| Inches | 0 to 100 inches | Bidirectional | 1 inch/sec |
| cm | 0 to 100 cm | Bidirectional | 1 cm/sec |

To specify the direction, domain, or speed explicitly, choose **Edit | Properties | Parameter** [38] or create an Animation button [149] and use Animate Properties [69].

## ▼ Motion Direction

The possible directions in which you can animate an object depends on the kind of object. When you start an animation using **Animate** [171] or the Motion Controller [263], Sketchpad uses the most common direction for the objects you animate. To access more advanced direction choices, create an Animation button [149].

- A point on a path can move forward, backward, bidirectionally, or randomly. (If the path is a circle, the choices are counter-clockwise and clockwise instead of forward and backward.) If you specify random motion on a path, each time the point is moved it's given a brand-new random position somewhere on the path.

- An independent point always moves randomly. Each time it moves, its new position depends both on its previous position and on its location within the window. A point moving slowly takes only small steps from its previous position, whereas a point moving quickly takes larger steps. If the point is near an edge of the window or outside the window, it's more likely to move toward the center of the window than away from it. In this way, randomly moving independent points usually remain visible.

- A parameter has a domain within which it is animated and can increase in value, decrease in value, change bidirectionally, or change randomly within that domain. To set the default domain for a parameter, use the Parameter Properties [38] panel. To set a parameter's domain and direction for an Animation button, use the Animate Properties [69] panel.

## ▼ Motion Speed

Use the Motion Controller, [263] the **Increase/Decrease Speed** [172] commands, or an Animation button[149] to set the speed of an animating object relative to the medium speed for that object.

You can set the ideal medium speed for points using System Preferences[284]. The actual motion speed depends both on this setting and on your computer.

The Motion Controller displays medium speed as speed 1.0. Sketchpad attempts to keep this speed constant, but if your sketch is complex or your computer is busy with other tasks, medium speed may be slightly slower than requested in System Preferences.

### ▼ Once-Only Motion

When you create an Animation button[149], you can specify that an object moves one time only. If the object is moving randomly, this means that one press of the button causes the object to move one time to a new random position. If the object is not moving randomly, once-only motion means that the object will stop once it returns to its starting position.

*See also:*
*Using the Motion Controller* [266]
*Parts of the Motion Controller* [265]

## 4.4 Text Palette

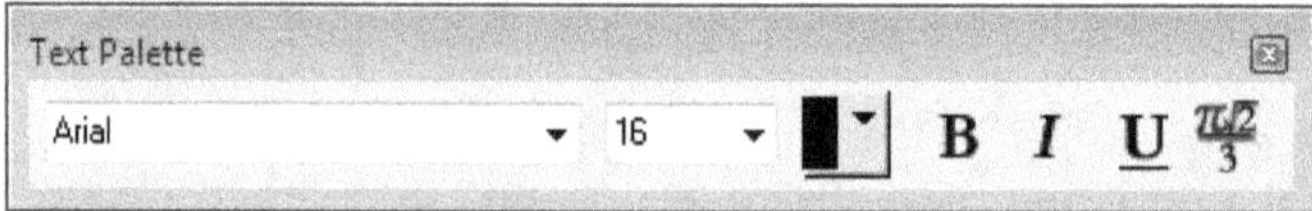


Use the Text Palette to format the font, size, style, and color[271] of labels[81], captions[56], measurements[36], and other text. You can also use the Text Palette to insert mathematical symbols and formatting[272] into captions.

The Text Palette can be used:

* when you have selected objects in your sketch and one or more of those objects either shows text or has a label. Use the Text Palette to change the appearance of the text or label of each selected object.

* when you're using the **Text**[120] tool to edit a caption and you have selected text in that caption. Use the Text Palette to change the appearance of the selected text.

### ▼ Showing, Hiding, and Moving the Text Palette

Normally the Text Palette appears automatically when you edit a caption. You can turn this behavior on or off using **Edit | Preferences | Text**[279]. To show or hide the Text Palette manually, choose **Display | Show Text Palette**[172] or **Display | Hide Text Palette**[172].

The keyboard shortcut for **Show/Hide Text Palette** is Shift+Ctrl+T (Windows) or Shift ⌘T (Mac).

If you're using a Windows computer, the Text Palette normally appears attached (or *docked*) to the bottom of the Sketchpad application window. You can drag it to a different position and leave it either floating or docked to the top or bottom of the window. If you're using a Macintosh computer, the Text Palette always floats above or beside your document and can be repositioned by dragging its title bar.

On either kind of computer, when the Text Palette is floating, you can hide it by clicking its Close button, and you can move it to a different position on the screen by dragging its title bar.

## 4.4.1 Using the Text Palette

The various parts of the Text Palette allow you to change the font, size, style, and color of labels 81 and text. You can also use the Text Palette to add mathematical notation 272 to your captions 56.

Choose **Show/Hide Text Palette** 172 to determine whether or not the Text Palette is visible.

To change the font, size, style, or text color of one or more labeled objects or objects displaying text, select those objects and use the Text Palette to make your desired modifications.

Pressing the down arrow triangles displays Font, Size, and Color pop-up menus.

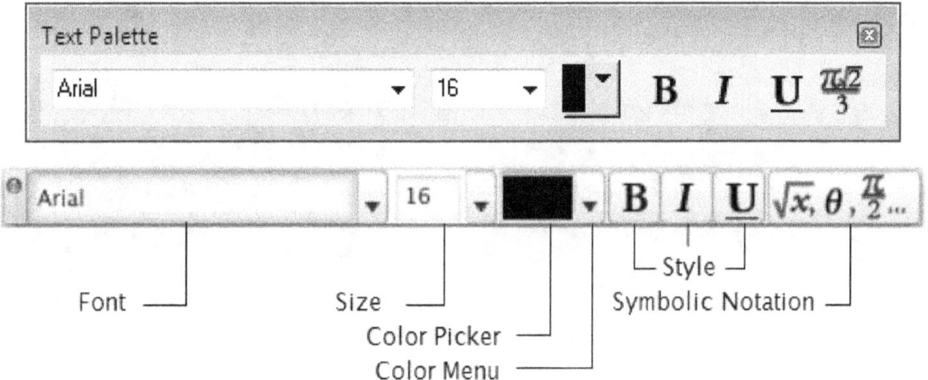


**Font:** Change text font by choosing a font from the pop-up menu. Recently-used fonts appear at the top of the menu.

**Size:** Change text size by typing a size or by choosing a size from the pop-up menu.

**Color Picker:** Change text color to any color your computer can display by clicking this button use your system's Color Picker 290 dialog box.

**Color Menu:** Change text color (to one of Sketchpad's default colors) by choosing a color from the pop-up menu.

When a Text Palette color is applied to selected geometric objects, it affects the color of those objects' labels. To change the color of the objects themselves, choose **Display | Color** 164. (You can change the available colors by holding the Shift key and choosing **Edit | Advanced Preferences | System** 284 **| Edit Color Menu.**)

**Style:** Change bold, italic, or underline text style by clicking a button. One click sets the style and depresses the button; a second click removes the style and releases the button.

**Symbolic Notation:** Display additional buttons for mathematical notation 272 by clicking here when you're editing a caption. Click a second time to hide the math formatting buttons.

When you change an object's font, size, or style, the default style for newly-created similar objects is updated to match. Press the Shift key, or uncheck **Update automatically when restyling existing objects** in Text Preferences 279, to preserve default settings for font, style, and size.

You can also use the **Display | Text** 166 submenu to set the style and font of labels and other text. The command shortcuts from this submenu are particularly convenient. There are shortcuts for **Bold,**

**Italic,** and **Underline** (Ctrl+B, Ctrl+I, and Ctrl+U in Windows, or ⌘B, ⌘I, and ⌘U on Mac) and for **Increase Size** and **Decrease Size** (Alt+> and Alt+< in Windows, or ⌘> and ⌘< on Mac).

*See also:*
*Display | Text submenu* 166
*Advanced Text Topics* 313
*Show/Hide Text Palette command* 172

## 4.4.2   Using Mathematical Notation in a Caption

While you're editing a caption 56, use the Text Palette 270 to enter mathematical symbols and mathematical formatting.

Press the Symbolic Notation button in the Text Palette to display additional notation buttons you can use to enter mathematical symbols and other formatting, like overbars, fractions, exponents, and grouping symbols.

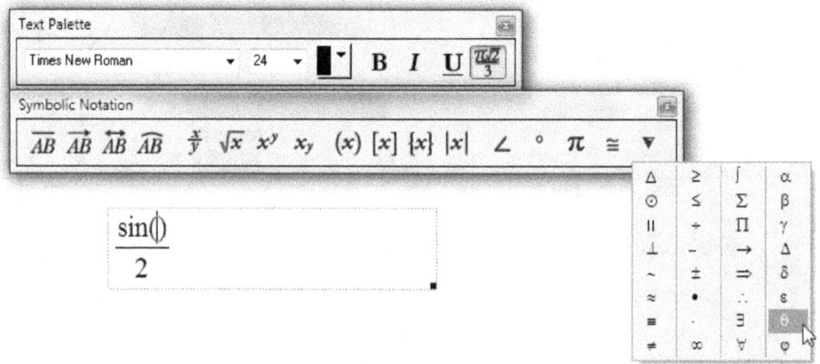

If you're using a Windows computer, you can drag these symbolic notation buttons so that they're floating or so that they're docked to any side of the Sketchpad window. If you're using a Macintosh computer, these symbolic notation tools appear as a second row of buttons in the Text Palette.

The symbolic notation buttons are divided into four groups.

| | |
|---|---|
| **Overbar Buttons:** Click one of the overbar buttons to apply a segment, ray, line, or arc overbar to text. If text is selected, the overbar is applied that text. If no text is selected, the overbar is applied to any new text you type without moving the insertion point. | $\overline{AB}\;\overrightarrow{AB}\;\overleftrightarrow{AB}\;\overset{\frown}{AB}$ <br><br> $\overline{AB}\perp\overleftrightarrow{CD}$ |
| **Operator Buttons:** Click one of the operator buttons to add a fraction, square root, superscript (exponent), or subscript to the selected text. If there is no selected text, the operator is inserted with ? placeholders in the places that you need to fill in. Use the Tab key to move from one ? placeholder to the next, or Shift-Tab to move from one ? placeholder to the previous one. | $\frac{x}{y}\;\;\sqrt{x}\;\;x^y\;\;x_y$ <br><br> $\sqrt{\dfrac{x^3}{5}}$ |
| **Grouping Buttons:** Click one of the grouping buttons to enclose the selected text in parentheses, square brackets, curly braces, or absolute value symbols. Unlike parentheses and brackets you type from the keyboard, mathematical grouping symbols always appear in pairs and resize automatically to fit whatever expression they enclose. If there is no selected text, the grouping symbols are inserted around a ? placeholder for you to fill in. Use the Tab key to move from one ? placeholder to the next, or Shift-Tab to move from one ? placeholder to the previous one. | $(x)\;[x]\;\{x\}\;\lvert x\rvert$ <br><br> $a\!\left(\dfrac{\lvert x\rvert+\lvert y\rvert}{2}\right)$ |
| **Symbol Buttons:** Click one of the symbol buttons to insert the angle symbol, the degree symbol, $\pi$, the congruence symbol, or any of the additional common symbols that appear in a pop-up menu when you press and hold on the last of these buttons. <br><br> To insert symbols and characters that aren't in this menu, use the Unicode input methods 314 available from your operating system. | $\angle\;\;{}^\circ\;\;\pi\;\;\cong\;\;\blacktriangledown$ <br><br> $\mathrm{m}\,\overset{\frown}{AB}=2\pi\,\dfrac{\mathrm{m}\angle AOB}{360^\circ}$ |

## 4.5　Properties

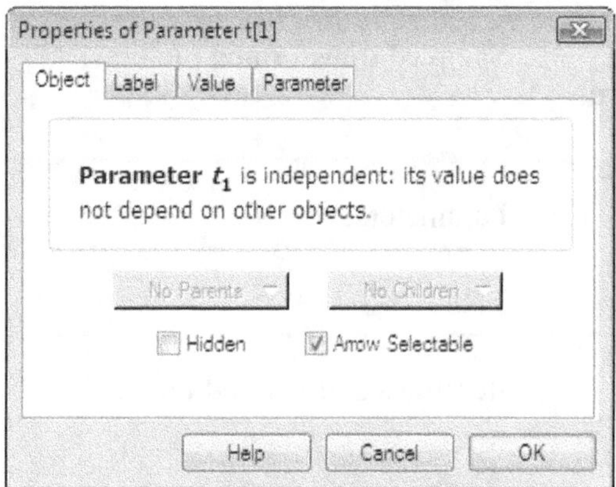

The Properties dialog box allows you to change various properties of a Sketchpad object.

To use the Properties dialog box, do one of the following:

- Select an object and choose **Edit | Properties** 158. The keyboard shortcut for **Properties** is Alt+? (Windows) or ⌘? (Mac).

- Select an object and choose **Properties** 158 from the Context 245 menu.

- Show the object's information balloon [124] and click the object's name in the balloon.

The Properties dialog box is arranged into separate panels of related properties. Switch from panel to panel by clicking on the tabs near the top of the dialog box. Which panels are available depends on the type of object selected.

While you're modifying one object's properties, you can switch to a different object by clicking that object in the sketch. (If the object to which you want to switch is obscured by the Properties dialog box, you'll first have to move the Properties dialog box aside to reveal the object.)

When you finish modifying an object's properties, click **OK** to make the changes permanent or click **Cancel** to leave the object with its original properties. When you click in the sketch to switch to a different object, your changes for the original object are made permanent before Sketchpad switches to the new object, just as if you'd clicked **OK** for that original object.

Keep in mind that for any specific object, only a few panels appear — not all of them!

Here are the possible Properties panels and the types of object for which each panel appears.

| Panel | Objects |
| --- | --- |
| Object [79] | All objects |
| Label [82] | All objects that can show labels [81] |
| Opacity [92] | All translucent objects [91] |
| Value [94] | All values [92] |
| Plot (Locus or Sampled Transformed Path) [95] | Loci [10] and sampled transformed paths [90] |
| Plot (Function Plot) [97] | Function plots [52] |
| Plot (Sampled Transformed Picture) [98] | Sampled transformed pictures [23] |
| Parameter [38] | Parameters [36] |
| Table [41] | Tables [39] |
| Iteration [35] | Iterations and iterated images [24] |
| Axis [44] | Axes [41] |
| Function [51] | Data-defined functions [51] |
| Marker [65] | Angle markers [60] and tick marks [63] |
| Hide/Show [67] | Hide/Show buttons [67] |

| | |
|---|---|
| Animate 69 | Animation buttons 69 |
| Move 71 | Movement buttons 71 |
| Presentation 72 | Presentation buttons 72 |
| Link 75 | Link buttons 75 |
| Scroll 76 | Scroll buttons 76 |

# 4.6    Preferences

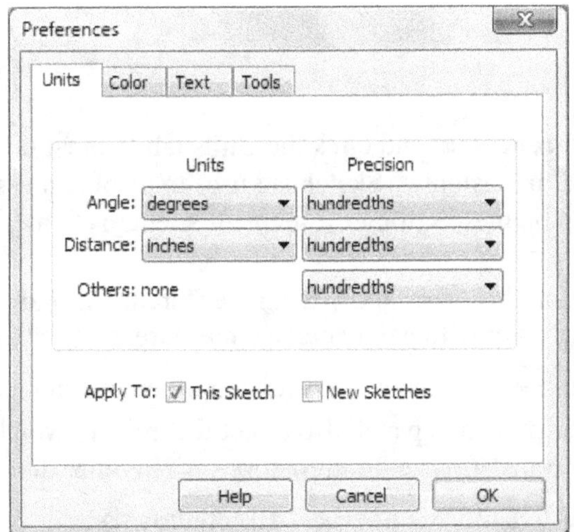


Choose **Edit | Preferences** 159 to open the Preferences dialog box and change a variety of settings that determine how Sketchpad works.

You can also choose **Preferences** from the Context 245 menu.

The Preferences dialog box includes four panels: Units 276, Color 277, Text 279, and Tools 280.

An Advanced Preferences 160 command allows you to control additional aspects of Sketchpad's operation that need to be changed rarely, if ever..

When you make changes in Sketchpad's Preferences, use the **Apply to** checkboxes at the bottom of the dialog box to decide whether your changes will apply to the current sketch only, to new sketches only, or to both the current sketch and new sketches. Choose **This Sketch** to have your changes affect newly constructed objects in the current sketch. Click **New Sketches** to have your changes affect all new sketches (including new blank pages you add to the current document).

If you want some changes to apply only to the current sketch and some to apply only to new sketches, you'll need to open the Preferences dialog box twice.

*Tools Preferences* `280`

## 4.6.1 Units Preferences

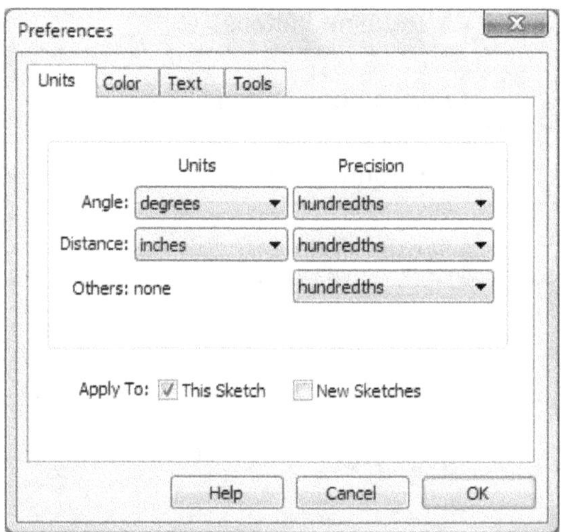

Choose **Edit | Preferences** `159` and click the Units tab to open this panel. The settings on this panel control the units and precision `277` Sketchpad uses to display measurements and calculations. `92` For example, depending on these settings, a segment's length `217` measurement may appear as "2.54 cm" or "1.0 in."

> The units set on this panel apply to all measurements and calculations in the sketch. The precision applies only to newly created measurements and calculations.

Use Value Properties `94` to change the precision of an existing measurement or calculation.

> The unit settings on this panel also affect the units in which you specify angles and distances for translations `200` and rotations `203` in other dialog boxes.

The choices for **Angle Units** are **degrees, directed degrees,** and **radians.**

> With directed degrees or radians, ∠ABC has the opposite measure as ∠CBA. With "plain" degrees, both of these angles have the same measure. Directed degrees are useful in transformational geometry, where the direction as well as the magnitude of an angle is important.

| | |
|---|---|
| Measurements in **degrees** are always positive and range from 0° to 180°. | m∠ABC = 20°<br>m∠CBA = 20° |
| Measurements in **directed degrees** are positive for counter-clockwise angles and negative for clockwise angles. Directed degrees range from −180° to 180°. | m∠ABC = 20°<br>m∠CBA = -20° |
| Measures of angles in **radians** are always directed and range from $-\pi$ to $\pi$. | m∠ABC = 0.11π radians<br>m∠CBA = -0.11π radians |

Set the **Distance Units** to **cm, inches,** or **pixels.** A pixel is a single dot on the computer screen and usually corresponds to about 1/72 inch on Macintosh and 1/96 inch in Windows.

Sketchpad's distance measurements are accurate when printed, but may not be exactly accurate on the screen, depending on the size of your monitor and on the resolution you've set in your computer's control panel. In the event that you require exactly accurate on-screen distance measurements, use the System Preferences `284` panel to set the precise number of pixels per inch or per centimeter on your screen. This panel is available only in **Advanced Preferences** `160`.

The choices for each of the three **Precision** settings are **units, tenths, hundredths, thousandths, ten-thousandths,** and **hundred-thousandths** (0, 1, 2, 3, 4, and 5 decimal places, respectively). These settings only affect how numbers are displayed, not how they are represented internally.

*See also:*
   *Accuracy vs. Precision* 277
   *Sketchpad's Internal Mathematics* 325

## 4.6.2    User Tip: Accuracy vs. Precision

The accuracy of a measurement 36 or calculation 39 in Sketchpad refers to how close the measured value is to the ideal, "correct" value. The precision of a measurement in Sketchpad refers to the number of decimal places used when the value is displayed on the screen.

> Don't confuse accuracy with precision. If you are displaying distances precise to the nearest tenth, Sketchpad can accurately represent the sum of two lengths measuring 1.4 as
>
> `1.4 + 1.4 = 2.8`
>
> If your precision is set to round displayed values to the nearest unit, however, the sum may seem nonsensical:
>
> `1 + 1 = 3`

The accuracy of Sketchpad's measurements and calculations is determined by the full computational power of your computer. Initial computations are usually accurate to about 15 significant digits. For instance, the square root of 2 is stored internally as 1.41421356237310. Calculations based on these initial values may be less accurate, because errors in the accuracy compound across calculations.

The precision of displayed measurements is determined by your choice in Units Preferences 276. Sketchpad rounds calculated values to your chosen precision when displaying them.

*See also:*
   *Sketchpad's Internal Mathematics* 325
   *Using Values* 92

## 4.6.3    Color Preferences

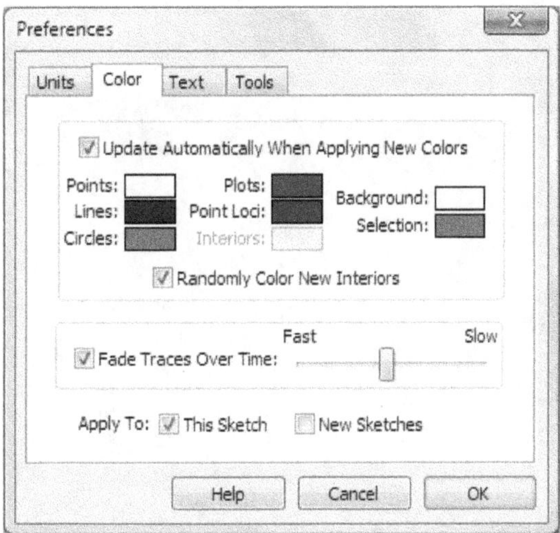

Choose **Edit | Preferences** 159 and click the *Color* tab to open this panel. The settings on this panel control the updating of color preferences, the default colors 86 of new objects you create, the background and selection 106 colors used in your sketch, and the fading behavior of traces 170.

**Update automatically when applying new colors:** When this box is checked, the preferences for your sketch are automatically changed whenever you color an object in the sketch. For example, if you color a line green and then construct a new segment, the new segment will also be colored green. Remove the check to have the preferences remain unchanged no matter how you modify the colors of objects in your sketch.

Even if *Update Automatically When Applying New Colors* is checked, you can prevent the preferences from being changed by holding down the Shift key while you change an object's color.

**Object** ⎡2⎤ **and sketch colors:** The first six color rectangles show the default colors for various types of objects in Sketchpad: points, lines (also including segments and rays), circles (also including arcs), interiors, point loci, and plots. New objects are assigned colors according to these settings. The last two color rectangles show the colors for selection markers and for the background of the sketch itself.

To change any of these colors, click the color rectangle you want to change. Your system's Color Picker ⎡290⎤ dialog box appears, allowing you to choose a new color.

**Randomly color new interiors:** Check this box to randomly assign a color when you construct a new interior ⎡8⎤, or leave it unchecked and use the Interiors color rectangle to determine the color for new interiors.

**Fade traces** ⎡170⎤ **over time:** Use this checkbox and speed control to determine whether or not traces fade and how fast they fade. When the box is checked, traces fade gradually over time, so recent traces are more vivid and older traces are fainter. If you remove this check, traces will never fade, and you'll have to remove them by using the Erase Traces ⎡171⎤ command (or by pressing the Esc ⎡294⎤ key). When fading is turned on, the speed control determines how quickly the traces fade.

Fading traces look best if your monitor is set to display more than 256 colors.

*See also:*
*Units Preferences* ⎡276⎤
*Text Preferences* ⎡279⎤
*Tools Preferences* ⎡280⎤

## 4.6.4 Text Preferences

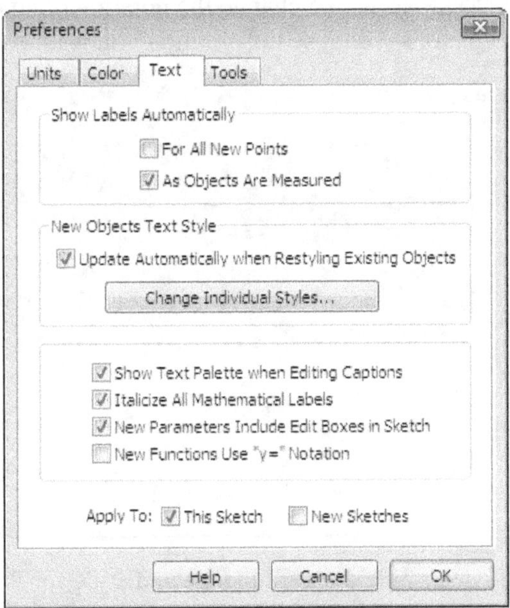

Choose **Edit | Preferences** 159 and click the Text tab to open this panel. These settings control when objects are labeled, and whether or not the Text Palette 270 appears automatically.

**Show Labels Automatically:** Check **For all new points** to show the labels of new points when they're created. Check **As objects are measured** to show labels when you measure an object. As you make a measurement, the labels needed to name the measurement will also appear. For example, if you measure the length of a segment between two unlabeled points, the segment endpoints' labels will also appear.

Even if you don't show labels automatically, you can show (or hide) individual objects' labels by using the **Text** 120 tool or choosing **Display | Show Labels** 168.

**New Objects Text Style:** Sketchpad maintains separate default text styles for newly created labels 81, captions 56, values 92, action buttons 66, tables 39, and axis tick numbers 41. Check **Update automatically when restyling existing objects** to automatically update one of these default styles when you change the style of an object. For instance, if automatic updating is checked and you change an action button's style to be Arial 14-point italic, the next action button you create will also be shown in Arial 14-point italic. Uncheck this box, or press Shift when applying a style, to prevent automatic updating. Click **Change Individual Styles** to open the Text Styles 280 dialog box and change the default text styles for the various categories of text.

**Show Text Palette when editing captions:** Check this box to make the Text Palette 270 appear automatically whenever you edit a caption and disappear when you finish editing. You can also display and hide the Text Palette by using the Show Text Palette 172 command.

**Italicize all mathematical labels:** Check this box to use automatic italics for most object labels, including geometric objects, measurements, parameters, and user-defined function names. Calculations and function definitions appear using mathematical italics in a way that is consistent with the most common standards for mathematical typesetting. Uncheck this box to turn off automatic italics.

**New parameters include edit boxes in sketch:** Check this box so that the values of newly created parameters are displayed with an edit box, and can be directly edited in the sketch.

**New functions use "y=" notation:** Check this box to display newly created functions in "$y=$" form rather than "$f(x)=$" form. To change the notation of an existing function, double-click it to edit it; in the Calculator 259 choose the notation you want from the Equation 259 pop-up menu.

### 4.6.4.1  Text Style Preferences

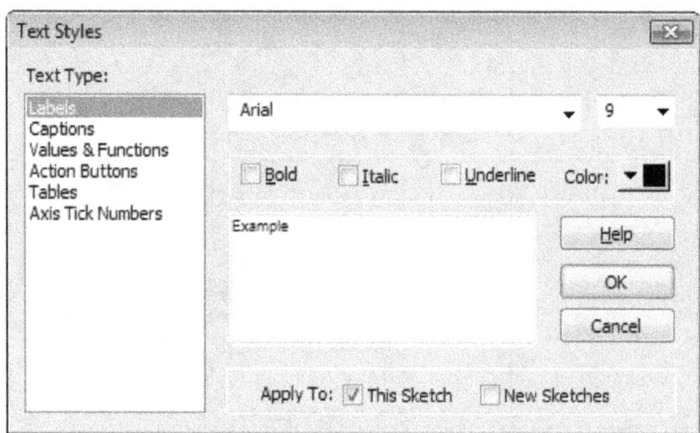

Use this dialog box to determine the default size and style for Sketchpad's labels and text objects.

To open this dialog box, choose **Edit | Preferences | Text** 279 and click **Set Text Style for New Text Objects.**

Use the list to choose a specific type of text, and use the other controls in the dialog box to change the style for new text of that type. You can change the style for labels 81, captions 56, values 92, action buttons 66, tables 39, or axis tick numbers 41.

These settings affect only new text; they have no effect on existing text. To change the style of existing objects or labels, select the objects and use the Text Palette 270.

## 4.6.5  Tools Preferences

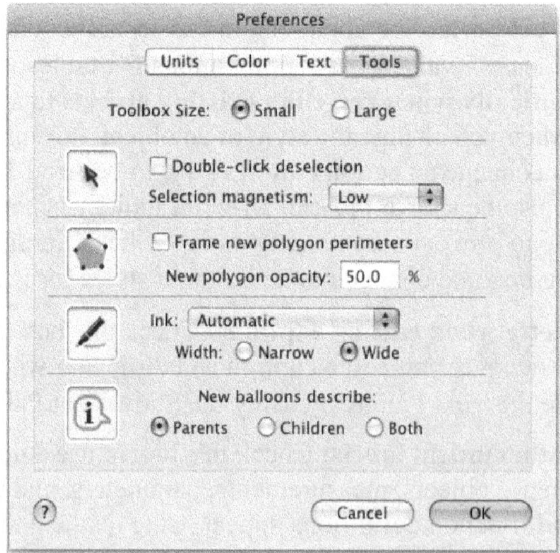

Choose **Edit | Preferences** 159 and click the Tools tab to open this panel.

This panel allows you control various aspects of how Sketchpad's tools work.

**Toolbox Size:** (Mac only) Click the appropriate radio button make the Toolbox icons larger or smaller.

**Arrow Tools:** Choose **Double-click deselection** to require a double-click in empty space to deselect all objects. (This reduces the chance of deselecting all objects by mistake.) Set the **Selection magnetism** to **Low, Medium,** or **High** to determine how close your pointer must be to an object in order to select it.

**Polygon Tools:** Check **Frame new polygon perimeters** to automatically show a frame when you create a new polygon, or uncheck it to show new polygons without a frame. Set **New polygon opacity** to determine how opaque or transparent new polygons will be. (Once a polygon is created, use the polygon's Opacity Properties 92 to control these settings for that polygon.)

**Marker Tool:** You can set the ink to **Brush Strokes** (which become thicker and thinner in response to pressure information), to **Smooth Curves** (which draw more smoothly), or to **Automatic** (which uses **Brush Strokes** when pressure information is available and uses **Smooth Curves** when it's not). In either case, you can set the ink to be **Narrow** or **Wide.**

**Information Tool:** Choose the appropriate radio button to determine whether an object's balloon shows only its parents, only its children, or both.

# 4.7    Advanced Preferences

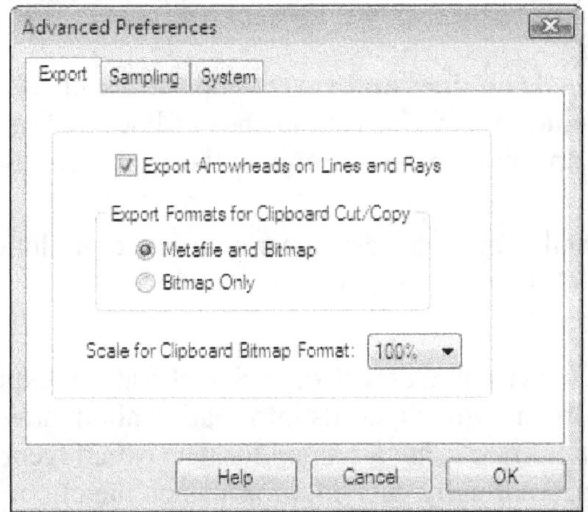


Hold the Shift key and choose **Edit | Advanced Preferences** 160 to open the Advanced Preferences dialog box. This dialog box controls advanced settings that determine how Sketchpad works. Change these settings only after careful consideration; most users never need to modify them or need to modify them only once.

Holding the Shift key changes the **Preferences** command to **Advanced Preferences.**

The Advanced Preferences dialog box includes three panels: Export 282, Sampling 283, and System 284.

## 4.7.1  Export Preferences

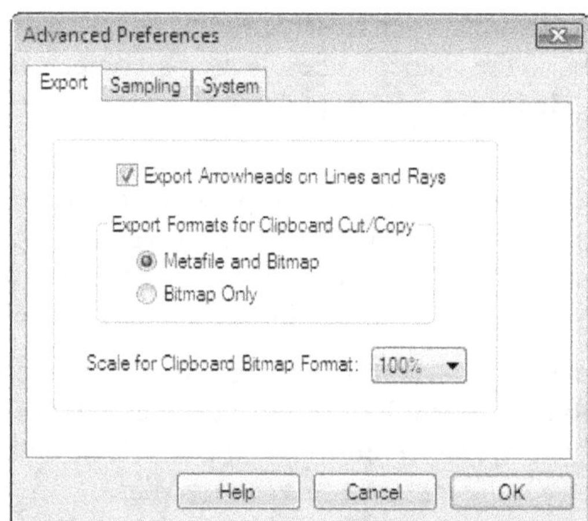

Hold the Shift key, choose **Edit | Advanced Preferences** 160, and click the Export tab to open this panel.

Holding the Shift key changes the **Preferences** 159 command to **Advanced Preferences** 160 .

The settings on this panel control how sketches are printed and how Sketchpad objects are copied 146 or cut 145 onto the clipboard.

**Export arrowheads on lines and rays:** If this box is checked, printouts and graphics copied to the clipboard include arrowheads at the ends of lines and rays. If this box is not checked, the ends of these objects appear on printouts and on the clipboard just as they do in the sketch — with no special markings.

To control whether arrowheads appear on a function plot or infinite point locus, select the object, choose Edit 144 | Properties 158 | Plot 95 , and check or uncheck the checkbox to show arrowheads.

**Export format for clipboard cut/copy:** Sketchpad can export a graphic image of your sketch using a vector format (which records information about how to draw objects like segments, circles, and captions) or using a bitmap format (which records the pixels that appear on the screen). Sketchpad normally puts both formats on the clipboard, leaving the decision to the program into which you are pasting. You can choose **Bitmap only** if special circumstances require it.

The vector format produces a smoother and more accurate image, and is usually the preferred format for word processors, graphics programs, and layout programs.

On Windows, the vector format is EMF+ (Enhanced Metafile Format); on Macintosh it's PDF.

**Scale for clipboard bitmap format:** This value determines the magnification used for clipboard bitmaps; use it to reduce the stair-step effect (called *aliasing*) that occurs with bitmaps. Set this to more magnification if you're copying pictures from Sketchpad to paste into a document in a word processor or page layout program that you intend to use for high-quality production. If you set the scale to more than 100%, the pasted picture will be large, and you'll need to scale it down in the word processing or page layout application into which you paste it. The result will be a higher-quality printout from the other application.

*See also:*
*Advanced Graphics Export* 320

## 4.7.2    Sampling Preferences

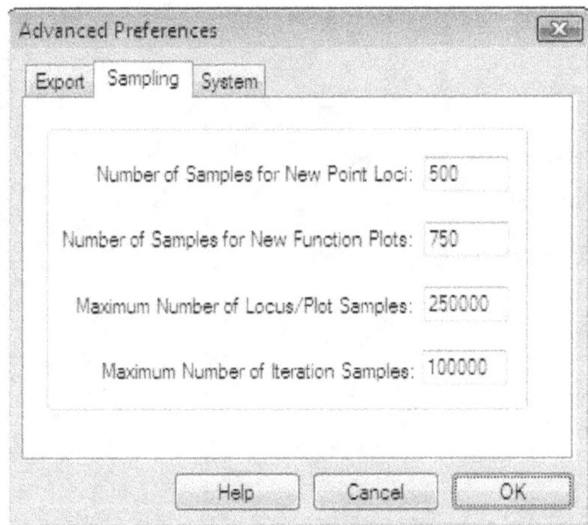

Hold the Shift key, choose **Edit | Advanced Preferences** 160, and click the Sampling tab to open this panel.

Holding the Shift key changes the **Preferences** 159 command to **Advanced Preferences** 160.

The settings on this panel control the number of samples used for loci 10, function plots 52, and iterations 24. In general, the more samples, the more accurate or detailed the object appears, but the slower it is for Sketchpad to compute and draw. If your computer is faster than most, you may want to increase the values of these settings; if it's slow to draw and recalculate loci, plots, and iterations you may want to decrease these values.

You can adjust the number of samples used in any specific locus or function plot by visiting its Plot Properties 95.

**Number of samples for new point loci** 10: This value determines the number of samples used in a newly created point locus. (The number of samples in a new locus of a nonpoint object is proportionate to, but smaller than, this number.)

**Number of samples for new function plots** 52: This value determines the number of samples used in a newly created function plot.

**Maximum number of locus/plot samples:** After you've created a locus or function plot, use **Properties** 95 to change the number of samples. This value determines the largest number you can use in the Plot panel 95 of the Properties dialog box 273 when you're changing the number of samples for such an object.

**Maximum number of iteration** 24 **samples:** This value limits the number of samples allowed in an iterated image. For an iteration using a single map, this is the maximum number of iterations. For an iteration using more than one map, this value limits the depth in such a way that the total number of iterated images of a single object never exceeds this number.

## 4.7.3    System Preferences

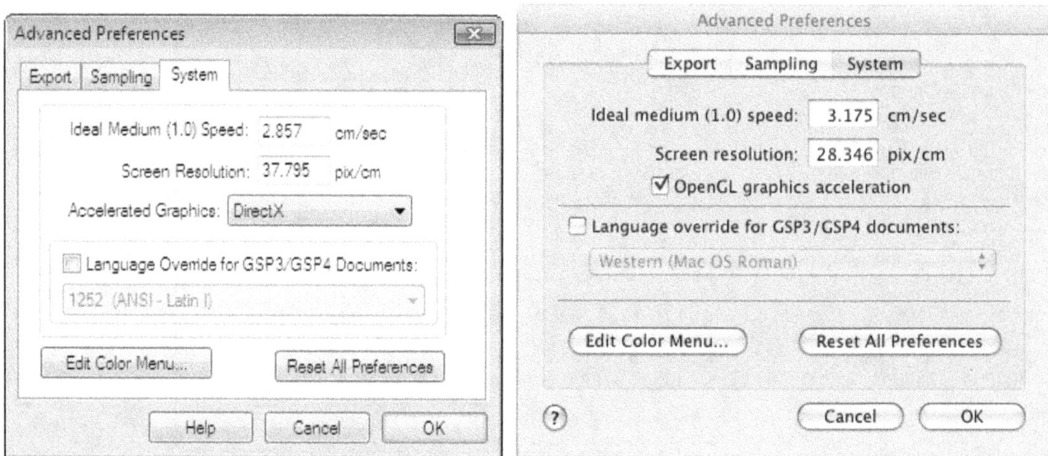

Hold the Shift key, choose **Edit | Advanced Preferences** 160, and click the System tab to open this panel.

Holding the Shift key changes the **Preferences** 159 command to **Advanced Preferences** 160.

The settings on this panel correlate Sketchpad's behavior with your particular computer, allow you to edit Sketchpad's color 164 menu, and allow you to change all preferences 159 back to their original values.

**Ideal medium speed:** This value sets the approximate speed represented by a value of 1.0 in the Motion Controller 263, or "medium" speed for a Movement button 71. You should modify the speed only if your computer is consistently too fast or too slow for your tastes when animating objects at a speed of 1.0 in the Motion Controller.

This value is "ideal" in the sense that Sketchpad attempts to reach it. If you're animating a particularly complex sketch or have a slower computer, animations may be slower than this.

**Screen resolution:** This value sets the correspondence between the pixels on your computer screen and real-world length units. The normal value is 96 pixels/inch (37.795 pix/cm) for Windows computers and 72 pixels/inch (28.346 pix/cm) for Macintosh. You should modify this value only if it's absolutely necessary that on-screen distances match Sketchpad's measured distances. Changing this value may result in discrepancies when you paste Sketchpad pictures into other applications, which assume your screen resolution is fixed at 96 or 72 pixels/inch.

**Graphics acceleration:** (This option may not available on some older computers). This setting determines the speed of the graphics displayed in your sketches. When it's enabled, Sketchpads uses available graphics software and hardware to optimize the speed of drawing on the screen.

If some objects are drawn incorrectly, try turning graphics acceleration off. If doing so eliminates the problem, first try to solve the problem by updating your computer's graphics software and/or firmware. If that fails, leave graphics acceleration turned off.

**Language override for GSP3/GSP4 documents:** If a Sketchpad 3 or Sketchpad 4 document was created in a different language, text may not appear correctly in Sketchpad 5, because older documents contain no information about the language in which they were created. (If Sketchpad is able to detect this situation, a dialog box 316 appears describing the problem.) To preserve the original text when you open such a document, use this pop-up menu to set the appropriate language override. Remember to set it back to your normal language when you're finished.

**Edit Color Menu:** This button displays either the Macintosh Color Picker[286] or the Windows Edit Color Menu[285] dialog box. These dialog boxes allow you to change the colors in the Color [164] menu. Use this feature to add your favorite shades of magenta, teal, or chartreuse. If you create a sketch with a dark or black background, you may want to add white to the Color menu.

**Reset All Preferences:** This button allows you to reset all Sketchpad preferences to their original values. Use it with care: you'll lose all the preferences values you've set in any other way.

*See also:*
*Motion Controller* [263]
*Animation Button* [149]
*Movement Button* [150]
*Color* [164]
*Color Picker* [290]
*Edit the Color Menu in Windows* [285]
*Edit the Color Menu on Macintosh* [286]

## 4.7.4    Edit the Color Menu for Windows

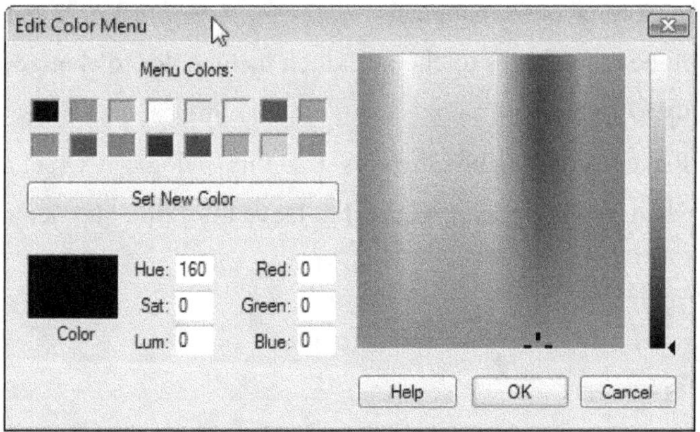


Hold the Shift key, choose **Edit | Advanced Preferences**[160], click the System tab, and click **Edit Color Menu** to open the Edit Color Menu dialog box.

Holding the Shift key changes the **Preferences**[159] command to **Advanced Preferences**[160].

Use this dialog box to set the colors that appear in Sketchpad's Color[164] submenu. The colors currently in the Color submenu appear under **Menu Colors.**

1. Click one of the **Menu Colors** boxes to choose which menu color to change.

2. Use the HSL controls, the RGB controls, or the color box and slider on the right to set a new color.

3. Click **Set New Color** to change the menu color to your new color.

4. Click another of the **Menu Colors** to choose a different menu color to change.

5. Click **OK** when you're done, or *Cancel* to leave the Color submenu unchanged.

*See also:*
*System Preferences* [284]
*Color Picker* [290]

### 4.7.5 Edit the Color Menu for Macintosh

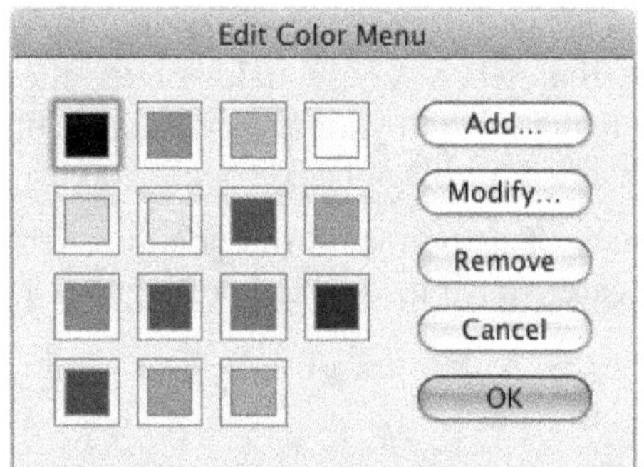

Hold the Shift key, choose **Edit | Advanced Preferences** [160], click the System tab, and click **Edit Color Menu** to open the Edit Color Menu dialog box.

Holding the Shift key changes the **Preferences** [159] command to **Advanced Preferences** [160].

Use this dialog box to set the colors that appear in Sketchpad's Color [164] submenu. The colors currently in the Color submenu appear.

1. Click one of the color boxes to choose which menu color to change.

2. Click Modify to change the color.

3. The Color Picker [291] dialog box appears. Use this dialog box to set a new color.

4. Click **OK** when you're done, or **Cancel** to leave the Color submenu unchanged.

*See also:*
*System Preferences* [284]
*Color Picker* [290]

## 4.8 Script View

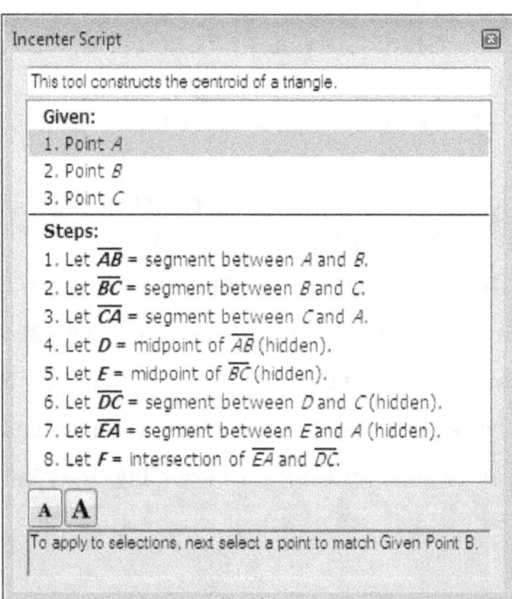

The Script View window displays a readable description of the mathematical construction performed by a custom tool 125. Use the Script View to review the given object requirements of a particular custom tool, or to investigate how a tool was originally defined. In the Script View, you can add or read comments about the tool written by the tool's author; change properties—such as color or line weight—of the objects that the tool creates; and even apply the tool's construction in a step-by-step fashion to objects in your sketch.

The Script View offers an "inside look" at the workings of any custom tool. You should be familiar with using custom tools before working with the Script View.

### ▼ Show the Script View

To display the Script View window, choose **Show Script View** 126 from the Custom Tools 126 menu or click the **Show Script View** checkbox in the Document Options 254 dialog box's view of custom tools. The Script View window sits on top of open document windows and displays the Script View of the most-recently chosen custom tool. To change the tool described by the Script View, choose a different custom tool from the available tools in the Custom Tools menu or in the Document Options dialog box.

Show Script View 126 is only available when one or more custom tools have been added to your Custom Tools menu.

You can reposition the Script View by dragging its title bar and resize it by dragging its resize area. The Script View remains visible until you close it by choosing **Hide Script View** 126 from the Custom Tools menu, or by clicking the window's close box.

To show the Script View at the same time as you choose a custom tool, hold the Shift key while you pull down the Custom Tools 126 menu.

### ▼ Change Script View Font Size

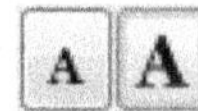

Use the Font Size buttons at the bottom of the Script View window to increase or decrease the font size used in the Script View window.

Smaller fonts are useful to see more steps; larger fonts are useful to project for students to see.

### ▼ Tool Comment

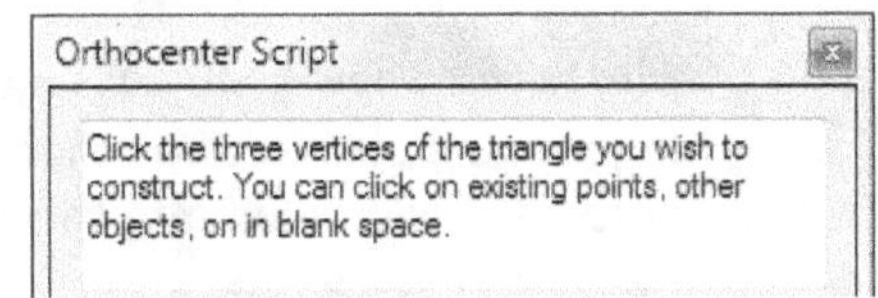


The Tool Comment appears at the top of the Script View window, and contains any comments about the custom tool's purpose or behavior that its author chose to add. Adjust the size of the Tool Comment by dragging the divider bar beneath it, and add to, or change the Tool Comment by typing. If you create new custom tools you plan to share with other Sketchpad users, use the Tool Comment to describe the tool's given objects, to identify yourself as the tool's author, and to explain how the tool should be used.

Any comments you add are saved with the tool, so you — or another Sketchpad user — can always refer to them when using the tool. When a Script View construction is printed, the comment is printed with the rest of the construction.

### ▼ Givens and Steps of a Tool

The Script View lists the given objects and steps that make up the custom tool 125, and allows you to change their attributes and rearrange the given objects. 288

### ▼ Use a Custom Tool with the Script View

If the Script View window is showing when you use a custom tool in a sketch, it provides visual feedback about the tool's use. Given objects 288 that you've already matched with the tool appear highlighted. Given objects that you haven't matched yet appear after those you've matched, and are not highlighted. The object that the tool is currently matching appears between the two.

> Use the Script View feedback to walk through a custom tool the first time you're trying a tool created by someone else.

### ▼ Use a Custom Tool as a Custom Command

Sometimes you may want to use a custom tool as a command 126 rather than as a tool, by selecting the objects first and then applying the command to the selected objects.

### ▼ Apply the Script View Step-by-Step

Sometimes you may want to apply a custom tool's construction to a sketch in a step-by-step fashion, 289 rather than use it as a tool.

### ▼ Print a Script View Construction

When the Script View is showing, you can print its Tool Comment and Object List. Right-click (Windows) or Ctrl-click (Macintosh) on any object in the script, and choose **Print Script View** from the Context 245 menu that appears.

*Subtopics:*


*See also:*


## 4.8.1    Givens and Steps of a Tool

The Script View's Object List describes, in a step-by-step fashion, all of the objects and constructed relationships that make up the custom tool 125. The object list is divided into two major sections: Givens and Steps.

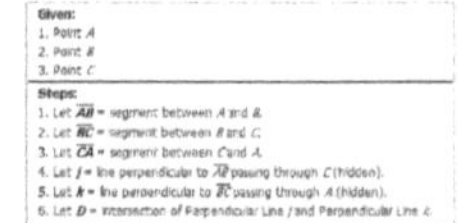

### ▼ Given Objects

The Given section shows all the given objects for the tool — all the objects that don't depend on any other objects in the tool. When using the tool in a sketch, these are the objects that you "match" in a sketch by clicking the tool. Given objects are listed in the order in which you must match them, which by default is the order chosen by the tool's author when the tool was created.

> You can change the order of given objects by dragging individual given objects up or down in Script View.

In some tools, the Given section may be divided into two parts, labeled **Assuming** and **Given.** The **Assuming** part contains assumed objects — given objects that are automatically matched to sketch objects having the same label when the tool is used in a sketch. The **Given** part contains the remaining given objects — the given objects that must be matched explicitly.

To change an object to be assumed, double-click the object and check **Automatically Match Sketch Object** in the Label panel 82.

Assumed objects don't need to be matched explicitly when you use the tool, unless the sketch does not contain objects to match with the same label. (By default, a tool's given objects are explicit, as opposed to assumed. But when you use a tool repeatedly in the same sketch, you may find it convenient to change some explicitly given objects into assumed objects.) For more information about assumed objects, see Automatically Match a Given Object 318.

### ▼ Steps

The Steps section shows all the objects that will be created from the tool's given objects. These include the tool's intermediate objects (which are not displayed when the tool is used in a sketch), and the tool's final results (which are displayed). Unlike given objects, a tool's steps cannot be reordered: they are determined by the tool's construction.

The script step for an intermediate object describes it as "hidden." To change the visibility of a tool object, double-click it and check or uncheck the **Hidden** checkbox in the Object Properties 79 panel.

### ▼ Working with the Object List

You can double-click any object in the Object List to change its properties. When you double-click, the Properties 158 dialog box appears. Use the various panels to make any desired changes to the properties. You can also use the Parents and Children 80 pop-up menus on the Object panel to view and change the properties of the parents or children of the object.

You can right-click (Windows) or Ctrl-click (Macintosh) any object in the list to display a Context 245 menu for the clicked object. Use this Context menu to change the color 164, point style 162, line style 163, or tracing 170 for the object on which you click. The Context menu also allows you to print the script 286.

*See also:*
*Advanced Tool Topics* 316

## 4.8.2 Apply the Script View Step-by-Step

Sometimes you may want to apply a custom tool's construction to a sketch in a step-by-step fashion, rather than use it as a tool. If you are reasoning through the construction described in the Script View, for example, it's convenient to watch the construction occur in your sketch one object at a time. Likewise, if you're demonstrating a particular construction to your teacher or students, it may be useful to advance object by object. When you use a custom tool directly in the sketch, its constructed steps happen all at once, as soon as you've matched given objects, and only the tool's final results are displayed. But when you apply a Script View construction step-by-step, objects appear one at a time, and the construction's intermediate objects are temporarily displayed (so you can see them) until all steps are complete (when Sketchpad hides them for you).

To apply Script View's construction step-by-step to a particular sketch:

1. Use the **Selection Arrow** 104 tool to select sketch objects that match the given objects displayed in Script View, in the order they appear in the Script View object list. For example, if the Script View lists three given points, select three matching points in your sketch.

If the Script View displays both Assumed and Given objects 288, select matching objects first for the Assumed objects and then for the Givens.

Assumed objects match automatically only when using custom tools directly, not when applying the Script View step-by-step.

As you select matching sketch objects, the Script View displays feedback about your selections. Given objects that match your selections appear highlighted; given objects for which you must still select matching sketch objects appear on a normal background.

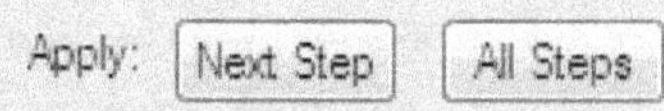


2. Once you have selected objects to match all of the Script View given objects, two buttons appear at the bottom of the Script View window.

Click **Next Step** to apply the first step of the construction. Click this button repeatedly to walk through the entire construction, step-by-step. As each step is applied, the corresponding object appears in your sketch. While stepping through the construction, you can drag objects in your sketch to investigate their relationships to other objects.

Click **All Steps** to finish the construction by immediately constructing all remaining steps.

3. When the last step of the construction has been completed, the construction's intermediate objects — which the Script View temporarily displays in your sketch during stepping — are hidden, the results are selected, and the **Next Step** and **All Steps** buttons disappear. The Script View returns to its normal appearance.

If the tool's results match the givens, you can apply the tool repeatedly to its own results simply by clicking the All Steps button.

To apply the Script View construction a second time, select new matching objects as described in step 1.

During stepping, you can press the mouse on any already-matched given object or already-constructed step in the Script View to highlight the corresponding sketch object to which it matches. If you want to stop step-by-step application of the construction without completing All Steps, press the Esc 294 key or close the Script View window.

*See also:*
  *Object List* 288
  *Advanced Tool Topics* 316

# 4.9    Color Picker

The Color Picker dialog box allows you to choose a specific color to be used for displaying a Sketchpad object, for displaying text in Sketchpad, or for changing any of the colors available on Sketchpad's Color menu.

The Color Picker appears when you choose the **Display | Color | Other** 166 command or when you click the Color Picker swatch on the Text Palette 270. It also appears when you click the Edit Color Menu button on the System panel 284 of the Advanced Preferences dialog box.

The Color Picker dialog box allows you to specify a color in several different ways. The most common methods involve RGB (Red-Green-Blue) values and HLS (Hue-Luminance-Saturation or Hue-Lightness-Saturation) values.

The actual Color Picker dialog box is provided by the particular Macintosh or Windows operating system installed on your computer, and may differ from the sample dialog boxes shown in the subtopics.

To specify a color using RGB, you set numeric values for the red, green, and blue components of the color. These numeric values are most commonly expressed either in percentages or in a range from 0 to 255, with higher numbers for a color corresponding to the presence of more of that color in the mix. Thus 255 for red and 0 each for green and blue specifies the purest possible red. Setting all three values to 0 specifies black (no color at all), and setting all three values to 255 specifies white (the brightest color, with the maximum amount possible of all three components).

The three values of HLS (Hue, Luminance, and Saturation) also allow you to specify a color numerically. Hue determines the color, such as red, blue, or green. Luminance determines how light or dark the colors are, with the maximum value corresponding to white and the minimum value (zero) corresponding to black. Saturation determines how much of the color is present; high saturation corresponds to vivid colors, and low saturation to pale colors. A saturation value of 0 specifies gray, with the shade of gray determined by the luminance. The most vivid pure colors correspond to the midpoint of the luminance scale and maximum saturation.

An alternative system called HSV (Hue-Saturation-Value) is similar to HLS, with the difference that the most vivid pure colors correspond to the maximum of the V scale (rather than the midpoint of the L scale) at maximum saturation. In this system white corresponds to zero saturation and maximum value.

*Subtopics:*


*See also:*


## 4.9.1 Macintosh Color Picker

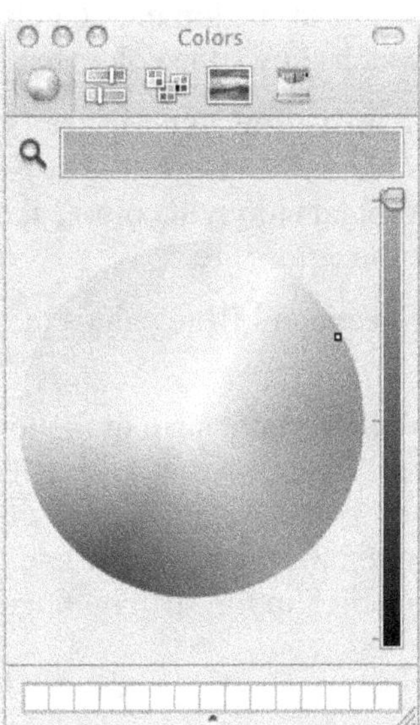


The top of the Macintosh Color Picker window lists several different ways in which you can pick a color, such as Color Wheel, Color Sliders, Color Palettes, Image Palettes, and Crayons.

If you use the Color Wheel, click anywhere in the color wheel to set the hue and saturation. Minimum saturation corresponds to the center of the wheel and maximum saturation to the edge of the wheel. Move the slider on the right to set the brightness.

*See also:*
  *Color* 164
  *Color Preferences* 277
  *Edit the Color Menu on Macintosh* 286

## 4.9.2   Windows Color Picker

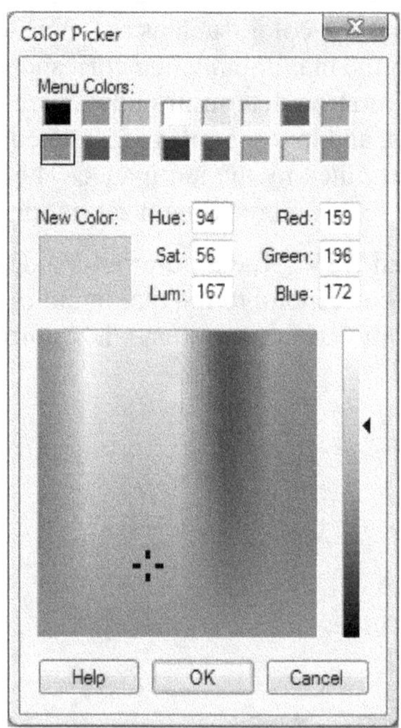

The Windows Color Picker shows at the top the 16 colors available from the **Display | Color** 164 submenu. The **New Color** swatch shows the color that will be chosen if you click the **OK** button.

**Hue, Saturation,** and **Luminance** allow you to specify a color numerically. The **Hue** value determines the color, such as red, blue, or green. The **Saturation** determines how much of the color is present; high saturation corresponds to vivid colors, and low saturation to pale colors. **Luminance** determines how light or dark the colors are.

Alternatively, use the **Red, Green,** and **Blue** values to specify a color based on how much of each of these primary colors is used.

Finally, use the rectangle and the vertical strip in the lower portion of the dialog box to specify the hue, saturation, and luminance. First click in the rectangle to choose the hue and saturation. Change the hue by moving horizontally in the rectangle, and change the saturation by moving vertically, with the minimum values at the bottom of the rectangle and the maximum values at the top. Once you've chosen the hue and saturation, click in the strip on the right to set the luminance.

*See also:*
  *Color* 164
  *Color Preferences* 277
  *Edit the Color Menu in Windows* 285

# Keyboard

# 5    Keyboard

As you gain experience in Sketchpad, you'll naturally seek ways to maximize your efficiency and productivity within the software environment. There are many keyboard techniques that will help you use Sketchpad productively.

## ▼ Menu Command Shortcuts

Many menu commands have shortcut keys, allowing you to access the menu command directly from the keyboard whenever that command is available. Most of these shortcuts require a modifier key (such as Alt or Ctrl on Windows or ⌘ on Mac) to be held down while you type the command's shortcut key. Shortcut keys (and their modifiers) are listed in the menu, directly across from the name of each command that can be accessed by a keyboard shortcut.

For example, in the Windows File menu the **Save** command lists Ctrl+S as its keyboard shortcut. In the Macintosh File menu it lists ⌘S as its keyboard shortcut. Thus, you can save the active document by typing the S key while holding down the Ctrl modifier key (in Windows) or the ⌘ modifier key (on Macintosh).

*See also:*
  *Menus* 136
  *Sketchpad's Menu Structure* 323

## ▼ The Esc Key

The Esc key provides a powerful general-purpose means to "escape" from your current activity. Depending on the activity you're presently engaged in, Sketchpad's response to the Esc key varies, but in general, Esc reverts Sketchpad to a less "special" state. Each press of the Esc key performs one of the following actions:

- If a caption 56 is being edited, Esc stops editing the caption 120.
- If a script view is being applied 289 step-by-step, Esc stops the execution without completing the remaining steps.
- If the **Arrow** 104 tool is not active, Esc activates it.
- If any object is selected 106, Esc deselects all objects.
- If any object is animating 171, Esc stops all animations 172.
- If any traces 170 are visible, Esc erases all traces 171.
- If information balloons 124 are showing, Esc hides them.
- If you're creating a drawing with the **Marker** 122 tool, Esc finishes the current drawing and allows you to start a new one.

Press Esc repeatedly to revert your document to a "normal" state, with no objects animating, no traces visible, no objects selected, and the **Arrow** tool active.

*See also:*
  *Special Keys* 295

## ▼ Special Keys

Click here for a list of special keys 295 that perform useful operations affecting selected objects, available commands, the active tool in the Toolbox 102, or the view of your document.

## ▼ Expressions in Dialog Boxes

Many of Sketchpad's dialog boxes permit or require you to enter various numeric quantities. Wherever you're required to enter a number, you can substitute an arithmetical expression — such as (1/3) or $2\pi$ — instead. Type expressions combining numbers, parentheses, addition (+), subtraction (−), multiplication (*), division (/), and exponentiation (^). Sketchpad evaluates your expression and uses the resulting value for the dialog box quantity.

To enter $\pi$ in a dialog box, type the letter p.

# 5.1 Special Keys

Keyboard shortcuts make it easy to perform special operations that affect the selected objects, available commands, your choice of the active tool in the Toolbox[102], or the view of your document.

In addition to the shortcuts listed here, you can use several caption-editing keys[297] when you're editing a caption[56], and you can use the command shortcuts listed next to many commands in the menus.

| *Key* | *Action* |
|---|---|
| Delete or Backspace | Clears[147] selected object(s) (same as **Edit**[144] \| **Clear**[147]). |
| ←, →, ↑, or ↓ | When objects are selected, drags the selected object(s) one pixel in the indicated direction. (Press and hold keys to drag longer distances.) |
| Shift+↑ or Shift+↓ | Changes the active tool to the next higher or lower tool in the Toolbox[102]. |
| Shift+← or Shift+→ | When the active tool is a **Selection Arrow**[104] tool, a **Straightedge**[115] tool, or a **Polygon**[118] tool, changes the active tool to the next or previous **Selection Arrow, Straightedge,** or **Polygon** tool in the Toolbox[102]. |
| + or − | When one or more loci[10] or function plots[52] are selected, increases (+) or decreases (−) the number of samples in those objects by a fixed percentage. |
| + or − | When one or more iterated images[24] are selected, increases (+) or decreases (−) the number of iterations[35] by one. |
| + or − | When one or more parameters[36] are selected, increases (+) or decreases (−) those parameters' values. (Press and hold keys to continue adjusting values; use Parameter Properties[38] to choose the keyboard adjustment amount for each parameter.) |
| Alt (Windows) or Option (Macintosh) | Temporarily invokes drag-scrolling. Press and drag in your document to scroll it in an arbitrary direction. Release the Alt key (Windows) or Option key (Macintosh) to resume the active tool. |

| | |
|---|---|
| Shift+menu | Menu command variations:<br><br>Changes **File \| Save As** [140] to [140]**Save As HTML** [140].<br><br>Changes **Edit \| Undo** [144] to [144]**Undo All** [144].<br><br>Changes **Edit \| Redo** [145] to [145]**Redo All** [145].<br><br>Changes **Edit \| Preferences** [159] to [159]**Advanced Preferences** [159].<br><br>Changes **Transform \| Iterate** [208] to [208]**Iterate to Depth** [208].<br><br>Changes **Graph \| Show/Hide Grid** [236] to **Show/Hide Coordinate System** [236].<br><br>Changes **Edit \| Split/Merge** [153] to **Merge Text To Point** [313].<br><br>Changes **Display \| Label Objects** [168] to **Reset Next Labels** [168] (provided no objects are selected).<br><br>Shows the Script View [286] at the same time as you choose a custom tool [126].<br><br>Changes **Choose Tool Folder** [133] to [133]**Forget Tool Folder** [133] in the Custom Tools [126] menu. |
| Shift+drag | Constrains **Straightedge** [115] tools.<br><br>Constrains the **Translate Arrow** [104] tool when dragging a control point of a straight object.<br><br>Maintains a picture's [18] aspect ratio while resizing with the [109]**Arrow** [109] tool [109]. |
| Shift+format | When changing object [86] or text appearance [162], preserves default settings for point style, line style, color, font, text style, and size. |
| Shift+Enter | Aligns [313] the selected text objects [99] or action buttons below the first selected text object. If the objects are already aligned, increases the vertical spacing between objects. |
| Shift+double-click table | Removes [229] most recently added row from the table [39]. |
| Shift+ **Information** [124] tool click | Keeps existing information balloons visible. |
| p | Types $\pi$ as part of a numeric value in a dialog box. |
| ! | Randomizes iterated points on paths [30] for the selected iteration [24]. |

## 5.2  Caption-Editing Keys

When you're editing a caption, use keyboard shortcuts to move the insertion point, to select text, or to change the caption's alignment.

| *Key* | *Action* |
|---|---|
| Delete or Backspace | Clears 147 selected text (same as **Edit | Clear** 147). |
| ←, →, ↑, or ↓ | Moves the insertion point by one character left or right, or by one line up or down. |
| Ctrl+← or Ctrl+→ | Moves the insertion point by one word left or right. |
| Shift+← or Shift+→ | Extends or contracts the currently selected text by one character left or right. |
| Shift+↑ or Shift+↓ | Extends or contracts the currently selected text by one line up or down. |
| Shift+Ctrl+← or Shift+Ctrl+→ | Extends or contracts the currently selected text by one word left or right. |
| Ctrl+I (Windows) or ⌘I (Mac) | Turns *italic* style on or off for the selected text. |
| Ctrl+B (Windows) or ⌘B (Mac) | Turns **bold** style on or off for the selected text. |
| Ctrl+U (Windows) or ⌘U (Mac) | Turns <u>underline</u> style on or off for the selected text. |
| Tab and Shift-Tab | When the caption includes "?" place-holders in fractions, square roots, superscripts and subscripts, and grouping symbols, the Tab key moves to the next "?" placeholder, and Shift-Tab moves to the previous one. |
| Alt+Arrow (Windows) Option+Arrow (Mac) | Sets text alignment when editing a caption. Use the left arrow for left alignment, the right arrow for right alignment, and either the up or down arrow for centered alignment. |

# Chapter 6

# What's New

# 6    What's New

The Geometer's Sketchpad Version 5 has many new features 300.

If you've used Sketchpad Version 4, you may want a quick overview of the many new features of Version 5, as well as notes on important changes in how you use Sketchpad. For details of the changes, see New Features in Version 5 300 and have a look at the Sample Documents 303 which illustrate Sketchpad's features.

*Subtopics:*
*New Features in Version 5* 300
*Reusing Your Version 4 Work* 302
*Sample Documents* 303

## 6.1    New Features in Version 5

The Geometer's Sketchpad Version 5 contains many improvements and new features. It's not only more powerful than previous versions, it's also easier to use and more expressive mathematically.

The most important changes are described here.

### ▼ Expressivity

- Display 86 points in a variety of sizes and paths using any of four widths and any of four patterns.

- Create angle markers 60 with the **Marker** 122 tool to identify angles, to mark congruent angles and right angles, and to measure angles.

- Create tick marks 63 with the **Marker** tool to identify path objects, to mark congruent segments, and to mark parallel lines.

- Show the frame of a polygon 8, and hide the interior 91 if you prefer.

- Set the transparency 91 and layer 91 of pictures 18, interiors 8, and loci 10 and iterations 24 of these objects.

- Display functions using either **y=** 45 notation 45 (for instance, $y = x^2$) or $f(x)$ notation (for instance, $f(x) = x^2$). Set the default notation for new functions by choosing **Edit | Preferences | Text** 279. Change the notation for a particular function using the Calculator's Equation pop-up menu 259.

- Display a value in radians 92 as a simple fraction of $\pi$ when appropriate and as a decimal value otherwise.

- Define a custom transformation 211 using any two points, one of which depends on the other, as an example, and apply the resulting transformation to almost any other object.

### ▼ Convenience

- Make points easier to select by making them larger 86 or by making them more magnetic 280.

- Use the Context menu 245 to easily access commands for specific objects, including commands to change the layer 247 of layered objects, to change the value 247 of parameters, to change the resolution 247 of loci and transformed pictures, and to change the depth 248 of iterations.

- Create a parameter 36 using a keyboard shortcut directly from the Calculator 257.

- Directly edit the value of a parameter [36] that has an edit box.

- Set the precision and keyboard adjustment increment for new parameters [226] easily, based on the number of decimal places you type when you create the parameter.

- Create two separate buttons — a Hide button and Show button — in a single step by holding the Shift key while choosing **Edit | Action Buttons | Hide & Show** [149].

- Open a sketch from a Sketchpad LessonLink activity directly from within Sketchpad. Choose **File | Sketchpad LessonLink** [138]®, enter your ClassPass, and choose the sketch you want.

### ▼ Powerful Text Features

- Use Hot Text [57] to include labels and values from your sketch as you create or edit a caption. Hot Text labels make it easier to include mathematical formatting in your captions, and Hot Text values allow you to display equations in forms like $y = 2x^2 - 5.4x + 2.3$.

- Display calculations, functions, and object labels using mathematical italics [81] consistent with the most common standards for mathematical typesetting. Turn mathematical italics on or off using **Edit | Preferences | Text** [279].

- Style text as bold, italic, or underline more easily by using keyboard shortcuts Ctrl+B, Ctrl+I, Ctrl+U (Windows) or ⌘B, ⌘I, ⌘U (Mac), or by using new commands on the **Display | Text** [166] submenu.

- Use Unicode [314] to include mathematical symbols and to write captions in various languages. Use shortcuts to include subscripts, Unicode symbols, and Greek letters [82] in labels.

- Set the default text style [280] easily for labels [81], captions [56], values [92], action buttons [66], tables [39], or axis tick numbers [41].

- When text affected by an iteration [24] is attached [153] to a point that's also affected by the iteration, the text is iterated along with the point.

### ▼ More and Better Tools

- Use the **Polygon** [118] tool to construct a polygon, a polygon with its edges, or only the edges.

- Use the **Marker** [122] tool to draw free-hand ink and to create angle markers [60] and tick marks [63]. Use drawn free-hand ink to define a function [232].

- Use the **Information** [124] tool to explore the relationship in a sketch, by showing objects' information balloons or exploring objects' properties [158].

- Use a custom tool as a command [126] by selecting the appropriate given objects and then holding the Shift key while choosing the tool from the Custom Tools menu.

- Choose a Tool Folder [133] easily, and automatically copy example tools to it.

### ▼ More and Better Objects

- Rotate and dilate pictures [23] and create custom transformations of pictures.

- Attach pictures [19] to one, two, or three points, enabling general affine transformations.

- Crop a picture to a polygon [152] to display only a particular portion of the picture. You can transform the resulting cropped picture.

- Save sketches with pictures more compactly with efficient compression techniques.

- Construct loci [10] and iterations [24] of other loci, of function plots, and of pictures.

- Construct a locus [10] based on a changing parameter.

- Create a Movement button [71] that moves a parameter's value to the value of some other parameter, measurement, or calculation.

- Create a Sound button [74] to play a sound defined by a mathematical function. Explore the properties of sound waves and the mathematics of oscillating functions and addition of waveforms.

### ▼ Easier and More Powerful Graphing

- Investigate the family of functions [10] $y = ax^2 + bx + c$ as parameter $a$ changes from -3 to 3.

- Measure the value of a point [221] on an axis [41] or on any other path [89]. Plot a value [237] on an axis, a line, or any other path.

- Construct an intersection [177] of a function plot with another function plot, an axis, or another geometric object.

- Use trigonometric numbering [235] on the horizontal axis of a rectangular or square coordinate system. Use Axis Properties [44] to use trigonometric numbering on a vertical axis.

- Use free-hand ink [122] or an imported picture [19] to define a function [232].

- Measure both the abscissa and the ordinate of one or more points in a single step by holding the Shift key while choosing **Measure | Abscissa & Ordinate** [223]. (If the coordinate system is polar, the command becomes **Measure | Polar Distance & Direction** [223].)

- Plot points using any combination of fixed and measured values by choosing Graph [233] | Plot Points [239] and providing values for $x$ and $y$ (or $r$ and $\theta$), either by typing fixed values or by clicking measured values in the sketch.

- Choose **Graph | Plot Parametric Curve** [241] to construct a parametric plot [54], with its $(x, y)$ or $(r, \theta)$ values determined by evaluating two functions.

### ▼ JavaSketchpad

- Export sketches as interactive HTML pages [307] that incorporate a wider variety of constructions, including functions [45] and function plots [52].

- Automatically copy **jsp5.jar** alongside a newly exported HTML document [140] so that you can preview the resulting JavaSketchpad web page [309].

## 6.2    Reusing Your Version 4 Work

The Geometer's Sketchpad Version 5 allows you to reuse all of the sketches and scripts you've created in Version 4 or Version 3.

### Using Version 4 Sketches

Version 4 sketches can be opened in Version 5 with all their contents intact, and with at most minor differences in formatting and appearance.

### Using Version 3 Sketches

Most Version 3 sketches can be opened in Version 5 with all their contents intact, and with at most minor differences in formatting and appearance. While some Version 3 sketches contain objects that are fundamentally unsupported in Version 5, even those documents can be partially opened, preserving all other elements of the sketch. (Unsupported objects include Macintosh publishers, imported objects such as imported graphics and sound, equations of lines and circles, and values derived from such equations.)

### Using Version 3 Scripts

Although Sketchpad Version 4 doesn't directly support Version 3 scripts, you can convert your Version 3 scripts into Version 4 custom tools by the following procedure:

1. Using Sketchpad Version 3, open the script file you want to convert.

2. Open a blank sketch.

3. Construct the required Given objects in the blank sketch.

4. Select the Given objects and play the script into the sketch.

5. Save the resulting sketch, and quit Sketchpad 3.

6. Using Sketchpad Version 5, open the sketch you just saved in Version 3.

7. Select all objects in the sketch.

8. Choose **Create New Tool** from the Custom Tools menu.

9. Give your new tool a name.

10. Save the sketch using an appropriate name (perhaps the same name as that of the tool).

Once the script has been saved as a custom tool in a Version 5 document, use that tool any time the document is open. You can also collect multiple tools into a single document, and you can place the document in Sketchpad's Tool Folder so that its tools are always available.

*See also:*
*Custom Tools* 125
*Tool Folder* 133

### Sketches and Scripts from before Version 3

Any sketch or script created by an earlier version of Sketchpad (Version 1 or Version 2) must be opened and resaved in Version 3 before you can use it in Version 5. Version 5 cannot directly open files from versions of Sketchpad earlier than Version 3.

## 6.3    Sample Documents

Sketchpad comes with many sample documents that explore ideas drawn from a wide range of mathematics; demonstrate potential Sketchpad applications; and contain tips, techniques, and examples for using Sketchpad effectively. Familiarizing yourself with their contents by browsing through them at least once can provide you another valuable Sketchpad learning resource. To view the sample documents, choose **Help** 244 | **Sample Sketches & Tools.**

# Chapter 7

# Appendix

# 7    Appendix

This Appendix contains a number of resources that are useful once you're familiar with the basics of Sketchpad.

| | |
|---|---|
| Special Constructions and Techniques 306 | Sketchpad users often ask about how to create particular constructions or use particular techniques. Here are answers to some of the most common questions. |
| JavaSketchpad and Web-Based Dynamic Geometry 307 | JavaSketchpad allows you to place simple sketches on web pages you publish on the Internet. Here's how to do it. |
| Advanced Topics 312 | You may be able to use various advanced techniques and benefit from information for experts. Find details here about text topics 313, tool topics 316, graphics export 320, menu structure 323, internal mathematics 325, and command-line flags 326. |
| Technical and InstallationSupport 327 | Having trouble? Here are possible solutions and contact information. |
| License 327 | The fine print. |
| Credits 331 | Who did what to make all this happen. |

## 7.1    Special Constructions and Techniques

Sketchpad users often ask about how to create particular constructions or use particular techniques. Here's a collection of some of the most common requests.

- Construct a Segment of Fixed Length 182
- Construct an Angle of Fixed Measure 204
- Construct Congruent Segments 184
- Construct Congruent Angles or Triangles 204
- Construct a Geoboard 44
- Zoom In and Out on a Sketch 111
- Show Just One Hidden Object 167
- Construct a Slider 222
- Make a Perpendicular Bisector Tool 132
- Use Split and Merge to Explore Constructions 156
- Construct a Sierpiński Gasket 33
- Use Signum to Construct a Piecewise Function 262

## 7.2    JavaSketchpad and Web-Based Dynamic Geometry

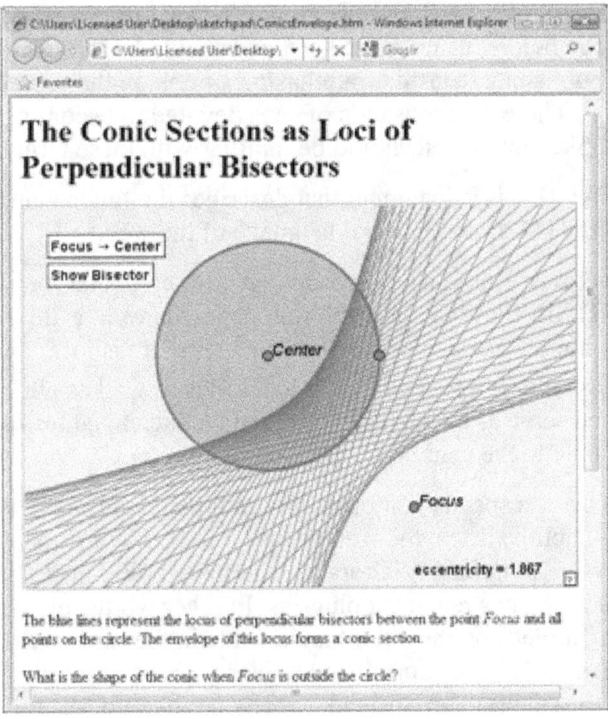


JavaSketchpad is a Sketchpad extension that allows you to place simple sketches inside web pages you publish on the Internet. These sketches appear as illustrations in your pages, and anyone who visits your web page can interact with them — by dragging points and pressing action buttons — directly from their web browsers, even if they don't have a copy of Sketchpad. If you create web pages, consider using JavaSketchpad as a way to enhance any mathematics — such as personal discoveries, journal articles, class syllabi, or assignments — that you post to the web.

> JavaSketchpad is intentionally small in comparison to The Geometer's Sketchpad so that it downloads quickly when someone visits your web page.

Not every sketch or activity you create in The Geometer's Sketchpad can be turned into an interactive JavaSketchpad illustration. Many of Sketchpad's advanced features require the full power of the program itself, rather than the slimmed-down version that visitors to your web page interact with. Also, while visitors can drag points and press action buttons in sketches you post as JavaSketchpad illustrations, they cannot draw or construct new objects. Despite these limitations, JavaSketchpad can enhance your web-publishing options significantly.

As an Internet technology, JavaSketchpad evolves rapidly, and the best source of information about it is on the web itself. The JavaSketchpad web site is the place to go for additional information; it includes sample applications, technical support, and a full description of the JavaSketchpad construction language that appears in your HTML files. You'll even find information about some features of JavaSketchpad that aren't available in The Geometer's Sketchpad. Visit the site at http://www.dynamicgeometry.com/JavaSketchpad.html.

*Subtopics:*

## 7.2.1 Web Publishing Overview

While the full intricacies of web publishing fall beyond the scope of this Reference Center, if you've created a web page before, using JavaSketchpad should be relatively straightforward. (If you've not yet started web publishing, consider purchasing a book on the subject or searching the web itself for more information. There are thousands of sites devoted to becoming a web author.) Before getting started with JavaSketchpad, you should be familiar with these terms.

- **HTML file:** This is a document that describes the fundamental appearance of a web page and is written in HTML, the internal language of the web.

- **Web directory:** This is a storage area on your computer, or a computer you have access to, that contains HTML files that describe web pages, as well as image files and other components that appear within a web page or set of web pages that visitors can see in their browsers.

  Because your browser can open local files as well as URLs, a folder on your own hard disk can serve as a web directory for JavaSketchpad testing purposes, even if it's not accessible to the general public via the Internet.

- **Applet:** This is a special computer program that resides in a web directory and that provides extended functionality to the web pages (HTML files) stored on that site. Applets are different from browser "plug-ins." Visitors have to install plug-ins to their browsers themselves, before they can access sites requiring plug-ins. In other words, plug-ins reside in the visitor's computer. Applets, on the other hand, reside in the same web directory as the HTML files that require them, so visitors don't need to worry about installing applets or reconfiguring their browsers to support them. Instead, applets are like HTML files: a visitor's browser accesses them from your web directory as needed to display your applet-enhanced web pages; the visitor doesn't have to worry about what happens "behind the scenes."

Putting this all together, JavaSketchpad is an applet that you can place in your web directory so that your HTML files present dynamic, interactive Sketchpad illustrations to your visitors.

*See also:*

*Essential JavaSketchpad Folder Structure* 308
*Create a JavaSketchpad Web Page* 309
*What Can Go Wrong* 311
*Modify and Publish Your Pages* 310

## 7.2.2 Essential JavaSketchpad Folder Structure

Two components work behind the scenes to provide a dynamic Sketchpad illustration in a web page. The HTML document contains information that describes the geometric construction to be visualized in a language that JavaSketchpad understands. The JavaSketchpad applet, **jsp5.jar,** provides the functionality that interprets this description, displays the figure in your visitors' browsers, and lets them interact with it. You can have many HTML files containing different illustrations that all use the same applet, just as on your local computer you can have many sketch documents that can all be opened by the same copy of Sketchpad.

When you save a web page containing a construction, Sketchpad actually saves two documents. The first is an HTML document describing your construction, and the second is a copy of the applet (**jsp5. jar**), enabling your browser to display the construction embedded in the HTML document.

If you later move the HTML document to a different location (for instance, to place it on a web server), you must also copy the **jsp5.jar** applet to the same location. By default, web browsers assume that **jsp5.jar** is located in the same place as your HTML file. Therefore, they'll only work if you keep a copy of **jsp5.jar** in the same folder as the HTML document.

**Experts:** If you don't want to store your HTML files in the same folder as the applet, specify a relative URL from the HTML file's base directory to the applet by modifying the <CODEBASE> parameter in your HTML file. See your HTML reference manual for more details.

For example, in the following illustration, the **Triangle.htm** web page has been stored in a folder (in this case, named **Web Folder**), and Sketchpad has automatically put a copy of **jsp5.jar** in the same folder. This is the correct relationship between the web page (**Triangle.htm**) and the applet (**jsp5.jar** ).

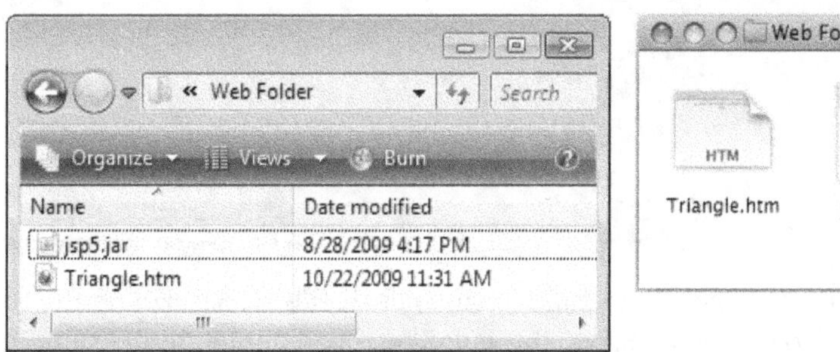


Proper folder structure (Windows and Macintosh)

*See also:*
*Web Publishing Overview* 308
*Create a JavaSketchpad Web Page* 309
*Modify and Publish Your Pages* 310
*What Can Go Wrong* 311

## 7.2.3    Create a JavaSketchpad Web Page

If this is your first time creating a JavaSketchpad illustration, start with a simple construction like a triangle.

1. Use Sketchpad to create or open a sketch containing the construction you want to show in a web page.

2. Resize the sketch window to be the size of your intended illustration on your web page. Adjust your construction so that it appears as you'd like your visitors to first see it.

   The size of the illustration in your web page is the same as the size of your sketch window at the moment you saved the HTML file. Be sure to decrease the size of the window before saving unless you want very large web page illustrations!

4. Choose **File | Save** 139 and save your document normally. This way you'll have a saved copy of the sketch from which you're about to create the web page.

5. Choose **File | Save As** 140.

   Shortcut: Hold down Shift while displaying the File menu to turn **Save As** into **Save As HTML.**

6. In the Save As dialog box, change the file format or type to **HTML/JavaSketchpad Document.** Click **Save.**

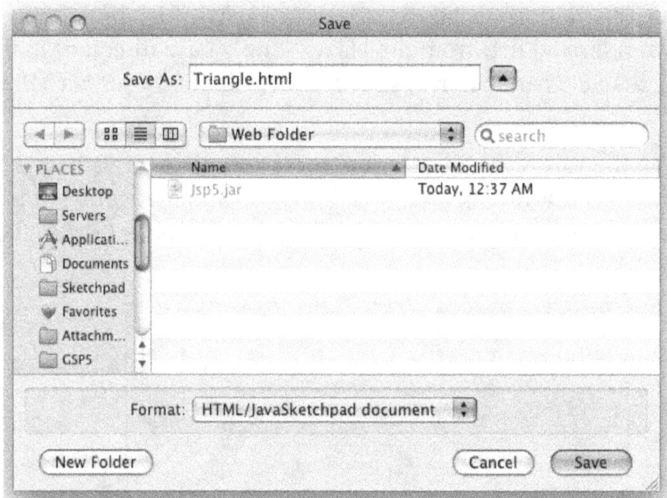

Sketchpad creates a new web page — an HTML file — describing the Sketchpad illustration. (Sketchpad also copies **jsp5.jar,** the JavaSketchpad applet, to the same folder it it's not there already.)

> Note: Sketchpad cannot open HTML documents, so saving your document as HTML does not save your document in a way that Sketchpad can reopen. This is the reason you did a normal save in step 4 above.

7. Assuming all goes well, Sketchpad will ask if you want to preview the web page in your browser. Click **Yes.**

8. Your browser opens and displays the web page. The first time your browser opens a page containing JavaSketchpad illustrations, it may take a few seconds to load.

9. Once your illustration appears, drag its independent points to explore your construction. You've successfully created your first Dynamic Geometry web page!

*See also:*
*Web Publishing Overview* 308
*Essential JavaSketchpad Folder Structure* 308
*Modify and Publish Your Pages* 310
*What Can Go Wrong* 311

## 7.2.4    Modify and Publish Your Pages

Once you've successfully previewed a JavaSketchpad web page, you'll find that although it may be exciting to have an interactive illustration, the rest of the page is rather dull. Sketchpad adds some default text to your illustration, but otherwise leaves the page blank.

Using your favorite HTML editor, you can replace the default text and add any new HTML content to the page that you want: a description, images, links, and so forth. When editing the HTML, be careful to preserve the large <APPLET>…</APPLET> block you'll find in the middle of the page. This block describes the JavaSketchpad illustration, and the illustration may no longer function if you alter any of the contents between the <APPLET> and the </APPLET> tags.

> You can even put multiple JavaSketchpad illustrations on the same page by copying an <APPLET> … </APPLET> block from one HTML file to another.

When you're ready to share your page on the Internet, copy it and the JSP applet 308 (**jsp5.jar**) to your web server. Remember to keep the applet in the same folder as your HTML file, even on your server. (You can store multiple HTML files in that folder and only one copy of the applet, but they must be in the same folder for JavaSketchpad to work.)

For information about other licenses, visit the JavaSketchpad web site.

Your license 327 for The Geometer's Sketchpad includes the use of JavaSketchpad on the Internet for noncommercial purposes. The purpose of this license is to allow you to post sketches that you or your students have created using The Geometer's Sketchpad. This license is granted provided your site can be freely visited by anyone on the Internet (that is, it's not password-protected or available only to subscribers) and that no direct or indirect profit is being made by having people visit your site.

*See also:*
*Web Publishing Overview* 308
*Essential JavaSketchpad Folder Structure* 308
*Create a JavaSketchpad Web Page* 309
*What Can Go Wrong* 311

## 7.2.5  What Can Go Wrong

Because of the many factors involved, it may take a few tries to get things right. Here are some common JavaSketchpad mishaps and steps you can take to avoid them.

More detailed information is available on the JavaSketchpad web site: http://www.dynamicgeometry. com/JavaSketchpad.html.

### ▼ Unsupported Objects

**Problem:** When you save, Sketchpad warns you that not all objects were successfully saved to JavaSketchpad format.

**Cause:** Because JavaSketchpad is smaller than "desktop" Sketchpad, it supports fewer ways of defining objects than you can use in the desktop application. If your sketch contains objects that JavaSketchpad doesn't support, Sketchpad warns you about them — and selects the unsupported objects and their children 80 in your sketch so that you can tell which ones were not supported. (Even when your sketch has unsupported objects, Sketchpad will save the objects that JavaSketchpad does support, so you can continue testing your web page.)

To work around unsupported objects, explore different ways of constructing the same illustration. For example, at present, JavaSketchpad does not support iterations or iterated images 24. If your sketch contains a construction that you've iterated using the **Transform | Iterate** 208 command, you may be able to replace it with one that you've iterated "manually" by actually constructing the first several iterations.

A complete list of objects supported by JavaSketchpad is available on the JavaSketchpad web site. Also, new versions of the applet—that support more and more Sketchpad objects — occasionally appear on the web site.

If your sketch or activity requires objects not supported by JavaSketchpad, you won't be able to share that sketch as an illustration. However, you can still post your original Sketchpad sketch — your **.gsp** file — as a downloadable file, so that visitors who use The Geometer's Sketchpad can download it and open it in Sketchpad, rather than in their browser.

### ▼ No Preview Offered

**Problem:** Rather than ask whether you want to preview your file in a browser, when you save, Sketchpad warns that you saved to a folder that does not contain a copy of **jsp5.jar.**

**Cause:** Sketchpad could not locate the **jsp5.jar** applet to copy it to the folder with your HTML page. The original copy of the applet may have been accidentally removed from its normal location in the Sketchpad folder. Reinstall Sketchpad to restore the applet.

### ▼ Non-Java Browsers

**Problem:** When you preview your web page in a browser, the page contains a message reading "Sorry, this page requires a Java-capable browser."

**Cause:** If you have an old web browser, it may not support the Java language and, therefore, won't work with JavaSketchpad. Contact your browser manufacturer for a newer version. Alternately, it may be that your browser supports Java, but that it's currently set to disable Java applets. Go to your browser's Preferences or Options to turn on support for Java applets.

### ▼ Java Exception or Error Occurs

**Problem:** When you preview your web page in a browser, a dialog box appears saying that a Java error or "exception" occurred.

**Cause:** If the message goes on to say that a "class" was not found — for example "GSP.class: class not found" — then **jsp5.jar** is either in the wrong location or has become corrupted. (For example, vital files within it may have been accidentally deleted or moved.) Try removing **jsp5.jar** from the folder and then resaving your document as HTML, so that Sketchpad can copy a fresh copy of **jsp5.jar** to the destination folder.

If the error message says something else, a different problem has occurred. While JavaSketchpad has been extensively tested on current versions of Internet Explorer, FireFox, and Safari, older browsers are erratic in their support of Java, and other browsers may have similar problems. (Java is a relatively new language and undergoes frequent modifications; different browsers support it to different degrees.) Check with your browser manufacturer to see if a more recent version of your browser is available. Errors or exceptions may also indicate a problem with JavaSketchpad itself. Visit the JavaSketchpad web site to see if a more recent version of the applet is available.

### ▼ Appearance Discrepancies

**Problem:** Your sketch saves correctly and previews in your browser, but certain details in the JavaSketchpad illustration — such as choice of fonts, size of exact position of text — do not match your original sketch.

**Cause:** These minor discrepancies are inevitable in Java applets, where less sophisticated graphic and text services are available to an applet than to a nonapplet program (like your browser or desktop Sketchpad). What you sacrifice in appearance flexibility you gain in generality: Applets like JavaSketchpad work well on a much wider variety of computers than are capable of running the full desktop version of Sketchpad.

*See also:*
*Web Publishing Overview* 308
*Essential JavaSketchpad Folder Structure* 308
*Create a JavaSketchpad Web Page* 309
*Modify and Publish Your Pages* 310

## 7.3 Advanced Topics

Once you become familiar with the basic ideas behind Sketchpad — the organization of its tools and menus, and the process of creating and exploring mathematical ideas through Dynamic Geometry constructions — you may want to explore some of the program's additional options. This section contains expert tips for using Sketchpad efficiently; instructions for using Sketchpad to create high-quality mathematical illustrations for use in other programs or in printed documents; overviews of the program's menu structure and internal mathematics; and special Windows command-line flags to control several Sketchpad settings.

*Subtopics:*


## 7.3.1 Advanced Text Topics

There are several advanced features you can use in connection with Sketchpad's text objects 99.

### ▼ Hot Text in Captions

While you're editing a caption, click an object in the sketch to insert a Hot Text 57 link showing its label or value.

### ▼ Unicode

Sketchpad fully supports Unicode 314, allowing you to insert a wide variety of mathematical symbols and to use multiple languages in captions and labels.

### ▼ Align Text Objects

Sketchpad's text-alignment shortcut allows you to line up selected text objects so that their left sides are aligned and they are spaced evenly beneath the first selected object.

When a sketch contains a number of text objects (captions, measurements, parameters, calculations, functions, and action buttons), you can line them up neatly to improve the appearance of the sketch. In the following examples, the aligned arrangement on the right looks neater than the unaligned arrangement on the left.

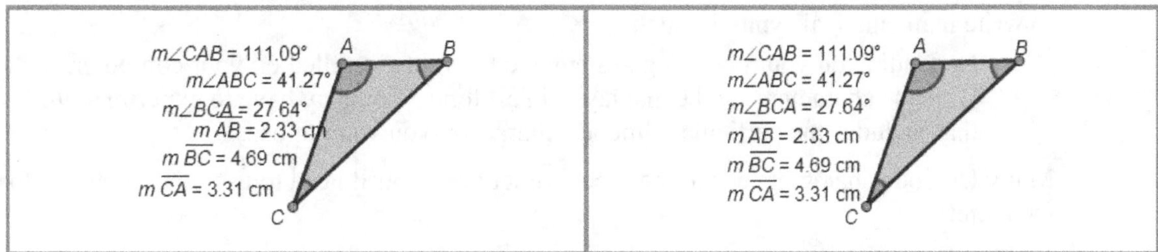


To align text objects, select them in the order in which you want them to appear, from top to bottom. Then hold down the Shift key and press the Enter key. The first selected object remains in its original position, and the remaining selected objects line up below the first.

To increase the spacing between the newly-aligned objects, make sure they are still selected. Then hold the Shift key and press Enter again. Each time you press Enter, a small amount of space is added between the objects. Continue holding Shift and pressing Enter until you have the vertical spacing you want.

If you accidentally create too much space, continue holding the Shift key and pressing the Enter key until the objects return to their minimum spacing.

## ▼ Label Objects with a Custom Sequence

Sketchpad allows you to modify the behavior of the Label Multiple Objects [168] dialog box by creating your own custom sequences.

When you use the **Display** [162] | **Label Objects** [168] command to label several objects, Sketchpad normally generates a sequence of labels by changing only the ending of whatever you type as the first label. Thus if you type $A_1$ as the first label, Sketchpad generates a sequence in which all the labels start with $A$, but with different numbers following the $A$: $A_1$, $A_2$, $A_3$, and so forth. Similarly, if you type *123a* as the first label, only the endings change: *123a*, *123b*, *123c*, and so forth.

To type a subscript such as the 1 in $A_1$, enclose the subscript in square brackets: "A[1]."

To generate a sequence in which the changing portion is not at the end, select the objects to label and choose **Display | Label Objects** [168]. In the Label Multiple Objects [168] dialog box, define a custom sequence by entering a special first label. This special custom sequence label must begin with an equal sign ("="), and contain "{...}" immediately following the part of the label that should change over the sequence.

For instance, you can type "=P{...}1" to generate the sequence starting with *P1, Q1,* and *R1.* Similarly, type "=A{...}[x]" to generate a sequence that begins with $A_x$, $B_x$, and $C_x$.

## ▼ Iterate Text Attached to a Point

When text affected by an iteration [24] is attached [153] to a point that's also affected by the iteration, the text is iterated along with the point.

*Subtopics:*
   *Unicode* [314]
   *Unrecognized Text Format* [316]

## 7.3.1.1   Unicode

Sketchpad fully supports Unicode, so you can use Unicode characters and mathematical symbols in captions [56], labels [81], and names of custom tools [125] and custom transformations [211]. You can write in multiple alphabets, such as живая математика, in Russian, or 动感几何, in Chinese. Use Unicode to write mathematical symbols, such as $\nabla f = \langle \partial f/\partial x, \partial f/\partial y \rangle$.

> Depending on your operating system and the fonts installed on your computer, not all Unicode characters can be displayed in all fonts. You may have to experiment to find a font that includes the particular Unicode characters you want to use.

Many Unicode characters cannot be typed directly, so you'll need to use other methods to enter these characters.

When you're editing a caption, you can insert a variety of Unicode characters using the Text Palette [270]. Press the Symbolic Notation button to show the mathematical notation choices. Then use the symbol buttons [272] on the right to enter various mathematical symbols and Greek letters.

When you're editing a label, you can insert certain Unicode characters using special codes, such as {angle} to enter the ∠ symbol. See Subscripts, Symbols, Greek Letters, and Unicode [82] for the list of codes you can use in labels.

If you want to type in a different language, it's most convenient to set your keyboard to switch between languages, so that you can type to enter the desired Unicode characters in either language.

If you want to enter other mathematical symbols or special characters, it may be most convenient to use your operating system's method for viewing and inserting Unicode characters from a table or palette.

### ▼ Set Your Keyboard to Type in a Different Language

Both Windows and Macintosh have methods for changing your keyboard so you can type in a variety of languages.

If you're using Windows Vista or Windows XP, choose **Start | Control Panel | Regional and Language Options.**

If you're using Mac OS X, choose **System Preferences | International.**

### ▼ Insert Unicode in a Caption or Label (Windows)

1. Open the Character Map by choosing **Start | All Programs | Accessories | System Tools | Character Map.**

2. In the Character Map, choose a font from the pop-up menu.

3. Highlight the character you want to insert into the document.

4. Click **Select,** and then click **Copy.**

5. Return to your Sketchpad document and position your cursor where you want the special character to appear.

6. Choose **Edit** [144] **| Paste** [147].

In Windows XP, support for Unicode characters is limited: many fonts are missing important Unicode characters. If you enter a Unicode character not supported by the current font, it will appear as a square. Windows Vista has much better Unicode support in default fonts.

### ▼ Insert Unicode in a Caption or Label (Macintosh)

1. Set up your Macintosh to display the Character Palette in the Input menu. Open **System Preferences.** In OSX 10.5, click the International icon, go to the Input Menu panel, and check the box for **Character Palette.** In OSX 10.6, click the Language & Text icon, go to the Input Sources panel, and check the box for **Keyboard & Character Viewer.** In either case, also check **Show input menu in menu bar.** (You only need to do this once.) The Input menu icon appears on the right side of your menu bar.

2. Choose **Show Character Palette** (OSX 10.5) or **Show Character Viewer** (OSX 10.6) from the Input menu on the menu bar. (If the Input menu does not appear, you may be using an application whose menu bar is too long to allow the Input menu to show. In this case switch to the Finder, which has a short menu bar, and then choose **Show Character Palette** or **Show Character Viewer** from the Input menu.)

3. In Sketchpad, click in the label or caption where you want to enter the special symbol.

4. In the Character Palette or Viewer, double-click the Unicode character or symbol that you want to insert into your Sketchpad document.

### 7.3.1.2  Unrecognized Text Format

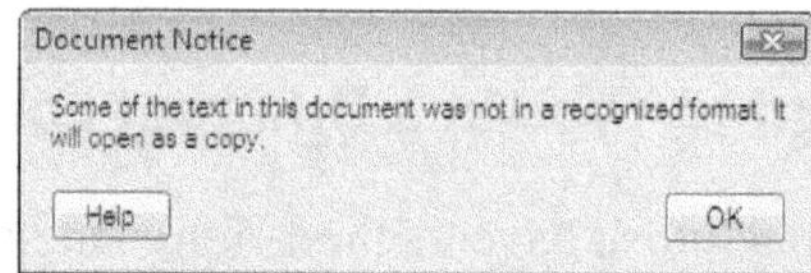

Documents that come from Sketchpad 4 and earlier versions of Sketchpad may not contain information about the language in which they were created. This presents no difficulty when you open an older document created in the same language as the one you use on your computer.

But if you try to open an older document that was created in a different language than the one your computer uses, Sketchpad may not be able to recognize the text because of the language differences. In this case, Sketchpad displays a message describing the problem.

If you know the language in which the original document was created, you may be able to read it successfully:

1. Hold the Shift key and choose **Edit | Advanced Preferences | System** 284.

2. Set the **Language override for GSP3/GSP4 documents** to the language in which the document was created.

3. Open the document.

4. Use **Advanced Preferences** 160 again to change the **Language override for GSP3/GSP4 documents** back to its original setting.

## 7.3.2  Advanced Tool Topics

There are several advanced options available to you when creating and using custom tools. You can:

- Use a custom tool as a command. 316

- Change the folder from which Sketchpad automatically loads custom tools when you start the program. 317

- Automatically match a tool's given object to a specific object in your sketch. 318

- Determine how objects produced by a tool are labeled. 319

*See also:*
   *Custom Tools* 125
   *Script View* 286
   *Document Options and Tool Options* 254

### 7.3.2.1  Use a Custom Tool as a Custom Command

Normally you use the Custom Tools 125 menu to choose a tool, and then employ the chosen tool by clicking in your sketch.

You can also use the Custom Tools menu to choose commands, called *custom commands,* that have the same effect as applying the tool to the selected objects.

To use a custom command:

1. Select the tool's required given objects 128, in the order listed in the custom tool's Script View 286. (There is no need to have previously chosen the desired tool or to have the Script View visible.)

2. Hold the Alt key (Windows) or ⌘ key (Mac) while pressing the **Custom** tool icon. The items listed

in the menu are unchanged, but with the Shift key down they represent custom commands rather than custom tools.

Custom commands that can be applied to the selected objects are enabled; all others are disabled.

3. Choose the desired command to apply it to the selections.

 Hint

If the meaning is not ambiguous, you can select multiple objects to satisfy the script's given objects multiple times. For example, if you have a **Perpendicular Segment** tool that constructs a perpendicular segment to a given line from a given point, you can select a point and a line to enable the **Perpendicular Segment** custom tool. You can also select, for example, five points and one line to enable the command, which now appears as **Perpendicular Segment (5x).** Similarly, you can select one point and three lines to enable the command as **Perpendicular Segment (3x).**

Using a custom command does not change your currently chosen custom tool.

## ▼ Missing Assumed Object for a Custom Command

A custom tool may be set to automatically match 318 one or more of its given objects to object(s) in the sketch with the same name. In the Script View 286, such an object appears in the Assuming section, before the Given section.

When you use such a tool, any assumed objects are matched immediately, without any action on your part. You only have to click to match objects in the Given section. If an assumed object is missing, then you must click to match that assumed object before matching the given objects.

When you use a custom command, assumed objects are similarly matched automatically (based on their names) to objects in the sketch, and the selected objects are matched to the given objects.

If the sketch does not contain a properly-named object of the correct type to match an assumed object, the command cannot be applied to the selected objects. In this case a dialog box appears, and you must name a sketch object properly to match the assumed object of the tool. (You may need to create the assumed object before naming it correctly.)

Depending on the type of object that's missing, you may be able to use the dialog box to tell Sketchpad to create such an object at random and name it appropriately.

## 7.3.2.2 Alternate Tool Folders

If you've designated a Tool Folder, Sketchpad checks that Tool Folder when it starts up, and puts every tool in the folder into the Custom Tools menu 126. (When you first install Sketchpad, there is no Tool Folder until you designate one by using the **Choose Tool Folder** command from the Custom Tools menu 126 menu.)

There are several possible ways in which students might use the Tool Folder.

**Students create and use their own tools:** In this scenario, the Tool Folder must be in a location, either on the local computer or on the network, where students have permission to save and modify documents. For instance, on a network students might use the **Choose Tool Folder** 133 command to designate their own Tool Folders within their personal Documents folders on the network. Though the location of this folder may vary depending on network configuration, a common location on the local hard drive is ~**\My Documents\Sketchpad\Tool Folder** (in Windows) or **Documents | Sketchpad | Tool Folder** (on Mac).

**All students use the same tools:** In this scenario, all students should have access to the same set of tools, located in the same Tool Folder. (For instance, you might have a group of geometry tools, or algebra tools, or calculus tools you want your students to use.) In this case, you can store the desired group of tools in a folder that's marked as read-only, or in a location on the network where you can save documents but students cannot. Using such a location prevents students from modifying the tools, or saving new tools, by mistake.

**Using a shortcut to set the Tool Folder on launch:** In Microsoft Windows, you can create a shortcut that students use to launch Sketchpad, and use a command-line flag 326 in conjunction with the shortcut to change the location of the Tool Folder. You can create multiple shortcuts with different command-line flags, so that each shortcut launches Sketchpad using a different Tool Folder.

### 7.3.2.3   Automatically Match a Given Object

When you create a custom tool, there may be a particular given object that you would like to always match to the same object in your sketch. Normally you must match that given object each time you use the tool, even though you're matching it to the same object each time.

To save the trouble of clicking the object each time and to make the tool easier to use, you can specify that the given object should be automatically matched to the same sketch object each time the tool is used. For example, suppose you create a tool that constructs one segment on a Poincaré disk model of the hyperbolic plane. Such a tool might have three given objects: a circle defining the Poincaré disk and two points defining the segment's endpoints. Since you'd like to use the same tool repeatedly to construct multiple segments on the same Poincaré disk, it's inconvenient to have to match the given circle to the same Poincaré disk each time you use the tool. You can change your tool to automatically match this given circle to the appropriate circle in your sketch, resulting in a tool that requires you to match only two points each time you use it. Since the Poincaré disk circle is matched automatically, this tool will always create segments on the same Poincaré disk.

To match a given object automatically:

1. Open the Properties dialog box 273 for the sketch object you want to match automatically. Use the Label panel 82 to assign a distinctive label to the object.

2. Choose *Show Script View* from the Custom Tools menu 126 to show the Script View 286 for the tool containing the given object to be matched automatically.

3. Double-click the given object 288 in the object list to open the Properties dialog box.

4. On the Label panel 82, assign the same label to the script object that you assigned to the sketch object.

5. Check the **Automatically match sketch object** checkbox and close the Properties dialog box.

When you use the tool, it will automatically match the tool's given object to the sketch object with the same label. Provided the appropriate checkbox is checked in Label Properties, a tool containing a given point labeled *Center* will automatically match the point labeled *Center* in your sketch. If no object in the sketch has a matching label, the tool will require you to match the given object manually.

> If all the given objects for a tool match automatically, the tool completes its construction as soon as you choose the tool. The active tool is then changed back to the **Selection Arrow** 104 tool.

When a given object in a tool is set to match automatically, it appears in the *Assuming* section of the script view, rather than the *Given* section.

(If a new tool has a coordinate system as one of its given objects, the coordinate system is automatically placed in the *Assuming* section of the script view, and the checkbox in the Label Properties dialog panel is **Automatically match marked coordinate system.**)

*See also:*


## 7.3.2.4    Generate Specific Labels

Normally when you use a custom tool 125, the results are assigned labels just the way they would be if they were constructed in any other way. But sometimes you may want to control the labels that are used for the results of a custom tool. There are two types of labels you can specify when you make the tool: constant labels and variable labels.

### ▼ Set a Constant Label

A result with a constant label is assigned the same label every time the custom tool is used. For example, a particular line produced by a tool might always be labeled *Mirror*.

To set a constant label for an object before you make the tool, use the Label panel 82 of the Properties dialog box 273 for the sketch object that should have the constant label. Enter the desired label and check **Use label in custom tools.**

To set a constant label for a tool object after you make the tool, use the Custom Tools menu 126 to show the tool's script view 286. Double-click the step corresponding to the resulting object that you want to have a constant label. On that object's Label panel, enter the desired label and check **Use label in sketches.**

### ▼ Set a Variable Label

When a tool has a result with a variable label, the label of the resulting object in the sketch is set based on the labels of other objects — objects that correspond to givens or other results of the tool. For instance, you can create a centroid tool that labels the centroid of a triangle based on the labels of the vertices of the triangle.

A variable label specifies how the labels of the givens and other results 288 of the tool should be used in constructing the desired label. To specify that a particular label should be used, start the label with an equal sign (=) and enclose in curly brackets the numeric index of the object whose label should be used. The number of an object is its position in the tool's script view, starting with 1 for the first given. If a tool has three givens, the givens are numbered from 1 to 3 and the number of the first step is 4.

For example, the following variable label combines the labels of the first and second given objects of the tool:

    ={1}{2}

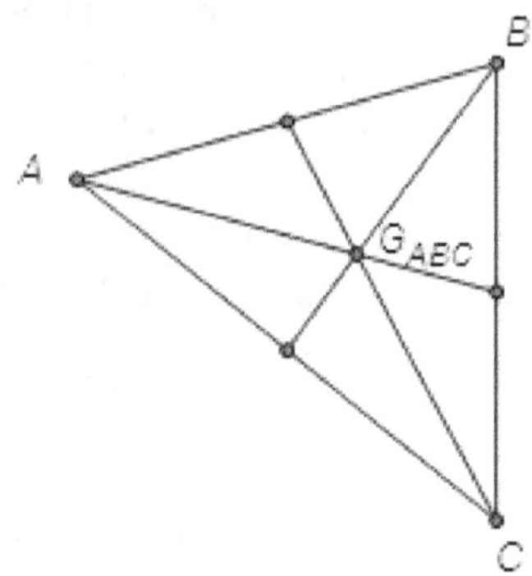

As a second example, if you create a tool to construct the centroid of a triangle, you can set the resulting centroid's label so that, if the tool is used on points $A$, $B$, and $C$, the centroid is labeled $G_{ABC}$. To accomplish this, set the centroid label in the tool to

=G[{1}{2}{3}]

> Square brackets in a label indicate that the portion of the label within square brackets should appear as a subscript.

To set a variable label for a tool object, show the tool's script view using the Custom Tools menu 126 . Double-click the step that should have the variable label to show its Properties 273 . On the Label panel 82 , enter the desired label and check **Use label in sketches.**

When you create a variable label for a resulting object, be sure to specify in the curly brackets only objects on which the labeled object depends. Otherwise it's possible that the labeled object will be produced before one of the objects that determines its label, and the variable label will not be created correctly.

## 7.3.3   Advanced Graphics Export

Use Sketchpad to create images to paste into other programs — word processors, illustration programs, page-layout programs, and programs that create web pages. And use Sketchpad to produce illustrations for handouts, tests, and quizzes for your class, to illustrate articles or books about geometry or algebra, or to decorate your personal web site or a web site for your class or school.

This topic describes several useful tips and techniques that can help you produce attractive, high-quality images for such purposes.

### ▼ Export Preferences

Hold the Shift key and choose **Edit** 144 | **Advanced Preferences** 160 | **Export** 282 to control several settings that affect graphics export.

**Arrowheads:** Decide whether to include arrowheads on lines and rays.

**Export Format:** Export to the clipboard in bitmap format only or in both bitmap and vector formats.

**Bitmap Scale:** Export bitmaps at 100%, 200%, 400%, or 800%.

### ▼ Vector Format and Bitmap Format

When you choose **Edit** 144 | **Copy** 146 , Sketchpad places a graphic image of the selected objects on the clipboard. The exported graphic on the clipboard can use a *vector* format or a *bitmap* format. Sketchpad normally puts both formats on the clipboard, allowing the program into which you are pasting to decide which format to use.

**Bitmap Format:** This format records the pixels that appear on the screen. In bitmap format, lines, circles, and text show stair-step effects (called aliasing); these effects are particularly visible when the graphic is enlarged, or when it's printed on a high-resolution printer. You can reduce the aliasing effect in a bitmap image by changing the scale for clipboard bitmap format.

Screen-capture programs also produce a bitmap format. This format is similar to Sketchpad's bitmap format, but the resulting graphic can also include objects such as window borders, menus, and the pointer.

On Windows, Sketchpad's bitmap format is DIB (Device-Independent Bitmap). For compatibility with some older programs, Sketchpad also creates a WMF (Windows MetaFile) which contains a DIB.

On Macintosh, Sketchpad's bitmap format is PICT.

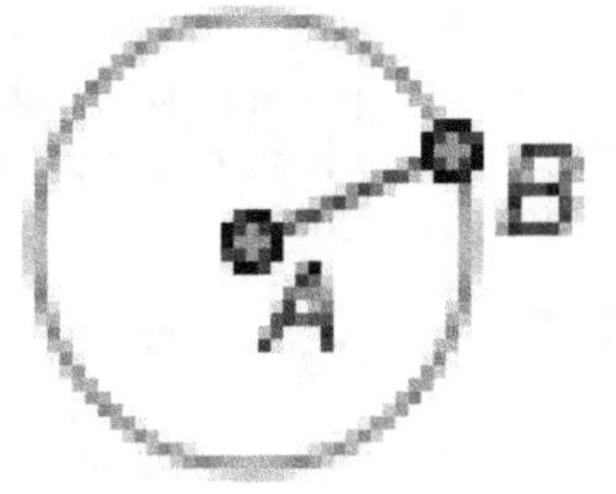

Bitmap Format Exported at
100% scale
(Magnified 4x)

**Vector Format:** This format records information about how to draw objects like segments, circles, and captions, and can produce smooth lines without aliasing. Vector graphics are smoother and more accurate, and are usually the preferred format for word processors, graphics programs, and layout programs.

On Windows, Sketchpad's vector format is EMF+ (Enhanced MetaFile).

On Macintosh, Sketchpad's vector format is PDF. Use the Preview program or other graphics software to convert PDF to other formats.

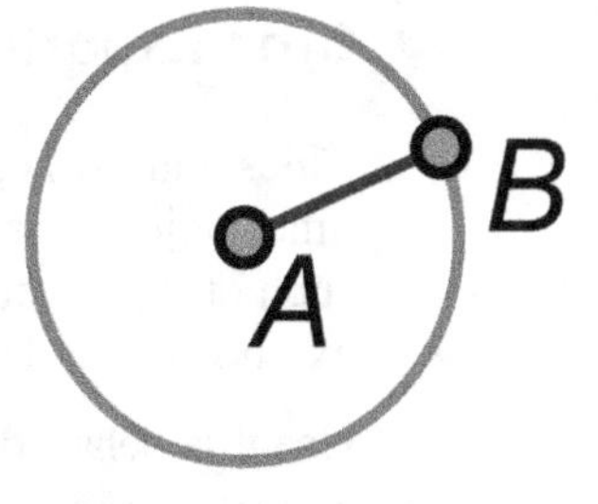

Vector Format
(Magnified 4x)

## ▼ Scaled Bitmap Export

When you choose **Edit** 144 | **Advanced Preferences** 160 | **Export** 282, you can set the scale used for bitmap export to 100%, 200%, 400%, or 800%. Use these settings to improve the quality of exported bitmap graphics when you're exporting to a program that doesn't recognize vector formats.

By using a higher scale, you can reduce the stair-step effect (called *aliasing*) that occurs with bitmaps. Set this to more magnification if you're copying bitmaps from Sketchpad to paste into a document in a word processor or page layout program that you intend to use for high-quality production. The result will be a larger graphic that includes more pixels. After you paste it into a word processing, page layout, or other application, you can scale it down to produce a higher-quality image of the desired size.

For instance, say you want use a 600 dpi printer to print a document containing a small Sketchpad graphic that is 50 pixels wide on the screen, and you'd like the printed graphic to be 0.2 inches wide. If you export the graphic at 100% and look closely at the document, you will be able to see the aliasing.

To improve the quality of the printed image, you can determine the appropriate scale for bitmap export. The number of pixels in the printed image will by 600 dpi * 0.2 in = 120 printer dots. Divide the number of desired printer dots by the number of screen pixels: 120 / 50 = 2.4. This corresponds to a scale of a little more than 200%, so you should use the Export Preferences 282 panel to set **Scale for clipboard bitmap export** to 200%. The images below show magnified views of the printed document using the original export scale of 100% and the calculated scale of 200%.

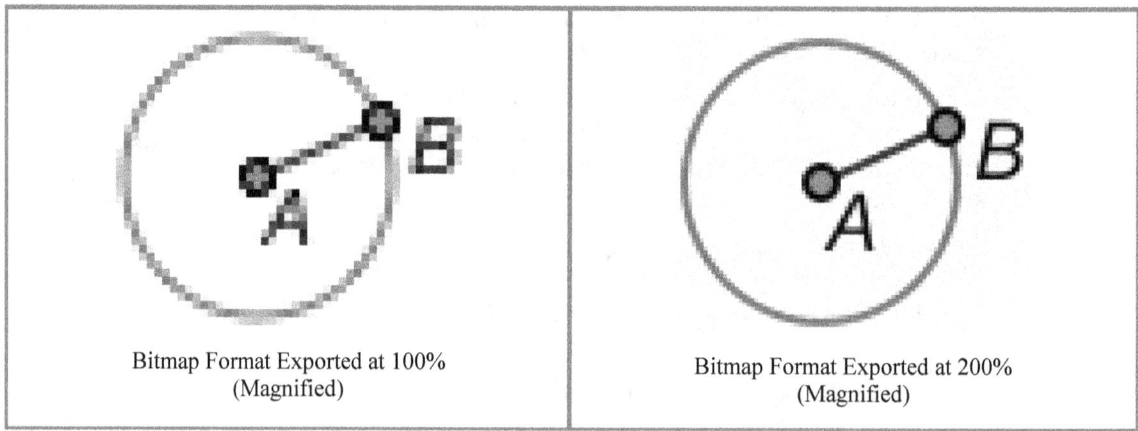


| | |
|---|---|
| Bitmap Format Exported at 100%<br>(Magnified) | Bitmap Format Exported at 200%<br>(Magnified) |

## ▼ Crop Exported Graphics

Lines and rays extend far beyond the limits of the screen. Depending on the specifics of a sketch, other objects, particularly loci and function plots, may have similar extents. But when you copy such objects, you don't want an image of infinite size. Accordingly, if an image being copied extends beyond the edges of the window, Sketchpad crops the image to the window dimensions.

Use this behavior to control the exact extent of the image copied to the clipboard. Before copying, resize 253 and scroll 295 your sketch window so that it shows the desired portion of the sketch. When you choose **Edit** 144 | **Copy** 146, the clipboard image will be correctly cropped.

## ▼ Graphics Document Export

Sketchpad normally exports graphics directly to the clipboard, for ease of pasting into word processing and page layout applications. If you like, you can preserve the exported graphic as a separate vector-graphics document on your computer, either using EMF+ (Enhanced Metafile) in Windows or PDF (Portable Document Format) on Macintosh.

**Windows:** Choose **File** 137 | **Save As** 140. In the Save As dialog box, choose **Enhanced Metafile (*.emf)** from the **Save as type** list. Then name your document and click **Save.** The resulting EMF+ document can be imported by many other Windows programs. The exported document includes all objects that appear within the window. Use the edges of the window to crop the exported graphics. Once you've saved the EMF+ document, there are a number of commercial graphics programs that can convert the EMF+ document to other graphics formats such as EPS or PDF. Depending on your printer driver, you may also be able to create a PDF document directly, by printing from Sketchpad and specifying PDF output in the Print dialog box.

If you want to export some but not all of the visible objects, use the **Copy** 146 command.

**Macintosh:** Select the objects you want to export and choose **Edit** 144 | **Copy** 146 to put a vector format (PDF) image on the clipboard. Then open the Preview application, and choose **File | New from Clipboard** to display the contents of the clipboard in a new PDF document. Save the document, either in its original PDF format or in any of the other formats Preview supports. You can also create PDF and Postscript (PS) documents by choosing **File** |137| | **Print** |143| and using the PDF button in the Print dialog box.

Once you've saved a PDF document, there are a number of graphics programs that can convert the PDF document to other graphics formats such as EPS.

### ▼ Screen Captures

You can create a bitmap graphic of the entire Sketchpad window, including even menus and the cursor, by using the screen capture capabilities built into Windows and Macintosh.

**Windows:** Press the Print Screen key to capture the entire screen, or Alt+Print Screen to capture the active window. The result is a bitmap on the clipboard; you can paste this bitmap into any program which recognizes bitmap files.

**Macintosh:** Press ⌘-Shift-3 to capture the entire screen, or ⌘-Shift-4 to capture part of the screen. The result is a picture on your desktop. Use the Preview program to change this picture into any common graphics format.

Hold the Ctrl key while pressing the other keys to put the captured picture on the clipboard instead of the desktop.

For more capture options on Macintosh, use the Grab program.

Screen captures are excellent for showing what the entire screen looks like and for including a menu or the cursor in your graphic. But a screen capture does a crude job on Sketchpad constructions. Diagonal lines and circles look jagged and blocky, text is of poor quality, and resizing the graphic may give unexpected results. For better results with Sketchpad constructions, use Sketchpad itself to copy the objects you want by selecting the desired objects and choosing **Edit** |144| | **Copy** |146|.

## 7.3.4 Sketchpad's Menu Structure

Sketchpad's menus are arranged thematically. While the File |137|, Edit |144|, and Display |162| menus contain commands relating to your Sketchpad documents and workflow, the other menus and tools are more mathematical in nature. Each of these menus presents a distinct mathematical viewpoint and commands appropriate to that viewpoint. Familiarizing yourself with their organizational structure can help you plan your approach to a given construction problem or mathematical challenge.

- **Toolbox** |102|**:** The **Compass** |113| and **Straightedge** |115| tools provide the fundamental tools of compass-and-straightedge Euclidean geometry.

  In that the **Compass** tool does not retain a fixed radius, it technically provides a collapsible, rather than a noncollapsible, compass.

- **Construct** |174|**:** This menu contains additional commands for working in compass-and-straightedge Euclidean geometry. (Many of the objects you create with this menu could be created with the Compass and Straightedge tools alone, though some would take many steps to create with those tools.) Where the **Construct | Circle by Center+Point** |182| command is equivalent to the **Compass** tool, **Construct | Circle by Center+Radius** |183| allows you to construct circles of a given radius, acting as a noncollapsible, rather than a collapsible, compass.

- **Transform** |192|**:** This menu contains commands drawn from the perspective of a metric transformational geometry. Use them to construct or explore symmetries and other transformational relationships. You may specify transformational parameters — such as angles of rotation or scaling factors of dilation — either geometrically, by referring to existing objects, or metrically, by entering numeric angles and lengths (or by referring to numerical values and calculations already defined in your sketch).

Since compass-and-straightedge geometry does not include tools for specifying lengths metrically, you cannot construct a segment of a given length 182 — say, 5.0 cm — using only the Construct menu. However, since the Transform menu's operations are metric in nature, you can use its commands to produce such a result.

- **Measure** 216 **:** This menu's commands continue the metric theme and offer a variety of ways to determine numeric relationships in your construction. The commands in the top part of this menu can be thought of as ruler-and-protractor operations: They measure distances, areas, and angles using the metric units you choose in Preferences 159. The commands that appear in the bottom part of the menu are analytic in nature and measure quantities in relationship to some (existing or newly-defined) coordinate system 41.

- **Number** 226 **:** This menu's commands allow you to create, calculate, and tabulate numerical values and functions, moving from the realm of number into algebra.

- **Graph** 233 **:** This menu's commands continue the analytic perspective, offering operations relating to coordinate systems and graphing.

While each of these menus reflects a unique mathematical perspective, in the course of any Sketchpad activity you may move back and forth between perspectives to focus on different aspects of your activity. In particular, move from a geometric or spatial visualization to a numeric perspective using commands from the Measure menu. (Think of these commands as "turning shapes into numbers.") Move from numbers back into geometric or spatial visualizations ("turn numbers into shapes") by using the Plot Points 239 and Plot Function 241 commands, or use Transform menu commands with marked 194 numeric values as transformational values.

Finally, in addition to the commands that produce or construct specific mathematical relationships in your sketch, each menu contains one command that produces a generalization of an arbitrary set of such relationships over some change.

- The Construct menu's **Locus** 190 command lets you visualize the position of a constructed object over a change in one point's position.

- The Transform menu's **Iterate** 208 command lets you visualize the orbit of one or more objects over some number of repetitions of a construction.

- The Measure menu's **Calculate** 227 command lets you express a general relationship arithmetically between two or more measured quantities.

- The Number menu's **Tabulate** 228 command lets you analyze the values of a set of measurements over time.

- The Graph menu's **Plot Function** 241 command lets you visualize a general function evaluated over a domain.

- The Toolbox's **Custom tool** 125 lets you generalize a set of relationships constructed between objects into a new tool that you can use to replicate that construction on a new set of objects.

Mastering these advanced commands allows you to move beyond the specific mathematical relationships, objects, tools, and commands that form Sketchpad's starting points and opens up a set of mathematical curves, shapes, and construction tools limited only by your imagination.

## 7.3.5 Sketchpad's Internal Mathematics

Sketchpad's internal mathematics determine how the program computes and represents numbers, geometric figures, functions, and other mathematical quantities. This in turn determines how these objects appear graphically, numerically, or symbolically.

At the numeric level, Sketchpad represents point coordinates and other quantities using 64-bit floating-point arithmetic. This standard representation for scientific computation allows your computer to represent a value with 14 to 16 significant digits of decimal precision over a wide range of magnitudes (roughly, as large as $\pm 10^{300}$ and as small as $\pm 10^{-300}$). While this is very precise, it is not exact in a mathematical sense. (For example, $\pi$ cannot be represented exactly with only 15 significant digits.) Sketchpad uses tuned algorithms to attempt to represent numbers as close to their exact value as possible, and to minimize the inevitable error introduced by calculating with only a finite number of significant digits. Nonetheless, you may witness numerical error effects in the least significant digits of numbers in sketches involving a lot of internal calculation. Regrettably, no computer or computer program can represent every number exactly: there are an infinite number of numbers, of course, and — at least today! — computers have only a finite amount of memory. Thus, while Sketchpad's numeric calculations are generally reliable and can serve as the basis of a convincing argument or conjecture, a Sketchpad result should never be mistaken for constituting a mathematical proof.

> Don't confuse the displayed precision of a value with its internal precision or accuracy [277]. When you display a value (such as a measurement), it appears only to the number of decimal places you choose in the Preferences [159] or Properties [158] dialog box. Internally, that number is represented to much greater precision, as described here.

At the graphical level, Sketchpad transforms its internal numeric representations into the shapes and positions that appear in your sketch window. For objects such as circles, points, and segments, the resulting images are as accurate as can be displayed on your computer screen. (When you print to a printer with higher resolution than your screen, you'll see the images are even more accurate than their on-screen representations.)

However, for plotted functions and loci, Sketchpad displays only a visual approximation of the curve's ideal mathematical shape. Primarily to maintain responsiveness when you're dragging objects, Sketchpad employs the same technique to plot these objects as a person might use if plotting them by hand: it evaluates the ideal curve at a number of different positions (called samples), then plots the curve by interpolating between these known samples. The samples themselves are very accurate, but the interpolations may or may not be, depending on the ideal shape of the mathematical object. In the Properties dialog box [273] for loci [95] and functions [97], the Plot panel [95] gives you control over how many samples Sketchpad uses to plot a function or a locus, as well as whether it displays that plot only as the (discrete) collection of accurate samples or by including the (continuous) interpolations between samples. If you consistently prefer a higher number of samples than Sketchpad uses by default, you can increase the default on the Sampling Preferences [283] panel. Also, while Sketchpad uses approximate interpolations for display purposes, it never relies on them for mathematical purposes. Thus, if you construct a point on a function plot or on a locus (or use the Calculator to evaluate a function at given values), that point's coordinates (or function's values) are always based on the exact curve or function and not on its visual approximation.

Finally, at the symbolic level, Sketchpad performs simple computer algebra to differentiate functions when you use the **Number | Define Derivative Function** [231] command. While these computed derivatives are generally reliable for graphing and evaluation purposes, they may not be exact. In particular, when differentiating intricate functions, Sketchpad may fail to simplify the result fully, introducing point discontinuities in the derivative; and Sketchpad does not compute domain restrictions on the derivative function. Use **Define Derivative Function** 231 to compute the slope of a function at an arbitrary point for graphing purposes or for mathematical constructions, but be sure to verify the result before using it as the basis of a mathematical proof.

### 7.3.6 Command-Line Flags for Sketchpad (Windows Version)

With the Windows version of Sketchpad you can set several command-line flags that determine Sketchpad's start-up behavior. Use command-line flags to maximize the Sketchpad application window, to maximize the frontmost document window, to specify a default document for Sketchpad to open, to determine the location of the Sketchpad Preferences file, to specify a folder to use as the Tool Folder 133, or to specify the folder Sketchpad will use when first opening or saving a file.

To set command-line flags, you must create a shortcut to the program itself. Windows automatically creates such a shortcut when you drag the Sketchpad icon to the Start button to install Sketchpad in the Start menu. You can also create such a shortcut on the desktop by right-clicking the Sketchpad icon and choosing **Send To | Desktop.** Consult your computer's Windows manual or online help to determine other ways to create a shortcut. Once you have created a shortcut, follow these steps to set its command-line flags.

1. If the shortcut is on the desktop, right-click the shortcut and choose **Properties.**

   If the shortcut is in the Start menu, activate the Start menu, right-click on Sketchpad, then choose **Properties.**

2. On the Shortcut panel of the Properties dialog box, click in the Target edit box and type the desired command-line flags at the end of the existing target. Here are examples of the command-line flags you can use.

   **-ma** Maximize the application window. The main Sketchpad window will fill your screen.

   **-md** Maximize the first document window. When you start Sketchpad, the document window will fill the application window.

   **"Read Me.gsp**

   **"** Open the sketch named **Read Me.gsp.**

   **-pref "h:\Sketchpad\Modified Preferences.dat"** Start Sketchpad using the preferences in the specified file. (Normally, Sketchpad uses the preferences in the file **Sketchpad Preferences.dat** from the Windows folder.)

   **-f "c:\Program Files\Sketchpad\Samples"** Sketchpad will use the Samples folder when it first opens or saves a sketch.

   **-t "Triangle Tools"** Start Sketchpad using the folder named **Triangle Tools** as the Tool Folder 133.

3. Click **OK.**

Your command-line flags will be in effect every time you run Sketchpad from the modified shortcut.

For example, if Sketchpad is installed in the folder **c:\Sketchpad,** the following target entry in the Shortcut Properties panel will start Sketchpad with the document maximized in the application window, using the folder named **My Tools** as the Tool Folder, and with the file named **Read Me.gsp** open:

```
"c:\Sketchpad\GSP 5.0.exe" -md "Read Me.gsp" -t "My Tools"
```

The following target entry in the Shortcut Properties panel will start Sketchpad with the file **GSP**

**Algebra Prefs.dat** as the preferences file:

```
"c:\Sketchpad\GSP 5.0.exe" -prefs "GSP Algebra Prefs.dat"
```

The following entry will start Sketchpad so that the first time the user chooses **File | Open** 138 or **File | Save** 139, the dialog box will show the folder containing the Geometry sample sketches:

```
"c:\Sketchpad\GSP 5.0.exe" -f "Samples\Sketches\Geometry"
```

# 7.4     Technical and Installation Support

## Web-Based Technical Support

A variety of technical support options are available at the Sketchpad Resource Center:

- Installation Questions: http://www.dynamicgeometry.com/Technical_Support/FAQ.html.

- Frequently Asked Questions: http://www.dynamicgeometry.com/Technical_Support.html

- Latest Product Updates: http://www.keypress.com/sketchpad/produpdates.html

- Technical Support Request Form: http://www.dynamicgeometry.com/Technical_Support/Support_Request.html

## E-Mail Technical Support

When you e-mail technical support, in addition to your request, please identify the type of CPU and operating system on which you use Sketchpad, and the full version number of the edition of Sketchpad you're using. (The full version number — for example, 5.00 — can be located in the program's About Sketchpad dialog box.)

Techsupport@keypress.com

## Telephone and Mail Technical Support

Single User or Site License customers within the USA may also access telephone-based or mail-based technical support. If you call, please have your customer or invoice number available, and try to call from a location where you can access your computer and Sketchpad.

Key Curriculum Press

1150 65th Street

Emeryville, CA 94608

1-510-595-7000

Outside the USA, please contact your local Sketchpad distributor for technical support.

# 7.5     License

## Key Curriculum Press Software License Agreement

### Definitions:

- Key means Key Curriculum Press, Inc., 1150 65th Street, Emeryville, CA 94608, USA.

- Software means The Geometer's Sketchpad Version 5 computer program that you received from any source.

- Documentation means the printed or electronic reference materials and other printed or electronic materials accompanying the Software.

- Product means Software and Documentation.

- Use means install, use, access, display, run, or otherwise interact with the Product.

- License means a Single-User License, a School/Institution License, a Student License, a Student 1-Year License, or any other license as defined by Key from time to time.

- Evaluation Mode means the restricted and time-limited state of the Software prior to the registration of a License Name and/or an Authorization Code as required for Use under a License and subject to the terms of use as stated in this Agreement.

- Licensee means an individual or institution that has legally obtained a License and that is legally able to Use the Product under the terms of that License, or any individual that Uses the Product under a Limited Preview License.

- Authorization Code means the unique code delivered to Licensee in conjunction with a License and that enables the Use of the Software under the terms of the License.

- Preview Mode means the restricted and time-limited state of the Software prior to the registration of the Software via an Authorization Code as required for Use under a License.

- Limited Preview License means a license to Use the Software in Preview Mode.

- Order Confirmation or License Confirmation means the printed or electronic copy of the invoice or invoice confirmation record received by a Licensee from Key or from one of Key's authorized educational dealers or distributors, or any other printed or electronic confirmation of a License received by Licensee from Key or displayed by the Software upon registration.

- License Term means the period of time associated with a License, if any.

## GRANT OF LICENSE

Key grants the Licensee a limited, non-exclusive license to Use the Product on one or more computers of the Licensee consistent with the conditions set forth below. If a Licensee is found to have violated these conditions of use, then Key may cancel the Licensee's License and terminate this Agreement. All other rights are expressly reserved by Key.

### Single-User Licenses

A Licensee with a Single-User License may Use the Product on up to three personal computers, provided only one copy of the Product is in use at one time.

### School/Institution Licenses

A Licensee with a School/Institution License may Use the Software or authorize Use of the Software on the number of computers as in the Order Confirmation or License Confirmation. Use by an instructor on a personal or home computer shall be considered one instance of use under a School/Institutional License.

### Student Licenses

A student Licensee with a Student License may Use the Product on up to three personal or family-owned computers, provided only one copy of the Product is in use at one time. A student who receives the Product and/or an Authorization Code from their school or educational institution in order to use the Software on a personal or family-owned computer shall be considered a Licensee and shall be bound by the terms of the Student License. A user of a Student License must be a student at an educational institution at the time the license was purchased or obtained.

**License Term**

If a License has a License Term, or if the Order Confirmation or License Confirmation indicates a License Term, Licensee may Use the Software for the specified License Term commencing on the day that the License was issued (the date on which the License Authorization Code was generated). The Software will not operate beyond the License Term and tampering with the Software in order to use it beyond the License Term is a violation of this Agreement.

## GRANT OF LIMITED PREVIEW LICENSE

Key grants each individual who installs the Product on a computer a limited, non-exclusive license to Use the Product in Preview Mode on the computer on which it is installed.

## LIMITATIONS ON USE

The Product is licensed as a single product and its component parts may not be separated for use on more than one computer or on the number of computers set out in the Order Confirmation or License Confirmation. The rights granted hereunder are personal to the Licensee. Neither the Product nor the rights granted hereunder may be resold, sub-licensed, assigned, leased, lent, or rented, whether for value or otherwise, except by Key pre-approved resellers or as noted in the terms of this Agreement. The Product shall not be Used as part of a time-share or service bureau arrangement. The Product may not be modified, reverse engineered, decompiled, or disassembled. The proprietary rights legends contained on and in the Product shall not be removed or obscured.

## ASSIGNMENT OF USE BY SCHOOLS

Schools or educational institutions that purchase either a School/Institution License or a Student License may assign or sell a sub-license in such Product for use by an instructor or a student enrolled in the school or educational institution. For the purposes of this License, such a faculty member or student is considered a Licensee subject to all of the conditions on Use of the Product as the original Licensee.

## REPRESENTATIONS OF LICENSEE

The Licensee represents that it has obtained all necessary consent and authority for the importation and Use of the Product in the jurisdiction in which the Licensee intends to Use the Product.

## INTELLECTUAL PROPERTY RIGHTS

Key and/or the rights holders named in the Product are the owners of and retain title to all proprietary and intellectual property rights in and to the Product, including copyrights, trade secrets, trademarks and know-how protected both by United States and Canadian copyright laws, and under the provisions of international treaties. Copying of the Product, other than as explicitly provided herein, constitutes an infringement of the rights holders' intellectual property rights. Licensee acknowledges the foregoing and agrees that it has no right, title or interest in the Product, except as specifically set forth herein, and that the Licensee has no rights in any trademarks identified as belonging to the rights holders.

## SUPPORT SERVICES

Key may provide the Licensee with support services related to the Product ('Support Services'). Use of Support Services is governed by Key's policies and programs described in the user software reference manual, in online documentation, and/or in other Key-provided materials. Any supplemental software provided to the Licensee as part of the Support Services shall be considered part of the Product and subject to the provisions of this Agreement. In the event that the Licensee provides technical information to Key pursuant to the delivery of Support Services, Key may use this information for its business purposes, including Product support and development.

## TERMINATION

Without prejudice to any other rights, Key may terminate this Agreement if the Licensee fails to comply with the terms and conditions hereof. In such event, the licensee agrees to remove all copies of the Product from any computers on which they were installed and destroy all such copies.

## LIMITED WARRANTY

THE PRODUCT IS PROVIDED "AS IS" WITHOUT WARRANTY OF ANY KIND, EITHER EXPRESSED OR IMPLIED, INCLUDING BUT NOT LIMITED TO IMPLIED WARRANTIES OF MERCHANTABILITY, FITNESS FOR A PARTICULAR PURPOSE, TITLE, AND NON-INFRINGEMENT, WITH REGARD TO THE PRODUCT, AND THE PROVISION OF OR FAILURE TO PROVIDE SUPPORT SERVICES. SOME JURISDICTIONS DO NOT ALLOW THE EXCLUSION OF IMPLIED WARRANTIES, SO THE ABOVE EXCLUSION MAY NOT APPLY TO YOU.

LIMITATION OF LIABILITY

IN THE EVENT THE EXCLUSION OF IMPLIED WARRANTIES DOES NOT APPLY AND IN THE EVENT OF A BREACH OF SUCH WARRANTIES, KEY'S AND ITS DEALERS' AND DISTRIBUTORS' ENTIRE LIABILITY AND YOUR EXCLUSIVE REMEDY SHALL BE, AT KEY'S OPTION, EITHER (A) RETURN OF THE PRICE PAID, IF ANY; OR (B) REPAIR OR REPLACEMENT OF THE PRODUCT RETURNED TO KEY WITH A PURCHASE RECEIPT. TO THE MAXIMUM EXTENT PERMITTED BY APPLICABLE LAW, IN NO EVENT SHALL KEY OR ITS SUPPLIERS BE LIABLE FOR ANY SPECIAL, INCIDENTAL, INDIRECT, OR CONSEQUENTIAL DAMAGES (INCLUDING, WITHOUT LIMITATION, DAMAGES FOR LOSS OF BUSINESS PROFITS, BUSINESS INTERRUPTION, LOSS OF BUSINESS INFORMATION, OR ANY OTHER PECUNIARY LOSS) ARISING OUT OF THE USE OF OR INABILITY TO USE THE PRODUCT OR THE PROVISION OF OR FAILURE TO PROVIDE SUPPORT SERVICES, EVEN IF KEY HAS BEEN ADVISED OF THE POSSIBILITY OF SUCH DAMAGES. AS SOME JURISDICTIONS DO NOT ALLOW THE EXCLUSION OR LIMITATION OF LIABILITY, THE ABOVE LIMITATION MAY NOT APPLY IN CERTAIN JURISDICTIONS.

## ENTIRE AGREEMENT

The Licensee agrees that this Agreement is the complete and sole statement of the agreement between Licensee, Key, and Key's distributors and dealers, and supersedes both all representations made in respect of the Product and all other agreements (whether written or oral) relating to the subject matter of this Agreement.

## PARTIAL ILLEGALITY

If any provisions of this Agreement shall be construed to be illegal or invalid, it shall not affect the legality or validity of any other provision thereof, and the illegal or invalid provisions shall be deemed stricken and deleted herefrom to the same extent and effect as if never incorporated herein, but all other provisions hereof shall continue in full force and effect.

## APPLICABLE LAWS

The rights and obligations of the parties under this Agreement shall not be governed by the United Nations Convention on Contracts for the International Sale of Goods. Instead, unless expressly prohibited by local law, the rights and obligations of the parties under this Agreement shall be governed by the State of California, and the laws of the United States applicable therein.

Key Curriculum Press 1150 65th Street Emeryville, CA 94608 Phone: 510-595-7000 Fax: 510-595-7040 Web: www.keypress.com

## 7.6 Credits

**Project Design**

Nicholas Jackiw

**Implementation**

Nicholas Jackiw, Scott Steketee, Matt Litwin

**Engineering Support**

Jon Brooks, Scott Johnson, Kirk Swenson

**Special Thanks To:**

Eugene Klotz

Kendra Lockman

Tawnia Litwin

Andres Marti

Vishakha Parvate

Steven Rasmussen

Doris Schattschneider

Daniel Scher

Nathalie Sinclair

and to all of our many field testers!

Portions of this work were funded by grants from the National Science Foundation to KCP Technologies, Key Curriculum Press, and the Visual Geometry Project at Swarthmore College.

The Geometer's Sketchpad® and Dynamic Geometry® are registered trademarks of KCP Technologies. The Sketchpad™ and Hot Text™ are trademarks of KCP Technologies. Sketchpad LessonLink™ is a trademark of Key Curriculum Press, Inc.

### ▼ Pthreads Win32

This software uses Pthreads-win32, the POSIX Threads Library for Win32.

Copyright © 1998 John E. Bossom

Copyright © 1999,2006 Pthreads-win32 contributors

The pthreads-win32 library and its use are covered under the terms of the GNU Library General

Public License.332

To dynamically link to your own version of the pthreads-win32 dll, place it next to the Sketchpad Application, and name it "pthreadVC2.dll".

## ▼ The GLee Library for OpenGL

This software uses the GLee Library for OpenGL.

Copyright (c)2009 Ben Woodhouse All rights reserved.

Redistribution and use in source and binary forms, with or without modification, are permitted provided that the following conditions are met: 1. Redistributions of source code must retain the above copyright notice, this list of conditions and the following disclaimer as the first lines of this file unmodified. 2. Redistributions in binary form must reproduce the above copyright notice, this list of conditions and the following disclaimer in the documentation and/or other materials provided with the distribution.

THIS SOFTWARE IS PROVIDED BY BEN WOODHOUSE ``AS IS AND ANY EXPRESS OR IMPLIED WARRANTIES, INCLUDING, BUT NOT LIMITED TO, THE IMPLIED WARRANTIES OF MERCHANTABILITY AND FITNESS FOR A PARTICULAR PURPOSE ARE DISCLAIMED. IN NO EVENT SHALL BEN WOODHOUSE BE LIABLE FOR ANY DIRECT, INDIRECT, INCIDENTAL, SPECIAL, EXEMPLARY, OR CONSEQUENTIAL DAMAGES (INCLUDING, BUT NOT LIMITED TO, PROCUREMENT OF SUBSTITUTE GOODS OR SERVICES; LOSS OF USE, DATA, OR PROFITS; OR BUSINESS INTERRUPTION) HOWEVER CAUSED AND ON ANY THEORY OF LIABILITY, WHETHER IN CONTRACT, STRICT LIABILITY, OR TORT (INCLUDING NEGLIGENCE OR OTHERWISE) ARISING IN ANY WAY OUT OF THE USE OF THIS SOFTWARE, EVEN IF ADVISED OF THE POSSIBILITY OF SUCH DAMAGE.

## 7.6.1 LG Public License

GNU LESSER GENERAL PUBLIC LICENSE

Version 2.1, February 1999

Copyright (C) 1991, 1999 Free Software Foundation, Inc.

51 Franklin Street, Fifth Floor, Boston, MA  02110-1301  USA

Everyone is permitted to copy and distribute verbatim copies of this license document, but changing it is not allowed.

[This is the first released version of the Lesser GPL.  It also counts as the successor of the GNU Library Public License, version 2, hence the version number 2.1.]

Preamble

The licenses for most software are designed to take away your freedom to share and change it. By contrast, the GNU General Public Licenses are intended to guarantee your freedom to share and change free software--to make sure the software is free for all its users.

This license, the Lesser General Public License, applies to some specially designated software packages--typically libraries--of the Free Software Foundation and other authors who decide to use it. You can use it too, but we suggest you first think carefully about whether this license or the ordinary General Public License is the better strategy to use in any particular case, based on the explanations below.

When we speak of free software, we are referring to freedom of use, not price. Our General Public Licenses are designed to make sure that you have the freedom to distribute copies of free software (and charge for this service if you wish); that you receive source code or can get it if you want it; that you can change the software and use pieces of it in new free programs; and that you are informed that you can do these things.

To protect your rights, we need to make restrictions that forbid distributors to deny you these rights or to ask you to surrender these rights. These restrictions translate to certain responsibilities for you if you distribute copies of the library or if you modify it.

For example, if you distribute copies of the library, whether gratis or for a fee, you must give the recipients all the rights that we gave you. You must make sure that they, too, receive or can get the source code. If you link other code with the library, you must provide complete object files to the recipients, so that they can relink them with the library after making changes to the library and recompiling it. And you must show them these terms so they know their rights.

We protect your rights with a two-step method: (1) we copyright the library, and (2) we offer you this license, which gives you legal permission to copy, distribute and/or modify the library.

To protect each distributor, we want to make it very clear that there is no warranty for the free library. Also, if the library is modified by someone else and passed on, the recipients should know that what they have is not the original version, so that the original author's reputation will not be affected by problems that might be introduced by others.

Finally, software patents pose a constant threat to the existence of any free program. We wish to make sure that a company cannot effectively restrict the users of a free program by obtaining a restrictive license from a patent holder. Therefore, we insist that any patent license obtained for a version of the library must be consistent with the full freedom of use specified in this license.

Most GNU software, including some libraries, is covered by the ordinary GNU General Public License. This license, the GNU Lesser General Public License, applies to certain designated libraries, and is quite different from the ordinary General Public License. We use this license for certain libraries in order to permit linking those libraries into non-free programs.

When a program is linked with a library, whether statically or using a shared library, the combination of the two is legally speaking a combined work, a derivative of the original library. The ordinary General Public License therefore permits such linking only if the entire combination fits its criteria of freedom. The Lesser General Public License permits more lax criteria for linking other code with the library.

We call this license the "Lesser" General Public License because it does Less to protect the user's freedom than the ordinary General Public License. It also provides other free software developers Less of an advantage over competing non-free programs. These disadvantages are the reason we use the ordinary General Public License for many libraries. However, the Lesser license provides advantages in certain special circumstances.

For example, on rare occasions, there may be a special need to encourage the widest possible use of a certain library, so that it becomes a de-facto standard. To achieve this, non-free programs must be allowed to use the library. A more frequent case is that a free library does the same job as widely used non-free libraries. In this case, there is little to gain by limiting the free library to free software only, so we use the Lesser General Public License.

In other cases, permission to use a particular library in non-free programs enables a greater number of people to use a large body of free software. For example, permission to use the GNU C Library in non-free programs enables many more people to use the whole GNU operating system, as well as its variant, the GNU/Linux operating system.

Although the Lesser General Public License is Less protective of the users' freedom, it does ensure that the user of a program that is linked with the Library has the freedom and the wherewithal to run that program using a modified version of the Library.

The precise terms and conditions for copying, distribution and modification follow. Pay close attention to the difference between a "work based on the library" and a "work that uses the library". The former contains code derived from the library, whereas the latter must be combined with the library in order to run.

TERMS AND CONDITIONS FOR COPYING, DISTRIBUTION AND MODIFICATION

0. This License Agreement applies to any software library or other program which contains a notice placed by the copyright holder or other authorized party saying it may be distributed under the terms of this Lesser General Public License (also called "this License"). Each licensee is addressed as "you".

A "library" means a collection of software functions and/or data prepared so as to be conveniently linked with application programs (which use some of those functions and data) to form executables.

The "Library", below, refers to any such software library or work which has been distributed under these terms. A "work based on the Library" means either the Library or any derivative work under copyright law: that is to say, a work containing the Library or a portion of it, either verbatim or with modifications and/or translated straightforwardly into another language. (Hereinafter, translation is included without limitation in the term "modification".)

"Source code" for a work means the preferred form of the work for making modifications to it. For a library, complete source code means all the source code for all modules it contains, plus any associated interface definition files, plus the scripts used to control compilation and installation of the library.

Activities other than copying, distribution and modification are not covered by this License; they are outside its scope. The act of running a program using the Library is not restricted, and output from such a program is covered only if its contents constitute a work based on the Library (independent of the use of the Library in a tool for writing it). Whether that is true depends on what the Library does and what the program that uses the Library does.

1. You may copy and distribute verbatim copies of the Library's complete source code as you receive it, in any medium, provided that you conspicuously and appropriately publish on each copy an appropriate copyright notice and disclaimer of warranty; keep intact all the notices that refer to this License and to the absence of any warranty; and distribute a copy of this License along with the Library.

You may charge a fee for the physical act of transferring a copy, and you may at your option offer warranty protection in exchange for a fee.

2. You may modify your copy or copies of the Library or any portion of it, thus forming a work based on the Library, and copy and distribute such modifications or work under the terms of Section 1 above, provided that you also meet all of these conditions:

* a) The modified work must itself be a software library.

* b) You must cause the files modified to carry prominent notices stating that you changed the files and the date of any change.

* c) You must cause the whole of the work to be licensed at no charge to all third parties under the terms of this License.

* d) If a facility in the modified Library refers to a function or a table of data to be supplied by an application program that uses the facility, other than as an argument passed when the facility is invoked, then you must make a good faith effort to ensure that, in the event an application does not supply such function or table, the facility still operates, and performs whatever part of its purpose remains meaningful.

(For example, a function in a library to compute square roots has a purpose that is entirely well-defined independent of the application. Therefore, Subsection 2d requires that any application-supplied function or table used by this function must be optional: if the application does not supply it, the square root function must still compute square roots.)

These requirements apply to the modified work as a whole. If identifiable sections of that work are not derived from the Library, and can be reasonably considered independent and separate works in themselves, then this License, and its terms, do not apply to those sections when you distribute them as separate works. But when you distribute the same sections as part of a whole which is a work based on the Library, the distribution of the whole must be on the terms of this License, whose permissions for other licensees extend to the entire whole, and thus to each and every part regardless of who wrote it.

Thus, it is not the intent of this section to claim rights or contest your rights to work written entirely by you; rather, the intent is to exercise the right to control the distribution of derivative or collective works based on the Library.

In addition, mere aggregation of another work not based on the Library with the Library (or with a work based on the Library) on a volume of a storage or distribution medium does not bring the other work under the scope of this License.

3. You may opt to apply the terms of the ordinary GNU General Public License instead of this License to a given copy of the Library. To do this, you must alter all the notices that refer to this License, so that they refer to the ordinary GNU General Public License, version 2, instead of to this License. (If a newer version than version 2 of the ordinary GNU General Public License has appeared, then you can specify that version instead if you wish.) Do not make any other change in these notices.

Once this change is made in a given copy, it is irreversible for that copy, so the ordinary GNU General Public License applies to all subsequent copies and derivative works made from that copy.

This option is useful when you wish to copy part of the code of the Library into a program that is not a library.

4. You may copy and distribute the Library (or a portion or derivative of it, under Section 2) in object code or executable form under the terms of Sections 1 and 2 above provided that you accompany it with the complete corresponding machine-readable source code, which must be distributed under the terms of Sections 1 and 2 above on a medium customarily used for software interchange.

If distribution of object code is made by offering access to copy from a designated place, then offering equivalent access to copy the source code from the same place satisfies the requirement to distribute the source code, even though third parties are not compelled to copy the source along with the object code.

5. A program that contains no derivative of any portion of the Library, but is designed to work with the Library by being compiled or linked with it, is called a "work that uses the Library". Such a work, in isolation, is not a derivative work of the Library, and therefore falls outside the scope of this

License.

However, linking a "work that uses the Library" with the Library creates an executable that is a derivative of the Library (because it contains portions of the Library), rather than a "work that uses the library". The executable is therefore covered by this License. Section 6 states terms for distribution of such executables.

When a "work that uses the Library" uses material from a header file that is part of the Library, the object code for the work may be a derivative work of the Library even though the source code is not. Whether this is true is especially significant if the work can be linked without the Library, or if the work is itself a library. The threshold for this to be true is not precisely defined by law.

If such an object file uses only numerical parameters, data structure layouts and accessors, and small macros and small inline functions (ten lines or less in length), then the use of the object file is unrestricted, regardless of whether it is legally a derivative work. (Executables containing this object code plus portions of the Library will still fall under Section 6.)

Otherwise, if the work is a derivative of the Library, you may distribute the object code for the work under the terms of Section 6. Any executables containing that work also fall under Section 6, whether or not they are linked directly with the Library itself.

6. As an exception to the Sections above, you may also combine or link a "work that uses the Library" with the Library to produce a work containing portions of the Library, and distribute that work under terms of your choice, provided that the terms permit modification of the work for the customer's own use and reverse engineering for debugging such modifications.

You must give prominent notice with each copy of the work that the Library is used in it and that the Library and its use are covered by this License. You must supply a copy of this License. If the work during execution displays copyright notices, you must include the copyright notice for the Library among them, as well as a reference directing the user to the copy of this License. Also, you must do one of these things:

  * a) Accompany the work with the complete corresponding machine-readable source code for the Library including whatever changes were used in the work (which must be distributed under Sections 1 and 2 above); and, if the work is an executable linked with the Library, with the complete machine-readable "work that uses the Library", as object code and/or source code, so that the user can modify the Library and then relink to produce a modified executable containing the modified Library. (It is understood that the user who changes the contents of definitions files in the Library will not necessarily be able to recompile the application to use the modified definitions.)

  * b) Use a suitable shared library mechanism for linking with the Library. A suitable mechanism is one that (1) uses at run time a copy of the library already present on the user's computer system, rather than copying library functions into the executable, and (2) will operate properly with a modified version of the library, if the user installs one, as long as the modified version is interface-compatible with the version that the work was made with.

* c) Accompany the work with a written offer, valid for at least three years, to give the same user the materials specified in Subsection 6a, above, for a charge no more than the cost of performing this distribution.

* d) If distribution of the work is made by offering access to copy from a designated place, offer equivalent access to copy the above specified materials from the same place.

* e) Verify that the user has already received a copy of these materials or that you have already sent this user a copy.

For an executable, the required form of the "work that uses the Library" must include any data and utility programs needed for reproducing the executable from it. However, as a special exception, the materials to be distributed need not include anything that is normally distributed (in either source or binary form) with the major components (compiler, kernel, and so on) of the operating system on which the executable runs, unless that component itself accompanies the executable.

It may happen that this requirement contradicts the license restrictions of other proprietary libraries that do not normally accompany the operating system. Such a contradiction means you cannot use both them and the Library together in an executable that you distribute.

7. You may place library facilities that are a work based on the Library side-by-side in a single library together with other library facilities not covered by this License, and distribute such a combined library, provided that the separate distribution of the work based on the Library and of the other library facilities is otherwise permitted, and provided that you do these two things:

* a) Accompany the combined library with a copy of the same work based on the Library, uncombined with any other library facilities. This must be distributed under the terms of the Sections above.

* b) Give prominent notice with the combined library of the fact that part of it is a work based on the Library, and explaining where to find the accompanying uncombined form of the same work.

8. You may not copy, modify, sublicense, link with, or distribute the Library except as expressly provided under this License. Any attempt otherwise to copy, modify, sublicense, link with, or distribute the Library is void, and will automatically terminate your rights under this License. However, parties who have received copies, or rights, from you under this License will not have their licenses terminated so long as such parties remain in full compliance.

9. You are not required to accept this License, since you have not signed it. However, nothing else grants you permission to modify or distribute the Library or its derivative works. These actions are prohibited by law if you do not accept this License. Therefore, by modifying or distributing the Library (or any work based on the Library), you indicate your acceptance of this License to do so, and all its terms and conditions for copying, distributing or modifying the Library or works based on it.

10. Each time you redistribute the Library (or any work based on the Library), the recipient automatically receives a license from the original licensor to copy, distribute, link with or modify the

Library subject to these terms and conditions. You may not impose any further restrictions on the recipients' exercise of the rights granted herein. You are not responsible for enforcing compliance by third parties with this License.

11. If, as a consequence of a court judgment or allegation of patent infringement or for any other reason (not limited to patent issues), conditions are imposed on you (whether by court order, agreement or otherwise) that contradict the conditions of this License, they do not excuse you from the conditions of this License. If you cannot distribute so as to satisfy simultaneously your obligations under this License and any other pertinent obligations, then as a consequence you may not distribute the Library at all. For example, if a patent license would not permit royalty-free redistribution of the Library by all those who receive copies directly or indirectly through you, then the only way you could satisfy both it and this License would be to refrain entirely from distribution of the Library.

If any portion of this section is held invalid or unenforceable under any particular circumstance, the balance of the section is intended to apply, and the section as a whole is intended to apply in other circumstances.

It is not the purpose of this section to induce you to infringe any patents or other property right claims or to contest validity of any such claims; this section has the sole purpose of protecting the integrity of the free software distribution system which is implemented by public license practices. Many people have made generous contributions to the wide range of software distributed through that system in reliance on consistent application of that system; it is up to the author/donor to decide if he or she is willing to distribute software through any other system and a licensee cannot impose that choice.

This section is intended to make thoroughly clear what is believed to be a consequence of the rest of this License.

12. If the distribution and/or use of the Library is restricted in certain countries either by patents or by copyrighted interfaces, the original copyright holder who places the Library under this License may add an explicit geographical distribution limitation excluding those countries, so that distribution is permitted only in or among countries not thus excluded. In such case, this License incorporates the limitation as if written in the body of this License.

13. The Free Software Foundation may publish revised and/or new versions of the Lesser General Public License from time to time. Such new versions will be similar in spirit to the present version, but may differ in detail to address new problems or concerns.

Each version is given a distinguishing version number. If the Library specifies a version number of this License which applies to it and "any later version", you have the option of following the terms and conditions either of that version or of any later version published by the Free Software Foundation. If the Library does not specify a license version number, you may choose any version ever published by the Free Software Foundation.

14. If you wish to incorporate parts of the Library into other free programs whose distribution conditions are incompatible with these, write to the author to ask for permission. For software which is copyrighted by the Free Software Foundation, write to the Free Software Foundation; we sometimes make exceptions for this. Our decision will be guided by the two goals of preserving the free status of all derivatives of our free software and of promoting the sharing and reuse of software generally.

NO WARRANTY

15. BECAUSE THE LIBRARY IS LICENSED FREE OF CHARGE, THERE IS NO WARRANTY FOR THE LIBRARY, TO THE EXTENT PERMITTED BY APPLICABLE LAW. EXCEPT WHEN OTHERWISE STATED IN WRITING THE COPYRIGHT HOLDERS AND/OR OTHER PARTIES PROVIDE THE LIBRARY "AS IS" WITHOUT WARRANTY OF ANY KIND, EITHER EXPRESSED OR IMPLIED, INCLUDING, BUT NOT LIMITED TO, THE IMPLIED WARRANTIES OF MERCHANTABILITY AND FITNESS FOR A PARTICULAR PURPOSE. THE ENTIRE RISK AS TO THE QUALITY AND PERFORMANCE OF THE LIBRARY IS WITH YOU. SHOULD THE LIBRARY PROVE DEFECTIVE, YOU ASSUME THE COST OF ALL NECESSARY SERVICING, REPAIR OR CORRECTION.

16. IN NO EVENT UNLESS REQUIRED BY APPLICABLE LAW OR AGREED TO IN WRITING WILL ANY COPYRIGHT HOLDER, OR ANY OTHER PARTY WHO MAY MODIFY AND/OR REDISTRIBUTE THE LIBRARY AS PERMITTED ABOVE, BE LIABLE TO YOU FOR DAMAGES, INCLUDING ANY GENERAL, SPECIAL, INCIDENTAL OR CONSEQUENTIAL DAMAGES ARISING OUT OF THE USE OR INABILITY TO USE THE LIBRARY (INCLUDING BUT NOT LIMITED TO LOSS OF DATA OR DATA BEING RENDERED INACCURATE OR LOSSES SUSTAINED BY YOU OR THIRD PARTIES OR A FAILURE OF THE LIBRARY TO OPERATE WITH ANY OTHER SOFTWARE), EVEN IF SUCH HOLDER OR OTHER PARTY HAS BEEN ADVISED OF THE POSSIBILITY OF SUCH DAMAGES.

END OF TERMS AND CONDITIONS

# Index

## - A -

# - D -